Lecture Notes in Mathematics

Volume 2256

This series reports on new developments in all areas of mathematics and their applications - quickly, informally and at a high level. Mathematical texts analysing new developments in modelling and numerical simulation are welcome. The type of material considered for publication includes:

1. Research monographs 2. Lectures on a new field or presentations of a new angle in a classical field 3. Summer schools and intensive courses on topics of current research.

Texts which are out of print but still in demand may also be considered if they fall within these categories. The timeliness of a manuscript is sometimes more important than its form, which may be preliminary or tentative.

More information about this series at http://www.springer.com/series/304

Bo'az Klartag • Emanuel Milman
Editors

Geometric Aspects of Functional Analysis

Israel Seminar (GAFA) 2017-2019
Volume I

 Springer

Editors
Bo'az Klartag
School of Mathematical Sciences
Tel Aviv University
Tel Aviv, Israel

Emanuel Milman
Department of Mathematics
Technion – Israel Institute of Technology
Haifa, Israel

Department of Mathematics
Weizmann Institute of Science
Rehovot, Israel

ISSN 0075-8434   ISSN 1617-9692   (electronic)
Lecture Notes in Mathematics
ISBN 978-3-030-36019-1   ISBN 978-3-030-36020-7   (eBook)
https://doi.org/10.1007/978-3-030-36020-7

Mathematics Subject Classification (2010): Primary: 46-XX; Secondary: 52-XX

This Springer imprint is published by the registered company Springer Nature Switzerland AG.
The registered company address is: Gewerbestrasse 11, 6330 Cham, Switzerland

Preface

Since the mid-1980s, the following volumes containing collections of papers reflecting the activity of the Israel Seminar in Geometric Aspects of Functional Analysis have appeared:

1983–1984 Published privately by Tel Aviv University
1985–1986 Springer Lecture Notes in Mathematics, vol. 1267
1986–1987 Springer Lecture Notes in Mathematics, vol. 1317
1987–1988 Springer Lecture Notes in Mathematics, vol. 1376
1989–1990 Springer Lecture Notes in Mathematics, vol. 1469
1992–1994 Operator Theory: Advances and Applications, vol. 77, Birkhäuser
1994–1996 MSRI Publications, vol. 34, Cambridge University Press
1996–2000 Springer Lecture Notes in Mathematics, vol. 1745
2001–2002 Springer Lecture Notes in Mathematics, vol. 1807
2002–2003 Springer Lecture Notes in Mathematics, vol. 1850
2004–2005 Springer Lecture Notes in Mathematics, vol. 1910
2006–2010 Springer Lecture Notes in Mathematics, vol. 2050
2011–2013 Springer Lecture Notes in Mathematics, vol. 2116
2014–2016 Springer Lecture Notes in Mathematics, vol. 2169

The first six were edited by Lindenstrauss and Milman, the seventh by Ball and Milman, the subsequent four by Milman and Schechtman, the subsequent one by Klartag, Mendelson, and Milman, and the last two by the present editors.

This is the first of two volumes from the years 2017–2019, the second volume is published in Springer Lecture Notes in Mathematics, vol. 2266. As in the previous Seminar Notes, these two volumes reflect general trends in the study of Geometric Aspects of Functional Analysis, understood in a broad sense. Two classical topics represented are the Concentration of Measure Phenomenon in the Local Theory of Banach Spaces, which has recently had triumphs in Random Matrix Theory, and the Central Limit Theorem, one of the earliest examples of regularity and order in high dimensions. Central to the text is the study of the Poincaré and log-Sobolev functional inequalities, their reverses, and other inequalities, in which a crucial role is often played by convexity assumptions such as log-concavity. The concept and properties of entropy form an important subject, with Bourgain's slicing problem and its variants drawing much attention. Constructions related to convexity theory are proposed and revisited, as well as inequalities that go beyond the Brunn–Minkowski theory. One of the major current research directions addressed is the identification of lower-dimensional structures with remarkable properties in rather arbitrary high-dimensional objects. In addition to functional analytic results, connections to computer science and to differential geometry are also discussed. All contributions are original research papers and were subject to the usual refereeing standards.

We are grateful to Vitali Milman for his help and guidance in preparing and editing these two volumes.

Tel Aviv/Rehovot, Israel   Bo'az Klartag
Haifa, Israel   Emanuel Milman

Jean Bourgain: In Memoriam

Our friend and mentor Jean Bourgain passed away on December 22, 2018, at the age of 64.

Jean Bourgain by Jan Rauchwerger. Courtesy of Vitali Milman

Jean Bourgain was one of the most outstanding mathematicians of our time. Bourgain changed the face of analysis; he revolutionized our understanding of analysis. He has introduced, mastered, and perfected many different methods in every corner of analysis, including a dozen of neighboring fields, and has left his mark in each of these directions. His achievements, vision, and insight united many distant and very diverse directions of mathematics into one enormously powerful and broad entity. When we say "Analysis" today we mean, besides classical directions, also ergodic theory, PDE, several directions of analytical number theory, geometry, and combinatorics (including complexity). This is undoubtedly the result of Bourgain's activity, unprecedented in its strength and diversity. The torrent of his achievements is difficult to grasp, the number of very long-standing problems Bourgain has solved can be counted in tens, perhaps approaching a hundred, and this would take a whole book to describe. It is almost impossible to believe: ∼550 hard analysis papers written over less than 40 years.

Jean's passing is a terrible loss for his family and friends, and a terrible loss to the mathematical world. We shudder at the thought of how many more theorems he could have proved and open problems he could have solved. He leaves behind an unbelievable legacy of results carrying his name, whose breadth is matched only by their depth.

In addition to being a mathematical giant, Jean has personally influenced us all. His influence was also well felt on the GAFA seminar notes. He has published a total of 50 papers (3 in the present volume) in **every** volume of the GAFA seminar notes since its inception (45 in Springer Math. Notes series, 2 in Birkhauser series, and 3 in Mathematical Sciences Research Institute (MSRI), Berkeley). We are proud that some of his last papers are published in this volume.

May he rest in peace.

Tel Aviv/Rehovot, Israel   Bo'az Klartag (Editor)
Haifa, Israel   Emanuel Milman (Editor)
Tel Aviv, Israel   Vitali Milman (Founding Editor)

Contents Overview for Volume II

Contents

Chapter 1
Gromov's Waist of Non-radial Gaussian Measures and Radial Non-Gaussian Measures

Arseniy Akopyan and Roman Karasev

Abstract We study the Gromov waist in the sense of t-neighborhoods for measures in the Euclidean space, motivated by the famous theorem of Gromov about the waist of radially symmetric Gaussian measures. In particular, it turns out possible to extend Gromov's original result to the case of not necessarily radially symmetric Gaussian measure. We also provide examples of measures having no t-neighborhood waist property, including a rather wide class of compactly supported radially symmetric measures and their maps into the Euclidean space of dimension at least 2.

We use a simpler form of Gromov's pancake argument to produce some estimates of t-neighborhoods of (weighted) volume-critical submanifolds in the spirit of the waist theorems, including neighborhoods of algebraic manifolds in the complex projective space.

For reader's convenience, in one appendix of this paper we provide a more detailed explanation of the Caffarelli theorem that we use to handle not necessarily radially symmetric Gaussian measures. In the other appendix, we provide a comparison of different variations of Gromov's pancake method.

The author Arseniy Akopyan was supported by the European Research Council (ERC) under the European Union's Horizon 2020 research and innovation programme (grant agreement No 716117).

The author Roman Karasev was supported by the Federal professorship program grant 1.456.2016/1.4 and the Russian Foundation for Basic Research grants 18-01-00036 and 19-01-00169.

A. Akopyan (✉)
Institute of Science and Technology Austria (IST Austria), Klosterneuburg, Austria

R. Karasev
Moscow Institute of Physics and Technology, Dolgoprudny, Russia

Institute for Information Transmission Problems RAS, Moscow, Russia
e-mail: r_n_karasev@mail.ru; http://www.rkarasev.ru/en/

© Springer Nature Switzerland AG 2020
B. Klartag, E. Milman (eds.), *Geometric Aspects of Functional Analysis*,
Lecture Notes in Mathematics 2256,
https://doi.org/10.1007/978-3-030-36020-7_1

1

1.1   Introduction

In [9] Mikhail Gromov proved the waist of the Gaussian measure theorem: For any continuous map $f : \mathbb{R}^n \to \mathbb{R}^k$ and a radially symmetric Gaussian measure γ, it is possible to find a fiber $f^{-1}(y)$ such that for any $t > 0$

$$\gamma(f^{-1}(y) + t) \geq \gamma(\mathbb{R}^{n-k} + t),$$

where $X + t$ denotes the t-neighborhood of a set X and $\mathbb{R}^{n-k} \subset \mathbb{R}^n$ denotes any $(n - k)$-dimensional linear subspace of $\mathbb{R}^n$. This statement is a very general version of the concentration phenomenon that was previously extensively studied for functions, the case $k = 1$ of the result.

In the recent papers [1, 2, 13] a Minkowski content version of such a result was considered, where one does not require the inequality for all t, but requires a lower bound on the asymptotics for $t \to +0$. In this simplified problem it turned out that γ can be replaced by other measures, such as the uniform measure on a centrally symmetric cube [13], the uniform measure on a Euclidean ball [1], the uniform measure on a ball in the spaces of constant curvature [2], and some other examples.

In this paper we address the question of extending the precise t-neighborhood waist theorem to measures other than the radially symmetric Gaussian measures, in the general understanding of [9, Question 3.1.A]. More precisely, for a given finite Borel measure μ in $\mathbb{R}^n$ we ask if the following is true: For any continuous $f : \mathbb{R}^n \to \mathbb{R}^k$ there exists a fiber $f^{-1}(y)$ such that for any $t > 0$

$$\mu(f^{-1}(y) + t) \geq \mu(\mathbb{R}^{n-k} + t).$$

Even restricting ourselves to radially symmetric measures, we find that the situation is not easy. A counterexample was given in [2, Remark 5.7] showing that for μ uniform in the Euclidean ball of sufficiently high dimension there is no such precise t-neighborhood waist theorem.

In Sect. 1.3 of this paper we give other counterexamples and discuss some positive results in the plane. In particular, Theorem 1.3.2 asserts that no such t-neighborhood waist theorem is possible for compactly supported radial measures and $2 \leq k < n$.

On the positive side, we have the following extension of Gromov's theorem to the case of not necessarily radially symmetric Gaussian measures:

Theorem 1.1.1 *Let $a_1 \geq a_2 \geq \cdots \geq a_n > 0$ and let γ be a Gaussian measure in $\mathbb{R}^n$ with density*

$$\rho = e^{-a_1 x_1^2 - a_2 x_2^2 - \cdots - a_n x_n^2}.$$

Then for any continuous map $f : \mathbb{R}^n \to \mathbb{R}^k$ *there exists a point* $y \in \mathbb{R}^k$ *such that for any* $t > 0$

$$\gamma(f^{-1}(y) + t) \geq \gamma(\mathbb{R}^{n-k} + t).$$

It is important that by $\mathbb{R}^{n-k} \subset \mathbb{R}^n$ *in this theorem we denote the coordinate subspace spanned by the first* $n - k$ *coordinates.*

The case of an arbitrary Gaussian measure with full-dimensional support is reduced to this one by translation and diagonalization of its covariance matrix, one only needs to care that a particular $(n - k)$-dimensional subspace is assumed in the right-hand side of the inequality.

To prove this theorem we use Gromov's original scheme, replacing the hard argument about choosing "centers" of the pancakes with an application of the Caffarelli theorem on 1-Lipschitz monotone transportation. This not only allows to extend the result to the non-radial case, but also simplifies the original argument, at least in our understanding.

After exhibiting, in Sect. 1.3, examples of radially symmetric measures, for which the neighborhood waist theorem does not hold, we address the question of choosing $y = 0$ in the waist theorems for odd maps in Sect. 1.4. It turns out that the neighborhood waist theorem holds true for all radially symmetric measures and odd maps, as Theorem 1.4.1 asserts.

In Sect. 1.5 we naturally pass to the following question: For which measure μ and a submanifold $X \subset \mathbb{R}^n$, or $X \subset \mathbb{S}^n$, can we claim the same neighborhood estimate as in the waist theorem? This question is inspired by the result of [14], and in particular we observe that the right neighborhood estimate holds for volume-critical submanifolds of $\mathbb{S}^n$ with the uniform measure, see Proposition 1.5.2, and that the neighborhood waist theorem turns out to be true for all radially symmetric measures and homogeneous holomorphic maps, see Theorem 1.5.4.

In one appendix we give the statement and an explanation of the Caffarelli theorem. In the other appendix we discuss Gromov's method of producing pancakes in Gromov's waist theorem.

1.2 Nonradial Gaussian Measures: Proof of Theorem 1.1.1

Let us start with establishing the linear case of the theorem. To do so, it is sufficient to consider the orthogonal projection $f : \mathbb{R}^n \to A$ for a linear subspace of dimension at most k, and for $B = A^\perp$ we need to find the lower bound on $\gamma(\nu_t(B))$. Here we use more explicit notation $\nu_t(X)$ for the t-neighborhood of a point or of a set X. We can also assume $\dim A$ to be precisely k, since further projections onto subspaces of smaller dimension may only improve the required estimate.

Note that the covariance form of the pushforward $f_*\gamma$ is obtained by restricting the covariance form of γ to A. It is a folklore fact that the restriction of a quadratic form with eigenvalues $\lambda_1 \leq \lambda_2 \leq \cdots \leq \lambda_n$ produces a quadratic form with eigenvalues $\kappa_1 \leq \ldots \leq \kappa_k$ such that

$$\kappa_1 \geq \lambda_1, \ldots \kappa_k \geq \lambda_k \quad \text{and} \quad \kappa_1 \leq \lambda_{n-k+1}, \ldots, \kappa_k \leq \lambda_n.$$

This fact is easily proven to induction, restricting the quadratic form to a codimension 1 subspace several times. Hence if we take the orthonormal coordinates $y_1, \ldots, y_k$ in A then the density of $f_* \gamma$ up to multiplication by a constant will be $e^{-b_1 y_1^2 - \cdots - b_k y_k^2}$ with

$$b_1 \geq a_{n-k+1}, \ldots, b_k \geq a_n$$

and therefore

$$
\begin{aligned}
\frac{\gamma(v_t(B))}{\gamma(\mathbb{R}^n)} &\geq \frac{\int_{v_t(0)} e^{-b_1 y_1^2 - \cdots - b_k y_k^2} \, dy_1 \ldots dy_k}{\int_{\mathbb{R}^k} e^{-b_1 y_1^2 - \cdots - b_k y_k^2} \, dy_1 \ldots dy_k} \\
&\geq \frac{\int_{v_t(0)} e^{-a_{n-k+1} x_{n-k+1}^2 - \cdots - a_n x_n^2} \, dx_{n-k+1} \ldots dx_n}{\int_{\mathbb{R}^k} e^{-a_{n-k+1} x_{n-k+1}^2 - \cdots - a_n x_n^2} \, dx_{n-k+1} \ldots dx_n},
\end{aligned}
\tag{1.2.1}
$$

the last inequality is done by the substitutions $t_i^2 = b_i x_i^2$ and $t_i^2 = a_{n-k+i} x_{n-k+i}^2$, after these substitutions the same expression in the two sides of the inequality is integrated over the two domains, one containing the other. Thus the theorem holds for linear maps with a simple choice $y = 0$.

For the general case, we follow Gromov's argument for the round Gaussian measure from [9] (see also the explanations in [13]). Let us state an appropriate version of the pancake decomposition theorem:

Theorem 1.2.1 (Essentially due to Gromov, [9]) *Let μ be a measure in the ball $B(R) \subset \mathbb{R}^n$ with density bounded from below and from above by positive numbers. Let $\mathcal{K}(B(R))$ be all convex bodies in $B(R)$ with nonempty interior and assume we have a map $F : \mathcal{K}(B(R)) \to \mathbb{R}^k$ continuous in the Hausdorff metric, for some $1 \leq k < n$.*

Then for any power of two $N = 2^l$ it is possible to produce a binary decomposition of $B(R)$ into convex parts $P_1, \ldots, P_N$ so that:

1) $\mu(P_1) = \cdots = \mu(P_N)$;
2) $F(P_1) = \cdots = F(P_N)$;
3) Given $\delta > 0$, for sufficiently large N depending on μ and δ, all the parts P_i will be δ-close to k-dimensional affine subspaces.

The last property is called the *k-pancake property.* The measure we are going to plug into this theorem is the restriction of γ to a big ball $B(R)$. The map F will be composed of f given in the waist theorem and a certain selection of a "center" $c(P_i)$ of a part with non-empty interior, which is the crucial part of our argument, different from the selection in [9, 13] and other previous works. Since the complement of $B(R)$ has arbitrarily small Gaussian measure, this will result in an arbitrary small error term in the estimate, which is handled by a compactness argument for possible values of y, the common value of $f(c(P_i))$. We give more explanations on the pancake decomposition in Sect. 1.7, outlining a version of the proof of Theorem 1.2.1.

The selection of the centers $c(P_i)$ of P_i is the crucial part of the argument, we need to do it so that the ratio

$$\frac{\gamma(\nu_t(c(P_i)) \cap P_i)}{\gamma(P_i)} \tag{1.2.2}$$

is bounded from below by a certain constant so that the summation of such lower bounds produces the required total estimate.

Consider the restriction $\mu_i = \gamma|_{P_i}$, we are interested in the estimate of (1.2.2) from below that now assumes the form

$$\frac{\mu_i(\nu_t(c(P_i)))}{\mu_i(\mathbb{R}^n)}. \tag{1.2.3}$$

Let us make the monotone transportation T_i of (an appropriately scaled) γ to μ_i. The measure μ_i is *more log-concave* than γ, that is μ_i can be expressed as the product of γ and a log-concave function, the characteristic function of P_i in our case. In this situation the Caffarelli theorem [6] (see also [15] and explanations in Sect. 1.6) implies that the monotone transportation T_i is 1-Lipschitz. Let us define the center $c(P_i) = T_i(0)$. By the stability property of the monotone transportation (see, for example, [5, Proposition 3.2]) the centers depend continuously on P_i in the Hausdorff metric while P_i keeps nonempty interior; thus we can plug this selection of the center into the Borsuk–Ulam type theorem for the configuration space of such partition (more details are in [9] and [13]) to ensure that

$$f(c(P_1)) = \cdots = f(c(P_N)) = y_N.$$

In order to have the total estimate it remains to bound (1.2.2) (or (1.2.3)) for a pancake P_i which is δ-close to an affine subspace A_i. For normalization, we translate P_i and μ_i so that the center $c(P_i)$ becomes the origin. Denote the translated μ_i by κ_i. The monotone transportation of (an appropriately scaled) γ into κ_i is the composition of the original transportation T_i of γ to μ_i and the translation, this can be seen from the fact that if we add a linear function to the potential of the monotone transportation then the transportation gets composed with a translation, below we work with potentials in more details. 1-Lipschitz property is also retained under a translation. Eventually we consider a measure κ_i obtained from (an appropriately scaled) γ by a 1-Lipschitz monotone transportation T_i (denoted by the same letter as before the translation) taking origin to the origin, from the pancake property the support of κ_i is δ-close to a linear subspace A_i (we keep the same letter for the translated A_i), and we need to have a lower bound for

$$\frac{\kappa_i(\nu_t(0))}{\kappa_i(\mathbb{R}^n)}. \tag{1.2.4}$$

Put for brevity $A = A_i$ and $B = A^\perp$, now they are linear subspaces, also denote by π_A and π_B the corresponding orthogonal projections. The map T_i is a monotone transportation, by definition it means that it has a convex potential

$$U : \mathbb{R}^n \to \mathbb{R}, \quad T_i(x) = \nabla U(x),$$

which must have the minimum at the origin, since we assume $T_i(0) = 0$. The convexity of U means that the Hessian is positive semidefinite, and the 1-Lipschitz property of T_i means that the difference

$$\text{Hess } U - dx_1^2 - \cdots - dx_n^2$$

is negative semidefinite. Here we use the fact that Hess U exists almost everywhere.

Now we are going to approximate U with another potential V satisfying the same property of convexity and the same bound on the Hessian, producing a monotone 1-Lipschitz transportation $S : \mathbb{R}^n \to A$, showing that T_i is also approximated by $S = \nabla V$.

Choose $\varepsilon > 0$ and then choose a radius R so that

$$\gamma(\nu_R(0)) \geq (1 - \varepsilon)\gamma(\mathbb{R}^n).$$

Let $X = \nu_R(A)$, then we also have

$$\gamma(X) \geq (1 - \varepsilon)\gamma(\mathbb{R}^n).$$

The assumption that the image of T_i lies in the neighborhood $\nu_\delta(A)$ means that

$$|\pi_B(\nabla U(x))| \leq \delta. \tag{1.2.5}$$

If we put $V(x) = U(\pi_A(x))$ then V is also convex, differentiable, and the inequality

$$\text{Hess } V - dx_1^2 - \cdots - dx_n^2 \leq 0$$

holds almost everywhere (here we again use the fact about the eigenvalues of the restriction of a quadratic form). Its gradient $S = \nabla V$ is 1-Lipschitz, and by integration (1.2.5) from $\pi_A(x)$ to x we have

$$|V(x) - U(x)| \leq (R + \varepsilon)\delta$$

on the neighborhood $v_\varepsilon(X)$. By our choice of coordinates both U and V have the minimum at the origin.

It is easy to see, that the estimate $|V - U| < 1/2\varepsilon^2$ on $v_\varepsilon(X)$ implies

$$|\nabla V - \nabla U| < \varepsilon$$

on X. Indeed, it is sufficient to prove this inequality in one-dimensional case, choosing a point x_0 where the inequality fails and the one-dimensional direction e along which

$$|(\nabla V - \nabla U) \cdot e| = \left| \frac{\partial V}{\partial e} - \frac{\partial U}{\partial e} \right| \geq \varepsilon.$$

In the one-dimensional case, we have without loss of generality

$$U(x_0) - V(x_0) \geq 0, \ U'(x_0) - V'(x_0) \geq \varepsilon.$$

For the second derivatives, we know that $0 \leq U''(x), V''(x) \leq 1$ almost everywhere, and therefore for the difference of functions and $x > x_0$ there holds the estimate:

$$U(x) - V(x) \geq \varepsilon(x - x_0) - \frac{1}{2}(x - x_0)^2.$$

Putting $x = x_0 + \varepsilon$ we have $U(x) - V(x) \geq 1/2\varepsilon^2$, a contradiction.

Therefore for any positive pancakeness parameter

$$\delta < \frac{\varepsilon^2}{4(R + \varepsilon)} \tag{1.2.6}$$

we have the estimate

$$|T_i(x) - S(x)| \leq \varepsilon$$

on the set X. The map S is 1-Lipschitz, it takes $v_t(B)$ to $v_t(0) \cap A$ and therefore T_i takes $v_{t-\varepsilon}(B) \cap X$ to $v_t(0)$. In view of (1.2.1) we establish

$$\frac{\gamma(v_t(c(P_i)) \cap P_i)}{\gamma(P_i)} = \frac{\kappa_i(v_t(0))}{\kappa_i(\mathbb{R}^n)} \geq \frac{\gamma(v_{t-\varepsilon}(B))}{\gamma(\mathbb{R}^n)} - \varepsilon \tag{1.2.7}$$

$$\geq \frac{\int_{v_{t-\varepsilon}(0)} e^{-a_{n-k+1}x_{n-k+1}^2 - \cdots - a_n x_n^2} \, dx_{n-k+1} \ldots dx_n}{\int_{\mathbb{R}^k} e^{-a_{n-k+1}x_{n-k+1}^2 - \cdots - a_n x_n^2} \, dx_{n-k+1} \ldots dx_n} - \varepsilon.$$

The rest of the proof is done like the argument in [9] or [13]. Now we choose a sequence $\varepsilon_N \to +0$, choose appropriate R_N and $\delta_N \to +0$ depending on ε_N

and satisfying (1.2.6). The Borsuk–Ulam argument ensures that the centers of the pancakes on Nth stage go to a single point y_N under the map f. For such y_N, the summation over the pancakes gives the bound

$$
\frac{\gamma\left(\nu_t(f^{-1}(y_N))\right)}{\gamma(\mathbb{R}^n)} \geq \frac{\int_{\nu_{t-\varepsilon_n}(0)} e^{-a_{n-k+1}x_{n-k+1}^2-\cdots-a_n x_n^2}\, dx_{n-k+1}\ldots dx_n}{\int_{\mathbb{R}^k} e^{-a_{n-k+1}x_{n-k+1}^2-\cdots-a_n x_n^2}\, dx_{n-k+1}\ldots dx_n} - \varepsilon_N.
$$

We may also assume that $y_N \to y$ from compactness considerations and for such y the above bound becomes precisely the required bound in the limit.

Remark 1.2 The given proof has a serious simplification compared to the original argument, achieved by using the Caffarelli theorem. It would be nice to have a version of the Caffarelli theorem on the sphere that would allow to simplify the proof of Gromov's waist of the sphere theorem (see e.g. [16]). At the moment we are not completely sure about the precise statement in the spherical case. To start with, we need to handle the following problem: Given a spherical convex set $C \subset \mathbb{S}^n$ and a hemisphere $H \subset \mathbb{S}^n$, find a 1-Lipschitz map sending the uniform measure on H to a multiple of the uniform measure on C. After that there remains the question of continuous dependence of the map on H and C.

Another possibly useful version would be to find a 1-Lipschitz map $f : \mathbb{S}^n \to C$ mapping the uniform measure of the sphere to a multiple of the uniform measure on C. The candidate for a centerpoint of C will be a point $y \in C$ whose preimage $f^{-1}(y)$ contains two opposite points of the sphere, like in the Borsuk–Ulam theorem. But with such a statement the continuous dependence of the centerpoint on C will also be problematic.

1.3 Existence of Waist Theorems for Some Radial Measures

Possibly the easiest case to consider is when μ is uniform on the unit sphere in $\mathbb{R}^n$. There is Gromov's waist of the sphere theorem [9, 16], but our question is not the same, because unlike that theorem the function is defined inside and outside the sphere in the Euclidean space and the distance we use is the Euclidean distance. To put it short, we have the following non-trivial observation:

Theorem 1.3.1 *A precise t-neighborhood waist theorem for μ distributed uniformly on the unit sphere and maps $f : \mathbb{R}^n \to \mathbb{R}^k$ is only possible for $n \leq 2$ or $n \leq k$.*

Proof The case $n \leq k$ is trivial, the origin is the fiber we need. Consider the case $n = 2$ and $k = 1$, that is we have a continuous function $f : \mathbb{R}^2 \to \mathbb{R}$.

Let us find the minimal T such that there exists a fiber $f^{-1}(y)$ whose T-neighborhood covers the whole circle $\mathbb{S}^1$ on which μ is supported, we evidently have $T \leq 1$ be considering the fiber passing through the origin.

For any $p \in \mathbb{S}^1$ consider the minimum and the maximum of f in the disk $B_p(T)$, they are continuous functions $m(p, T)$ and $M(p, T)$ of p and T. By the choice of y and T we have

$$\forall p \in \mathbb{S}^1, \quad m(p, T) \leq y \leq M(p, T).$$

The assumption of the minimality of T shows that both equalities must become equalities

$$m(p, T) = y = M(q, T),$$

for some $p, q \in \mathbb{S}^1$, since otherwise the intersection of all segments $[m(p, T), M(p, T)]$ consisted of a nonzero segment (because we are dealing with continuous functions on a compactum) and we could decrease T keeping this intersection non-empty by the uniform continuity of m and M.

Now we have two disks $B_p(T)$ and $B_q(T)$ with the property that $f \geq y$ on the former and $f \leq y$ on the latter. For any $0 < t < T$ the circle $C = \mathbb{S}^1(\sqrt{1 - t^2})$ intersects both of them.

Then by the intermediate value theorem there exist at least two points $a, b \in C$ such that $f(a) = f(b) = y$ and such that a and b cannot be simultaneously covered by the interior of a $B_{p'}(T)$ for $p' \in \mathbb{S}^1$, see Figs. 1.1 and 1.2. Therefore the arcs $\mathbb{S}^1 \cap B_a(t)$ and $\mathbb{S}^1 \cap B_b(t)$ are disjoint and they together show that $f^{-1}(y) + t$ intersects $\mathbb{S}^1$ by a sufficiently big set.

Now we are going to describe the counterexamples for $n \geq 3$ by a modification of [2, Remark 5.7]. We will have two essentially different cases.

Case $k = 1$: We choose $p \in \mathbb{S}^{n-1}$ and start from the function $g(x) = |x - p|$. This function has the t-neighborhood property for $t = 1$ satisfied on the level $G = \{g = 1\}$ only. Now we modify g near the intersection $G \cap \mathbb{S}^{n-1}$ so that the new fiber $F = \{f = 1\}$ becomes orthogonal to $\mathbb{S}^{n-1}$ in their intersection and F still remains the only fiber satisfying the 1-neighborhood property, the latter is guaranteed by

Fig. 1.1 The case of one-dimensional sphere

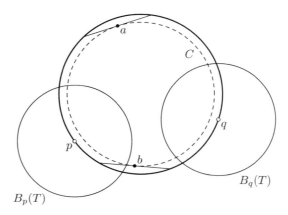

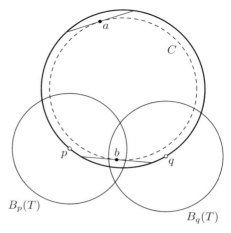

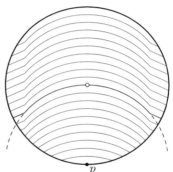

choosing $f \leq g$. In fact, this can be made in two-dimensions and extended to higher dimensions by rotation invariance about the $0p$ axis, see Fig. 1.3.

Now we test small $t > 0$ against the constructed f and its fiber F. The orthogonality condition provides that for $n = 2$ the asymptotics of $(F + t) \cap \mathbb{S}^1$ is correct. For $n \geq 3$ the intersection $I = F \cap \mathbb{S}^{n-1}$ is an $(n - 2)$-dimensional sphere of radius strictly less than 1 and from the orthogonality it follows that the asymptotics in $t \to +0$ of the surface area of $(F + t) \cap \mathbb{S}^{n-1}$ is the same as of $(I + t) \cap \mathbb{S}^{n-1}$, which is insufficient to match $(\mathbb{R}^{n-1} + t) \cap \mathbb{S}^{n-1}$, just because I has smaller $(n - 2)$ volume than $\mathbb{R}^{n-1} \cap \mathbb{S}^{n-1}$.

Case $2 \leq k < n$: Let us build a counterexample as follows. The components $f_1, \ldots, f_{k-1}$ will be just the linear coordinates. Considering the value $t = 1$ we readily see that the fiber presumably satisfying the t-neighborhood waist theorem must lie in

$$\mathbb{R}^{n-k+1} = \{f_1 = \cdots = f_{k-1} = 0\},$$

otherwise the t-neighborhood would not reach some of the points with $f_i = \pm 1$.

Fig. 1.4 Level lines for one
of the map coordinates

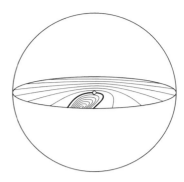

Moreover, for any definition of the last coordinate $f_k : \mathbb{R}^n \to \mathbb{R}^k$ the fiber $F = f^{-1}(y)$ we can presumably take must pass through the origin. Indeed, it already has to lie in $\mathbb{R}^{n-k+1}$ and in order to reach a point p with $f_1 = \pm 1$ with $t = 1$ it must pass through the origin since the origin is the only point of $\mathbb{R}^{n-k+1}$ at distance at most 1 from p.

Now we pass to $\mathbb{R}^{n-k+1}$ noting that $n - k + 1 \geq 2$. It remains to build a function $f_k : \mathbb{R}^{n-k+1} \to \mathbb{R}$ whose fiber passing through the origin is very close to a radius segment $[0, p]$, $|p| = 1$, see Fig. 1.4. Then the fiber F that we have to choose assuming the t-neighborhood waist theorem will be close to a half of $\mathbb{R}^{n-k}$ instead of the full $\mathbb{R}^{n-k}$ producing losses in the intersection $(F + t) \cap \mathbb{S}^{n-1}$ for $0 < t < 1$.

$\square$

In fact the last counterexample extends to arbitrary radially symmetric and compactly supported measure.

Theorem 1.3.2 *Let μ be a radially symmetric compactly supported measure in $\mathbb{R}^n$ different from the delta-measure at the origin. Then there is no t-neighborhood waist theorem for μ and maps $f : \mathbb{R}^n \to \mathbb{R}^k$ when $2 \leq k < n$.*

Proof By scaling we assume that the support of μ is precisely the unit ball. Then the last construction in the previous proof works, considering 1-neighborhoods we need to choose the fiber passing through the origin and arbitrarily close to a half of $\mathbb{R}^{n-k}$. But this fiber does not work for intermediary $0 < t < 1$.

$\square$

So far we only have positive examples in the plane, but it is also possible to make a counterexample.

Example 1.3.3 Take μ so that half of it is the delta-measure at the origin and the other half is uniformly distributed on the unit sphere. Let the function $f : \mathbb{R}^2 \to \mathbb{R}$ be just $f(x) = |x|$. Assuming the t-neighborhood waist property for this measure and considering $t \to +0$ we see that y has only two possibilities, $y = 0$ and $y = 1$.

In both cases considering $t \to 1-0$ exhibits a discontinuous jump of $\mu(f^{-1}(y) + t)$ to its maximal value, which contradicts the inequality

$$\mu(f^{-1}(y) + t) \geq \mu(\mathbb{R} + t),$$

since the right hand part has no jump.

There remain low-dimensional questions, which we still cannot answer:

Question 1.3.4 Does the uniform measure on the two-dimensional Euclidean disk have the t-neighborhood waist property? The same question for the uniform measure on the three-dimensional ball.

Note that the counterexample in [2, Remark 5.7] starts working from dimension 4.

1.4 Waist for Radial Measures and Odd Maps

The waist theorems estimate the measure of $f^{-1}(y) + t$, but it is not clear which choice of y is good in every particular situation. There is an easy case when we can choose a particular y, the case of odd maps. In [3] it was noted that for a continuous odd map $f : \mathbb{S}^n \to \mathbb{R}^k$ the fiber $f^{-1}(0)$ intersects every k-dimensional subsphere $S \subset \mathbb{S}^n$ at least twice; this is sufficient to invoke Crofton's formula and conclude that the $(n - k)$-dimensional volume (in a certain sense) of the fiber is sufficiently large. But in order to estimate the n-dimensional volume of the neighborhood $f^{-1}(0) + t$ we need a trickier argument:

Theorem 1.4.1 *Let $f : \mathbb{S}^n \to \mathbb{R}^k$ be an odd continuous map and let $t > 0$. Then*

$$\mathrm{vol}(f^{-1}(0) + t) \geq \mathrm{vol}(\mathbb{S}^{n-k} + t).$$

This theorem is an improved version of a particular case of [17, Theorem 5.1.2], here we provide a simpler proof of it.

Proof The proof of Gromov and Memarian [9, 16] mostly works in this case; the thing that needs an adjustment is the Borsuk–Ulam-type theorem for partitions of the sphere with a binary tree hierarchy of cuts.

If we want a partition into $N = 2^\ell$ parts $P_1, \ldots, P_N$, use hyperplanes with the choice of the normal from an m-dimensional sphere ($m = k + 1$) each time, then the configuration space $\mathcal{M}$ of the binary partitions with the choice of the normals has dimension $m(N - 1)$. This is sufficient to satisfy the constraints

$$f(c(P_1)) = \cdots = f(c(P_N)), \quad \mathrm{vol}\, P_1 = \cdots = \mathrm{vol}\, P_N.$$

If we count the constraints then we see that we have precisely $m(N - 1)$ constraints. The free group action on $\mathcal{M}$ in this Borsuk–Ulam-type result is the group of symmetries of the graded binary tree, which is $\Sigma_N^{(2)}$, the 2-Sylow subgroup of the permutation group.

Now, for centrally symmetric partitions, we have a choice of the normal at the root of the tree, and after this we only have to choose the normals at one of the two subtrees below the root. Thus the configuration space $\mathcal{M}$ gets reduced to the product of $N/2$ m-dimensional spheres, giving a configuration space $\mathcal{M}_0$ of dimension $mN/2$. Some of the constraints are also satisfied automatically, we only need to solve the following equations:

$$f(c(P_1)) = \cdots = f(c(P_{N/2})) = 0,$$
$$\operatorname{vol} P_1 - \operatorname{vol} \mathbb{S}^n/N = \cdots = \operatorname{vol} P_{N/2} - \operatorname{vol} \mathbb{S}^n/N = 0.$$

The number of constraints written in this way is also $mN/2$. In fact, we have written some redundant volume constraints, but writing the constraints this way shows that we can interpret them as m sets of $N/2$ equations with the same action of the symmetry group inside each set. The flip at the root, for example, will change the sign of all the constraints. The symmetry group this time is $G = \mathbb{Z}/2 \times \Sigma_{N/2}$, the first factor flips the root, the second is responsible of its subtrees.

The Borsuk–Ulam-type statement for the modified problem is established by the *test map scheme*, see also [13]. We consider a linear projection $\pi : \mathbb{S}^m \to \mathbb{R}^m$ and replace, for each i, the pair

$$\big(f(c(P_i)), \operatorname{vol} P_i - \operatorname{vol} \mathbb{S}^n/n\big)$$

with the projection $y_i = \pi(v_i^0 + s v_i^1 + \cdots + s^\ell v_i^\ell)$, where s is a small number and v_i^j is the normal that participates in the building of the part P_i on level j of the tree. It is easy to see that such a collection of $(y_1, \ldots, y_N) \in (\mathbb{R}^m)^N$ keeps the symmetries of $(f(c(P_i)), \operatorname{vol} P_i - \operatorname{vol} \mathbb{S}^n/n)_{i=1}^N$ under the action of $\Sigma_N^{(2)}$ and its subgroup G.

It is only possible to satisfy the constrains $y_1 = \cdots = y_{N/2} = 0$ over $\mathcal{M}_0$ by choosing v_i^j from the pairs of points with $\pi(v_i^j) = 0$. Similar to the original problem, this describes a unique non-degenerate orbit of zeroes of the test G-equivariant map $\mathcal{M}_0 \to \mathbb{R}^{nN/2}$. From the parity considerations, for any other G-equivariant map $\mathcal{M}_0 \to \mathbb{R}^{mN/2}$ maps something to zero. This establishes the needed Borsuk–Ulam-type result for mapping the centers of the pancakes into zero. The rest of the proof proceeds as in [9, 16]. $\qquad\square$

It turns out that for radially symmetric measures in $\mathbb{R}^n$ and odd maps we have a very general result:

Theorem 1.4.2 *Let μ be a radially symmetric measure in $\mathbb{R}^n$ and $f : \mathbb{R}^n \to \mathbb{R}^k$ be an odd continuous map. Then*

$$\mu(f^{-1}(0) + t) \geq \mu(\mathbb{R}^{n-k} + t).$$

Proof Put $Z = f^{-1}(0)$, it suffices to consider measures with a radially symmetric density ρ, since the general case can be obtained by weak approximation of any measure by measures with density.

Consider the intersection of $Z + t$ and a sphere $\mathbb{S}_r^{n-1}$ of radius r. We will estimate this slice from below by the t-neighborhood (in the Euclidean metric) of $Z \cap \mathbb{S}_s^{n-1}$

with $s = \sqrt{r^2 - t^2}$. In terms of the spherical geometry, we consider $Z \cap \mathbb{S}_s^{n-1}$ in $\mathbb{S}_s^{n-1}$, consider its α-neighborhood in spherical geometry with $\alpha = \arcsin t/r$ and then inflate this neighborhood r/s times to put it onto $\mathbb{S}_r^{n-1}$, this will be precisely the t-neighborhood of $Z \cap \mathbb{S}_s^{n-1}$ in the Euclidean metric intersected with $\mathbb{S}_r^{n-1}$.

From Theorem 1.4.1 we have an estimate for the α-neighborhood of $Z \cap \mathbb{S}_s^{n-1}$. Then we multiply it by r/s, multiply by $\rho(r)$, and integrate over r to have an estimate from below for $\int_{Z+t} \rho(x)\, dx$. There is no need to write down the explicit formulas since for the case $Z = \mathbb{R}^{n-k}$ we always have a strict equality in all steps of this estimate, which we just put in the right hand side of the total estimate. □

Another result was established in [2, Theorem 5.2] using Gromov's version of the Borsuk–Ulam theorem for the images of the $(1 - t)$-scaled centers of the pancakes: Suppose $K \subset \mathbb{R}^n$ is a convex body, μ is a finite log-concave measure supported in K, and $f : K \to Y$ is a continuous map to a $(n - k)$-manifold Y. Then for any $t \in [0, 1]$ there exists $y \in Y$ such that $\mu(f^{-1}(y) + tK) \geq t^{n-k}\mu K$.

Using the modifications in the Borsuk–Ulam-type argument from the proof of Theorem 1.4.1, we readily obtain its version for centrally symmetric body and measure and an odd map (compare also with [13, Theorem 5.7]):

Theorem 1.4.3 *Suppose $K \subset \mathbb{R}^n$ is a centrally symmetric convex body, μ is a finite centrally-symmetric log-concave measure supported in K, and $f : K \to \mathbb{R}^{n-k}$ is an odd continuous map. Then for any $t \in [0, 1]$*

$$\mu(f^{-1}(0) + tK) \geq t^{n-k}\mu K.$$

1.5   Neighborhoods of Critical Submanifolds

Let us develop the ideas of Sect. 1.4 about t-neighborhoods of given submanifolds. Bo'az Klartag in [14] established another result of this kind, where a neighborhood volume estimate

$$\gamma(f^{-1}(0) + t) \geq \gamma(\mathbb{C}^{n-k} + t)$$

for a holomorphic map $f : \mathbb{C}^n \to \mathbb{C}^k$ with $f(0) = 0$. We try to understand such phenomena in view of the fact that the variety in that case $f^{-1}(0)$ is a critical point of the volume functional under local perturbations.

1.5.1   Neighborhoods of Submanifolds in the Euclidean Space

Consider a k-dimensional smooth submanifold $X \subset \mathbb{R}^n$; note that we have interchanged k and $n - k$ compared to the statement of the waist theorem, but this will be convenient in our argument. Let X be *properly embedded*, which effectively means that X is a closed submanifold without boundary. Let $P_X : \mathbb{R}^n \to X$ be the metric projection, and let μ be a measure in $\mathbb{R}^n$. We assume μ has a log-concave smooth density ρ.

The map P_X is defined almost everywhere and is not necessarily continuous. Still, it induces a fiber-wise decomposition of μ

$$\int_{\mathbb{R}^n} f \, d\mu = \int_X \left(\int_{P_X^{-1}(x)} f \, d\mu_x \right) d \operatorname{vol}_k(x) \qquad (1.5.1)$$

where μ_x is supported in $P_X^{-1}(x)$ for every $x \in X$, so its support has dimension $n - k$.

First let us note that every μ_x is log-concave. Indeed, the decomposition (1.5.1) can be approximated by the following finite decompositions. Take a sufficiently dense discrete subset $S \subset X$ and build its Voronoi diagram, assigning to every $s \in S$ the convex set

$$V_s = \{x \in \mathbb{R}^n : \forall s' \in X \ |x - s| \leq |x - s'|\}.$$

When taking the limit over more and more dense S the measures $\mu|_{V_s}$ approach μ_x if $s \to x$. This is quite informal, but can be made precise similar to the argument with passing to infinite partitions in [9, 16]. The Voronoi diagram of S is built by choosing the closest point in the set S to a given point x'; in the limit we obtain the presentation of the whole measure μ as its disintegration into μ_x, that is we want to integrate a function over μ, we first integrate it over every μ_x and then integrate over $x \in X$.

What is seen from the Voronoi diagram picture, is that since every restriction $\mu|_{V_s}$ is log-concave, the limit measures μ_x are also log-concave. If μ is more log-concave than a Gaussian density $e^{-A|x|^2}$ then μ_x is also more log-concave than the same density, let us call this situation *strongly log-concave*.

Now, if we want to estimate $\mu(X + t)$, where $X + t$ is the t-neighborhood of X, following [9, 16], we have to estimate $\mu_x B_x(t)$ for every t, if we know that

$$\mu_x B_x(t) \geq C_t \mu_x \mathbb{R}^n \qquad (1.5.2)$$

from some strong log-concavity assumption, then we integrate to obtain

$$\mu(X + t) \geq C_t \mu \mathbb{R}^n.$$

In order to have a working estimate (1.5.2), it is preferable to have the situation where x is a point of maximum density of μ_x, apart from the strong log-concavity of μ. Here we consider the density ρ_x of μ_x in its $(n - k)$-dimensional convex support, which coincides with the normal $(n - k)$-dimensional subspace to X at x

locally. More precisely, in a tubular neighborhood of X the density ρ_x is smooth as a function of $y \in P_X^{-1}(x)$ and is also smooth as a function of x.

Another way to consider ρ_x is to say that this is the Jacobian of the exponential map from the normal bundle of X to $\mathbb{R}^n$, which establishes a diffeomorphism of an open neighborhood U_X of X in its normal bundle with almost all of $\mathbb{R}^n$, the rest of $\mathbb{R}^n$ is called the *cut locus*. This is seen from the representation of μ_x as the disintegration of μ under the metric projection onto X. Therefore the value ρ_x may be considered as the density of the pull-back of the volume in $\mathbb{R}^n$ to U_X.

The condition that ρ_x has a maximum of density at x in the direction orthogonal to X is purely local, in view of its log-concavity, and must have a local expression. In case of the constant density of μ this is definitely related to the trace of the second fundamental form of $X \subset \mathbb{R}^n$. Arguing geometrically, when μ has constant density near x, try to deform X along a vector field v supported in a neighborhood of x in X and orthogonal to X, we may note that the derivative of ρ_x at points $x' \in \operatorname{supp} v$ in the direction of v, averaged over x', vanishes if and only if the first variation of $\operatorname{vol}_k X$ in the direction v vanishes.

It seems plausible that for the case of not necessarily uniform μ with density ρ, the condition on the central point of μ_x being x seems to mean vanishing of the first variation of the ρ-weighted k-dimensional Riemannian volume of X. Then a sufficient property of X in order to have a good estimate for $\mu(X + t)$ is that X is a critical point of the ρ-weighted Riemannian k-volume, say ρ-*critical* for short. Let us state what the above argument proofs:

Proposition 1.5.1 *Let a k-dimensional properly embedded submanifold $X \subset \mathbb{R}^n$ be ρ-critical for a Gaussian density γ centered at the origin with density ρ. Then for all $t > 0$*

$$\gamma(X + t) \geq \gamma(\mathbb{R}^k + t).$$

Returning to Klartag's theorem [14] we observe that a zero set of holomorphic functions X is volume-critical (see Sect. 1.5.2) but it is not necessarily ρ-critical for the Gaussian density. Therefore our observation is insufficient to reprove Klartag's theorem, but we can prove something useful in the sphere $\mathbb{S}^n$ with its intrinsic Riemannian structure and the Riemannian volume.

1.5.2 Neighborhoods in the Sphere and the Complex Projective Space

A similar to the above argument works in the sphere $\mathbb{S}^n$ with its uniform Riemannian volume as μ. In this case instead of log-concavity we need another notion, expressing the fact that the pancake measure μ_x can be approximated by the uniform measure restricted to convex subsets of $\mathbb{S}^n$, that is the notion of $\sin^k$-concave measures, introduced in [16]. From the results of [16] we only have to know that once a measure μ_x is $(n-k)$-dimensionally supported $\sin^k$-concave and is centered at x in terms of the maximum density, there is an estimate

$$\mu_x(B_x(t)) \geq c_{n,k}(t)\mu_x\mathbb{S}^n,$$

where $c_{n,k} = \mu(\mathbb{S}^k + t)/\mu\mathbb{S}^n$ for a standardly embedded $\mathbb{S}^k \subset \mathbb{S}^n$.

For a volume-critical k-dimensional submanifold $X \subset \mathbb{S}^n$ and its metric projection $P_X : \mathbb{S}^n \to X$ we again have a decomposition of the spherical Riemannian volume

$$\int_{\mathbb{S}^n} f \, d\,\mathrm{vol} = \int_X \left(\int_{P_X^{-1}(x)} f d\mu_x \right) d\,\mathrm{vol}_k(x), \tag{1.5.3}$$

with $\sin^k$-concave measures μ_x (as shown by approximating this decomposition with a Voronoi partition). The "center at x" assumption is satisfied for μ_x by the variational argument, since otherwise the perpendicular to X derivative of the density of μ_x would be non-zero and it would be possible to variate X with a linear order change of its volume. Hence we have the estimate and integrating the estimate we obtain:

Proposition 1.5.2 *A volume-critical smooth properly embedded submanifold $X \subset \mathbb{S}^n$ of dimension k satisfies, for $t > 0$,*

$$\mathrm{vol}(X + t) \geq \mathrm{vol}(\mathbb{S}^k + t),$$

where $\mathbb{S}^k \subset \mathbb{S}^n$ is standardly embedded.

We can also push forward the spherical observation to the complex projective space:

Theorem 1.5.3 *Let $X \subset \mathbb{C}P^n$ be an algebraic submanifold with $\dim_{\mathbb{C}} X = k$. If $\mathbb{C}P^n$ is considered with its standard Fubini–Study metric then, for $t > 0$,*

$$\mathrm{vol}(X + t) \geq \mathrm{vol}(\mathbb{C}P^k + t),$$

where $\mathbb{C}P^k \subset \mathbb{C}P^n$ is standardly embedded.

Proof Note that [10, Section I] (also noted previously in [7, 18], [8, §4], [4]) that X is critical in $\mathbb{C}P^n$ because for any $2k$-dimensional $X' \subset \mathbb{C}P^n$ we have the *calibration inequality*

$$\mathrm{vol}_{2k} X' \geq \int_{X'} \frac{\omega^k}{k!},$$

where ω is the Fubini–Study symplectic form of $\mathbb{C}P^n$. And at the same time for a complex subspace X we have

$$\mathrm{vol}_{2k} X = \int_X \frac{\omega^k}{k!}.$$

Since the form ω^k is closed, the right hand side of the estimates does not change under the deformations of X to X' and then X is not only volume-critical, but also volume-minimal.

Now consider the Hopf map $H : \mathbb{S}^{2n+1} \to \mathbb{C}P^n$ and $Y = H^{-1}(X)$. From [11, Theorem 2] we know that Y is also volume-critical.

From Proposition 1.5.2 we have a lower bound for the volume of the t-neighborhood of Y, with equality holding for the case $X = \mathbb{C}P^k$. The Hopf map is a quotient map of Riemannian manifolds with all fibers of length 2π, hence for the t-neighborhood of X we will have the same estimate divided by 2π. And again, the resulting estimate must be attained for $X = \mathbb{C}P^k$, thus completing the proof. □

The above observations allow to establish a particular case of the result of [14] (for homogeneous maps, generalized to non-Gaussian radial measures):

Theorem 1.5.4 *Let Z be the zero set of a homogeneous holomorphic map $f : \mathbb{C}^n \to \mathbb{C}^k$; let μ be a radially symmetric measure on $\mathbb{C}^n$. Then for any $t > 0$*

$$\mu(Z + t) \geq \mu(\mathbb{C}^{n-k} + t).$$

Proof By generically perturbing the coordinates of the map f we may assume that every intersection of Z with a sphere $\mathbb{S}_s^{2n-1}$ of radius centered at the origin is a smooth submanifold of the sphere, all such intersections for different s are similar to each other from homogeneity. By the argument from the proof of Theorem 1.5.3 every such intersection $Z \cap \mathbb{S}_s^{2n-1}$ is volume-critical and therefore has an appropriate lower bound on the volume of its t-neighborhood for every t. The rest of proof is the same as the proof of Theorem 1.4.2. □

The following opposite estimate (an upper bound on the volume of the neighborhood) for algebraic manifolds was prompted to us by Alexander Esterov in private communication:

Theorem 1.5.5 *Let $X \subset \mathbb{C}P^n$ be an algebraic submanifold with $\dim_{\mathbb{C}} X = k$ and degree d. If $\mathbb{C}P^n$ is considered with its standard Fubini–Study metric then, for $t > 0$,*

$$\mathrm{vol}(X + t) \leq d \, \mathrm{vol}(\mathbb{C}P^k + t),$$

where $\mathbb{C}P^k \subset \mathbb{C}P^n$ is standardly embedded.

Proof We again lift everything to $\mathbb{S}^{2n+1}$ and make estimates in every pancake. Note that in every pancake the measure μ_x is $\sin^{2k+1}$-concave and has maximum density ρ_x in the point $x \in X'$. Its $\sin^{2k+1}$-concavity property means that for the ratio

$\frac{\mu_x B_x(t)}{\rho_x(x)}$ is bounded from above by the similar ratio for the test case $X = \mathbb{C}P^k$, $X' = \mathbb{S}^{2k+1}$, see [16, Lemma 4.6] for example.

The integral of $\rho_x(x)$ over X' is just the $(2k+1)$-volume of X', this follows from the Minkowski volume formula. So integrating the estimate

$$\mu_x B_x(t) \le \rho_x(t) \frac{\text{vol}(\mathbb{S}^{2k+1} + t)}{\text{vol}_{2k+1} \mathbb{S}^{2k+1}}$$

we obtain

$$\text{vol}(X' + t) \le \frac{\text{vol}_{2k+1} X' \cdot \text{vol}(\mathbb{S}^{2k+1} + t)}{\text{vol}_{2k+1} \mathbb{S}^{2k+1}}.$$

The volume $\text{vol}_{2k+1} X' = 2\pi \text{vol}_{2k} X$, by Crofton's formula or the calibration equality, is the volume of $\mathbb{S}^{2k+1}$ multiplied by the degree of X, so we have eventually

$$\text{vol}(X' + t) \le d \, \text{vol}(\mathbb{S}^{2k+1} + t),$$

which is equivalent to the required estimate. $\qquad\square$

Remark 1.5.6 As was communicated to us by Sergei Ivanov, this result can be obtained by analyzing the Riccati equation for the second fundamental form of the hypersurface $H_t = \partial(X' + t)$,

$$\text{II}'_t = -(\text{II})^2 - R,$$

with the quadratic form R obtained by plugging the normal vector of H_t into the Riemann curvature form in the last and the first position. This equation shows that the trace of II in case of X' decreases quicker than in case of $\mathbb{S}^{2k+1}$, and so does the logarithm of the volume of the neighborhood, while for small t both volumes have the same asymptotic behavior. This works for smooth H_t, but possible non-smoothness of the boundary after some t may only improve the estimate.

Question 1.5.7 Is it true that, for any closed n-dimensional Riemannian manifold M with sectional curvature bounded from below by 1, any volume-critical smooth closed k-dimensional $X \subset M$, and any $t > 0$, we have

$$\frac{\text{vol}(X + t)}{\text{vol} \, M} \ge \frac{\text{vol}(\mathbb{S}^k + t)}{\text{vol} \, \mathbb{S}^n}?$$

The pancake approach does not work in this case, but it seems plausible that the answer follows from the investigation of the evolution of the volume and the second fundamental form of the boundary of $(X + t)$ when t varies from 0 to a certain value.

There is another question about real projective spaces:

Question 1.5.8 Does this result generalize to the estimate

$$\text{vol}(X + t) \geq \text{vol}(\mathbb{R}P^k + t)$$

for k-dimensional submanifolds $X \subset \mathbb{R}P^n$ homologous to $\mathbb{R}P^k \subset \mathbb{R}P^n$?

1.6 Appendix: Explanation of the Caffarelli Theorem

For the reader's convenience we give a more detailed argument from [15] explaining the Caffarelli theorem.

Theorem 1.6.1 (Caffarelli) *Assume a measure μ_Q has a smooth density e^{-Q} on the whole $\mathbb{R}^n$, while another measure μ_P has a smooth density e^{-P} on an open convex set, we assume P convex. Assume also that for every $x \in \mathbb{R}^n$, y in the domain of μ_P, and a nonzero vector v we have*

$$D_v^2 P(y) - D_v^2 Q(x) \geq 0,$$

where D_v^2 denotes the second derivative in the direction of v and also assume that $D_v^2 P(y)$ is positive definite at every y as function in v. Then the monotone transportation taking μ_Q to μ_P is 1-Lipschitz.

Proof Under the smoothness and positivity assumptions, the potential $U : \mathbb{R}^n \to \mathbb{R}$ of the transportation map from μ_Q to μ_P satisfies the equation:

$$\det D^2 U(x) = e^{P(DU(x))-Q(x)},$$

where DU is the derivative of U at a point x that gives the transportation map $x \mapsto DU(x)$, and $D^2U(x)$ is the Hessian quadratic form at a point x. If we are interested in the first and the second derivatives in a given direction v we write $D_v U$ or $D_v^2 U$.

Taking the logarithm we obtain

$$\ln \det D^2 U(x) = P(DU(x)) - Q(x).$$

Applying additional translations (that do not change the assumptions of the theorem) assume we consider the situation at the origin and the image of the origin under the transportation is again the origin, thus $DU(0) = 0$. In this case the second order of the right hand side in v near the origin is the quadratic form

$$D_{\Delta_0(v)}^2 P(0) - D_v^2 Q(0),$$

where $D^2 P$ and $D^2 Q$ are regarded as quadratic forms, whose values are taken at the vectors $\Delta_0(v)$ and v respectively, where Δ_0 is the derivative of the transportation map, that is $D^2 U(0)$ regarded as a linear operator. If we take v to be an eigenvector of Δ_0 and assume its eigenvalue λ is greater than 1, then we see that the value of this quadratic form on v becomes

$$\lambda^2 D_v^2 P(0) - D_v^2 Q(0) = (\lambda^2 - 1) D_v^2 P(0) + D_v^2 P(0) - D_v^2 Q(0),$$

which is positive by the assumption of the theorem. We are going to show that there is a contradiction, thus showing that $D^2 U$ cannot have eigenvalues greater than 1 and thus the transportation map is 1-Lipschitz.

In order to have a contradiction we need to show that the quadratic term of $\ln \det D^2 U(v)$ in its expansion in $v \to 0$ is non-positive, this will be established under a certain choice of a point x (the one we translate to origin) and a vector v in which we expand this quantity. Let again $\Delta_0 = D^2 U(0)$, viewed as a symmetric positive definite matrix and let us check how the expression $D^2 U(x)$ changes when we change x, putting $x = tv$ with some fixed nonzero vector v and variable $t \in \mathbb{R}$. We are interested in the first and the second order, so we assume

$$D^2 U(tv) = \Delta_0 + \Delta_1 t + \Delta_2 t^2 + o(t^2),$$

where we consider

$$\Delta_1 = D^2 D_v U(0), \quad \Delta_2 = \frac{1}{2} D^2 D_v^2 U(0)$$

as symmetric matrices, and Δ_0 can be assumed positive definite in the considered smooth transportation case. In order to simplify the formulas put $A = \Delta_0^{-1/2} \Delta_1 \Delta_0^{-1/2}$, $B = \Delta_0^{-1/2} \Delta_2 \Delta_0^{-1/2}$ and write

$$\det(\Delta_0 + \Delta_1 t + \Delta_2 t^2 + o(t^2)) = \det \Delta_0 \cdot \det(I + At + Bt^2 + o(t^2)),$$

where I is the unit matrix. A simple calculation produces the expansion of the logarithm of this matrix expression

$$\ln \det(\Delta_0 + \Delta_1 t + \Delta_2 t^2 + o(t^2)) = \ln \det \Delta_0 + \operatorname{tr} At$$
$$+ \left(\operatorname{tr} B + \operatorname{tr} \wedge^2 A - 1/2 (\operatorname{tr} A)^2 \right) t^2 + o(t^2),$$

where $\operatorname{tr} \wedge^2 A$ is the second invariant polynomial of A. If A is diagonalized with eigenvalues $s_1, \ldots, s_n$ then we have

$$\operatorname{tr} \wedge^2 A - 1/2 (\operatorname{tr} A)^2 = \sum_{i<j} s_i s_j - 1/2 \left(\sum_i s_i \right)^2 = -1/2 \sum_i s_i^2 \leq 0.$$

Hence we obtain a useful inequality for logarithms of determinants of matrices

$$\ln \det(\Delta_0 + \Delta_1 t + \Delta_2 t^2 + o(t^2)) \leq \ln \det \Delta_0 + \operatorname{tr} At + \operatorname{tr} Bt^2 + o(t^2)$$

Now assume we choose the unit vector v and the point x for which the second directional derivative $f(x) = D_v^2 U(x)$ is maximal, it can be assumed that the maximum is attained in the situation when the support of μ_P is compact, in this case $f(x)$ will tend to zero as $|x| \to +\infty$. Other cases are reduced to this by going to the limit with the usage of the stability of the transportation map. So we have x and v producing a maximal $D_v^2 U(x)$, as before, we translate this optimal x and its transportation image $DU(x)$ to the origin to simplify the calculations.

At this point the matrix $\Delta_2 = 1/2 D^2 f = 1/2 D^2 D_v^2 U$ is negative semidefinite from the maximality assumption; hence $B = \Delta_0^{-1/2} \Delta_2 \Delta_0^{-1/2}$ is also negative semidefinite. The maximality assumption on the vector v means that the quadratic form $D_v^2 U(0)$ attains its maximum in v; hence v has to be the eigenvector of the corresponding linear operator Δ_0 (the same matrix as $D^2 U(0)$) with the maximal possible eigenvalue λ, which we assumed to be greater than 1 somewhere and hence greater than 1 in the situation where it is maximal. In this situation the second derivative of the expression

$$\ln \det D^2 U(tv)$$

equals to the trace of a negative semidefinite matrix $B = \Delta_0^{-1/2} \Delta_2 \Delta_0^{-1/2}$ and is non-positive, while the second derivative of

$$P(DU(tv)) - Q(tv) = D_{\Delta_0(tv)}^2 P(0) - D_{tv}^2 Q(0) + o(t^2)$$

$$= t^2 \left(\lambda^2 D_v^2 P(0) - D_v^2 Q(0) \right) + o(t^2)$$

is positive at $t = 0$ by the assumption of the theorem. This is a contradiction. $\square$

1.7 Appendix: Explanation of the Pancake Decomposition

In [13] the method of producing pancakes for Gromov's waist theorem was different compared to the original work [9]. Some of the readers told us that because of this they have an impression that the method to produce the pancake decomposition in [9] was incorrect. That is why we have decided to give more explanations here about the argument from [9].

Let us assume we are proving Theorem 1.2.1, although we will sometimes speak about the extension of the argument to its spherical version, where the Euclidean ball of radius R is replaced by the sphere $\mathbb{S}^n$ and the binary partition is made by hyperplanes through the center of the sphere.

Let us choose a sequence of uniformly distributed linear subspaces $\{L_i\}$ in $\mathbb{R}^n$ of dimension $n - k - 1$ each. Note that we only consider the essential case $n > k$, hence the dimension is at least 0. In fact we do not need a uniform distribution in the Grassmannian $G_{n-k-1}(\mathbb{R}^n)$, we will be quite satisfied if the sequence L_i visits any open subset of the Grassmannian infinitely many times.

1.7.1   General Version of the Equipartition Argument

After that we build a binary decomposition of $\mathbb{R}^n$ into $N = 2^l$ parts, on ith stage of the decomposition we use hyperplane cuts parallel to L_i. Every individual cut (including the cut at infinity) is parameterized by a sphere S^{k+1}, the total hierarchy of cuts is parameterized by $(S^{k+1})^{N-1}$. Similar to what is happening for the standard ham sandwich theorem, the generalized Borsuk–Ulam type theorem (from [9]) for $\mathfrak{S}_N^{(2)}$-equivariant (the 2-Sylow subgroup of the permutation group) maps

$$\left(S^{k+1}\right)^{N-1} \to \left(\mathbb{R}^{k+1}\right)^N / \Delta(\mathbb{R}^{k+1}), \quad \text{where} \quad \Delta(x) = \underbrace{(x, \ldots, x)}_{N}$$

allows us to find a binary partition with equal measures of parts and equal images $F(P_i)$. The proof of this generalization of the Borsuk–Ulam theorem is essentially a parity counting argument, establishing that generically the number of solution $\mathfrak{S}_N^{(2)}$-orbits is odd, as explained in [13], for example.

For empty or degenerate parts the values $F(P_i)$ are undefined and we need to deal with it in order to apply the Borsuk–Ulam-type theorem. Let Z be the closed subset of $(S^{k+1})^{N-1}$ corresponding to the partitions with equal measures of all parts. Evidently, the values $F(P_i)$ are well-defined on Z and in a neighborhood of it. After that we may modify the functions $F(P_i)$ so that they remain the same over Z and get extended to the whole $(S^{k+1})^{N-1}$ continuously, to achieve this, it is sufficient to multiply them by a continuous function supported in the neighborhood of Z. Then the Borsuk–Ulam type theorem is applied to the values $\mu(P_1), \ldots, \mu(P_N)$, and the k-dimensional vectors $F(P_1), \ldots, F(P_N)$, now continuously defined over the whole configurations space. Once we obtain a solution from the Borsuk–Ulam-type theorem, the obtained equality $\mu(P_1) = \cdots = \mu(P_N)$ guarantees we are on Z, and hence on the set where $F(P_i)$ are originally defined.

1.7.2   Modification of the Equipartition Argument for the Spherical Waist Theorem

The pancake decomposition argument is also used in Gromov's waist of the sphere theorem [9, 16], and in its generalization [12] for maps to smooth manifolds $f :$

$\mathbb{S}^n \to M^k$. In this case we use the binary partition of the sphere $\mathbb{S}^n$ by cuts through the origin, we again choose a uniformly distributed sequence of linear subspaces $L_i \subseteq \mathbb{R}^{n+1}$ of dimension $n - k - 1$ each (assuming the nontrivial case $n > k$). The possible cuts on stage i are made by hyperplanes through L_i, and in every node of the binary tree the possible cuts are again parametrized by spheres S^{k+1}, the unit spheres of the orthogonal complements of L_i.

The corresponding version of the Borsuk–Ulam theorem in this case is about $\mathfrak{S}_N^{(2)}$-equivariant maps (not stated explicitly in [12], but proven there)

$$\left(S^{k+1}\right)^{N-1} \to \left(M^k \times \mathbb{R}\right)^N / \Delta(M^k \times \mathbb{R}), \quad \text{where} \quad \Delta(x) = \underbrace{(x, \ldots, x)}_{N},$$

which for $M^k = \mathbb{R}^k$ is the same as above, the factor $\mathbb{R}$ corresponds to the requirement to equipartition the volume.

In this case we need more care to continuously extend the map $F(\cdot)$ from its original domain, since the target space M^k need not be contractible. Using the fact that $F = f \circ c$, it is sufficient to continuously extend the center map $c(P_i)$ assigning a "center point" to a part in the sphere, and then compose the extension of c with f. Moreover, since we are going to plug F into the Borsuk–Ulam-type theorem, we may consider a part P_i as a member of the binary partition and let $c(P_i)$ depend continuously on the position of P_i in the partition, minding that the collection of $c(P_i)$ must keep some equivariance under the $\mathfrak{S}_N^{(2)}$-action. Then we may impose the restriction that the extended center map $c(P_i)$ will also depend on the first partition stage, and the extended $c(P_i)$ has to go to the open hemisphere of the first sphere partition stage to which a possibly empty or degenerate P_i is assigned, this restriction is valid wherever $c(P_i)$ is defined originally. This way the map extension problem is to extend a continuous section of a fiber bundle with contractible fibers, such an extension problem always has a solution.

What *seems important to us* is that the center point selection procedure for spherical convex sets in [9, 16] still remains somewhat complicated and we do not see a Caffarelli-type simplification for the spherical case at the moment.

1.7.3 Simplified Version of the Equipartition Argument

In fact, the application if the Borsuk–Ulam-type theorem in the proof of Theorem 1.2.1 can be simplified, although this simplification seems to be not suitable in the spherical case.

Consider *only* binary decompositions with equal measures of parts. It means that on ith stage we cut all parts into equal halves by hyperplanes parallel to L_i. Every such cut is parameterized by unit vectors $u \perp L_i$, just because after we choose the direction of a hyperplane we in fact find a unique hyperplane that cuts a given part in two parts of equal measure. Since every such u is chosen from a copy of S^k, the total hierarchy of equipartition cuts is parameterized by $(S^k)^{N-1}$. The generalized Borsuk–Ulam type theorem for $\mathfrak{S}_N^{(2)}$-equivariant maps

$$\left(S^k\right)^{N-1} \to \left(\mathbb{R}^k\right)^N / \Delta(\mathbb{R}^k), \quad \text{where} \quad \Delta(x) = \underbrace{(x, \ldots, x)}_{N}$$

is then applied to the map that sends a binary partition to the collection of values

$$F(P_1), \ldots, F(P_N)$$

and gives a configuration with all the $F(P_i)$ equal.

1.7.4 Proof that the Parts Are Pancakes

Having a binary partition into parts of equal measure and with coinciding $F(P_i)$, we need to show that for any given δ we can choose sufficiently large N so that almost all the parts are pancakes.

The contrary to the needed pancake property is that a part P_i contains a $(k + 1)$-dimensional disc D of radius δ, and this happens for arbitrarily large N. This can be seen, following [13], from the fact that every convex body in $\mathbb{R}^n$ can be approximated by its John ellipsoid up to scaling by n, and if an ellipsoid is not δ/n-close to the affine subspace spanned by its k principal exes (which would imply P_i is δ-close to the same k-dimensional affine subspace) then the ellipsoid does contain a $(k + 1)$-dimensional disk D of radius δ/n. Thus we assume the contrary and denote δ/n by δ for brevity.

An elementary observation shows that if a convex body $K \subset \mathbb{R}^n$ is cut into equal halves by a hyperplane h then h divides the corresponding directional width of K in ratio at least $1 - 2^{-n} : 2^{-n}$. Since we have fixed the dimension n, this is just some small constant. Considering our measure instead of the volume, whose density is in the range $[m, M]$ we will have the same estimate for the width ratio with a rough constant $c_\mu = \frac{m}{M}(1 - 2^{-n})$.

Now we track the origin of the part that contains the disk D, let the affine full of D be A. During the production of this part, by the assumption, we have made arbitrarily large number of cuts almost parallel to the $(n - k - 1)$-dimensional orthogonal complement of A, just because many L_i were close to this complement. Those cuts produced the convex parts of the big ball,

$$Q_1 \supset Q_2 \supset \cdots \supset Q_I = P_i,$$

each Q_{i+1} produced from Q_i by an equipartition of its μ-measure. The orthogonal projection $\pi_A(Q_i)$ was evidently inclusion-decreasing. Moreover, for cuts with L_i close to $A^\perp$ there was a direction u in A such that the width $w_u(\pi_A(Q_{i+1}))$ was bounded from above

$$w_u(\pi_A(Q_{i+1})) \leq (1 - c_\mu/2)w_u(\pi_A(Q_i)).$$

From this it follows that, assuming sufficiently large number of such width-decreasing cuts we get a contradiction with the inclusion $\pi_A(Q_I) \supseteq D$. One way to argue is that we may also assume that the number of cuts with approximately same u was also large and resulted in the decrease of w_u from R to a number smaller than δ. The other way is to note that the $(k + 1)$-dimensional volume of $\pi_A(Q_i)$ under cuts with L_i close to $A^\perp$ also had a guaranteed decrease

$$\mathrm{vol}_{k+1}\, \pi_A(Q_{i+1}) \leq \left(1 - (c_\mu/2)^{k+1}\right) \mathrm{vol}_{k+1}\, \pi_A(Q_i),$$

which gives a decrease from $\frac{\pi^{(k+1)/2}}{((k+1)/2)!} R^{k+1}$ to $\mathrm{vol}_{k+1}\, D = \frac{\pi^{(k+1)/2}}{((k+1)/2)!}\delta^{k+1}$ in a finite number of steps, and gives a contradiction so sufficiently large N.

If we are interested in the spherical version, then the local situation is essentially the same up to small curvature. Having a $(k + 1)$-dimensional disk D of radius δ (for small δ it is very close to its Euclidean analogue) in a single part will contradict the fact that many of the cuts were almost passing through the $(n - k - 1)$-dimensional orthogonal complement in $\mathbb{R}^{n+1}$ to the $(k + 2)$-dimensional cone over D, that is many cuts were essentially perpendicular to D and had to decrease its width in some direction.

Acknowledgements The authors thank Alexey Balitskiy, Michael Blank, Alexander Esterov, Sergei Ivanov, Bo'az Klartag, Jan Maas, and the unknown referee for useful discussions, suggestions, and questions.

References

1. A. Akopyan, R. Karasev, A tight estimate for the waist of the ball. Bull. Lond. Math. Soc. **49**(4), 690–693 (2017). arXiv:1608.06279. http://arxiv.org/abs/1608.06279
2. A. Akopyan, R. Karasev, Waist of balls in hyperbolic and spherical spaces. Int. Math. Res. Not. (2018). arXiv:1702.07513. https://arxiv.org/abs/1702.07513
3. A. Akopyan, A. Hubard, R. Karasev, Lower and upper bounds for the waists of different spaces. Topol. Methods Nonlinear Anal. **53**(2), 457–490 (2019)
4. M. Berger, Quelques problèmes de géométrie Riemannienne ou deux variations sur les espaces symétriques compacts de rang un. Enseignement Math. **16**, 73–96 (1970)
5. Y. Brenier, Polar factorization and monotone rearrangement of vector-values functions. Commun. Pure Appl. Math. **XLIV**(10), 375–417 (1991)
6. L.A. Caffarelli, Monotonicity properties of optimal transportation and the FKG and related inequalities. Commun. Math. Phys. **214**(3), 547–563 (2000)
7. G. de Rham, *On the Area of Complex Manifolds. Notes for the Seminar on Several Complex Variables* (Institute for Advanced Study, Princeton, 1957–1958)
8. H. Federer, Some theorems on integral currents. Trans. Am. Math. Soc. **117**, 43–67 (1965)

9. M. Gromov, Isoperimetry of waists and concentration of maps. Geom. Funct. Anal. **13**, 178–215 (2003)
10. R. Harvey, H.B. Lawson Jr., Calibrated geometries. Acta Math. **148**(1), 47–157 (1982)
11. W.-Y. Hsiang, H.B. Lawson Jr., Minimal submanifolds of low cohomogeneity. J. Differ. Geom. **5**, 1–38 (1971)
12. R. Karasev, A. Volovikov, Waist of the sphere for maps to manifolds. Topol. Appl. **160**(13), 1592–1602 (2013). arXiv:1102.0647. http://arxiv.org/abs/1102.0647
13. B. Klartag, Convex geometry and waist inequalities. Geom. Funct. Anal. **27**(1), 130–164 (2017). arXiv:1608.04121. http://arxiv.org/abs/1608.04121
14. B. Klartag, Eldan's stochastic localization and tubular neighborhoods of complex-analytic sets (2017). arXiv:1702.02315. http://arxiv.org/abs/1702.02315
15. A.V. Kolesnikov, Mass transportation and contractions (2011). arXiv:1103.1479. https://arxiv.org/abs/1103.1479
16. Y. Memarian, On Gromov's waist of the sphere theorem. J. Topol. Anal. **03**(01), 7–36 (2011). arXiv:0911.3972. http://arxiv.org/abs/0911.3972
17. N. Palić, *Grassmannians, measure partitions, and waists of spheres*, PhD thesis, Freie Universität Berlin (2018). https://refubium.fu-berlin.de/handle/fub188/23043
18. W. Wirtinger, Eine determinantenidentität und ihre anwendung auf analytische gebilde und Hermitesche massbestimmung. Monatsh. Math. Phys. **44**, 343–365 (1936)

Chapter 2
Zhang's Inequality for Log-Concave Functions

David Alonso-Gutiérrez, Julio Bernués, and Bernardo González Merino

Abstract Zhang's reverse affine isoperimetric inequality states that among all convex bodies $K \subseteq \mathbb{R}^n$, the affine invariant quantity $|K|^{n-1}|\Pi^*(K)|$ (where $\Pi^*(K)$ denotes the polar projection body of K) is minimized if and only if K is a simplex. In this paper we prove an extension of Zhang's inequality in the setting of integrable log-concave functions, characterizing also the equality cases.

2.1 Introduction

Given a convex body (compact, convex with non-empty interior) $K \subseteq \mathbb{R}^n$, its *polar projection body* $\Pi^*(K)$ is the unit ball of the norm given by

$$\|x\|_{\Pi^*(K)} := |x||P_{x^\perp}K|, \qquad x \in \mathbb{R}^n$$

where $|\cdot|$ denotes both the *Lebesgue measure* (in the suitable space) and the Euclidean norm, and $P_{x^\perp}K$ is the *orthogonal projection* of K onto the hyperplane orthogonal to x. The *Minkowski functional* of a convex body K containing the origin

The first and second authors are partially supported by MINECO Project MTM2016-77710-P, DGA E26_17R and IUMA, and the third author is partially supported by Fundación Séneca, Programme in Support of Excellence Groups of the Región de Murcia, Project 19901/GERM/15, and by MINECO Project MTM2015-63699-P and MICINN Project PGC2018-094215-B-I00.

D. Alonso-Gutiérrez (✉) · J. Bernués
Área de Análisis Matemático, Departamento de Matemáticas, Facultad de Ciencias, Universidad de Zaragoza, IUMA, Zaragoza, Spain
e-mail: alonsod@unizar.es; bernues@unizar.es

B. González Merino
Departamento de Didáctica de las Ciencias Matemáticas y Sociales, Facultad de Educación, Universidad de Murcia, Murcia, Spain
e-mail: bgmerino@um.es

© Springer Nature Switzerland AG 2020
B. Klartag, E. Milman (eds.), *Geometric Aspects of Functional Analysis*,
Lecture Notes in Mathematics 2256,
https://doi.org/10.1007/978-3-030-36020-7_2

is defined, for every $x \in \mathbb{R}^n$, as $\|x\|_K := \inf\{\lambda > 0 \mid x \in \lambda K\} \in [0, \infty]$. It is a norm if and only if K is centrally symmetric.

The expression $|K|^{n-1}|\Pi^*(K)|$ is affine invariant and the extremal convex bodies are well known: *Petty's projection inequality* [10] states that the (affine class of the) n-dimensional Euclidean ball, B_2^n, is the only maximizer and *Zhang's inequality* [13] (see also [8] and [1]) proves that the (affine class of the) n-dimensional simplex Δ_n is the only minimizer. That is, for any convex body $K \subseteq \mathbb{R}^n$,

$$\frac{\binom{2n}{n}}{n^n} = |\Delta_n|^{n-1}|\Pi^*(\Delta_n)| \leq |K|^{n-1}|\Pi^*(K)| \leq |B_2^n|^{n-1}|\Pi^*(B_2^n)|$$

$$= \frac{\pi^{\frac{n}{2}}\Gamma\left(1 + \frac{n-1}{2}\right)^n}{\Gamma\left(1 + \frac{n}{2}\right)^n}.$$

In recent years, many relevant geometric inequalities have been extended to the more general context of *log-concave functions*, i.e., functions $f : \mathbb{R}^n \to [0, \infty)$ of the form $f(x) = e^{-v(x)}$ with $v : \mathbb{R}^n \to (-\infty, \infty]$ a convex function. The set of all log-concave and integrable functions in $\mathbb{R}^n$ will be denoted by $\mathcal{F}(\mathbb{R}^n)$. This family of functions contains the set of convex bodies via the natural injections $K \to \chi_K$ (where χ_K denotes the characteristic function of K) or $K \to e^{-\|\cdot\|_K}$ (when K is a convex body containing the origin). We refer the reader to [9] or [7] and the references therein for a quick introduction on this topic.

The aim of this paper is to provide an extension of Zhang's inequality for every $f \in \mathcal{F}(\mathbb{R}^n)$. For that matter, for every $f \in \mathcal{F}(\mathbb{R}^n)$ we consider the (centrally symmetric) *polar projection body of* f, denoted $\Pi^*(f)$, which was defined in [2] by

$$\|x\|_{\Pi^*(f)} = 2|x| \int_{x^\perp} P_{x^\perp} f(y) dy, \quad x \in \mathbb{R}^n$$

where $P_{x^\perp} f : x^\perp \to [0, \infty)$ is *the shadow of* f, which was defined in [9] by $P_{x^\perp} f(y) := \max_{s \in \mathbb{R}} f\left(y + s\frac{x}{|x|}\right)$.

Petty's projection inequality was extended by Zhang, see [14], to compact domains and it was shown to be equivalent to the so called *affine Sobolev inequality*, which for log-concave functions in the suitable Sobolev space takes the following form: For every $f \in \mathcal{F}(\mathbb{R}^n) \cap W^{1,1}(\mathbb{R}^n) = \left\{ f \in \mathcal{F}(\mathbb{R}^n) : \frac{\partial f}{\partial x_i} \in L^1(\mathbb{R}^n), \forall i \right\}$

$$\|f\|_{\frac{n}{n-1}}|\Pi^*(f)|^{\frac{1}{n}} \leq \frac{|B_2^n|}{2|B_2^{n-1}|},$$

with equality if and only if f is the characteristic function of an ellipsoid.

Remark 2.1 We point out that the factor 2 in the definition of $\|\cdot\|_{\Pi^*(f)}$ appears so that for every $f \in \mathcal{F}(\mathbb{R}^n) \cap W^{1,1}(\mathbb{R}^n)$ the norm equals

$$\|x\|_{\Pi^*(f)} = \int_{\mathbb{R}^n} |\langle \nabla f(y), x \rangle| dy,$$

which was the expression of the norm that was considered in [14].

Our extension of Zhang's inequality has the following form:

Theorem 2.1.1 *Let $f \in \mathcal{F}(\mathbb{R}^n)$. Then,*

$$\int_{\mathbb{R}^n} \int_{\mathbb{R}^n} \min\{f(y), f(x)\} dy dx \leq 2^n n! \|f\|_1^{n+1} |\Pi^*(f)|.$$

Moreover, if $\|f\|_\infty = f(0)$ then equality holds if and only if $\frac{f(x)}{\|f\|_\infty} = e^{-\|x\|_{\Delta_n}}$ for some n-dimensional simplex Δ_n containing the origin.

Remark 2.2 The inequality in Theorem 2.1.1 is affine invariant, i.e., the inequality does not change under compositions of f with affine transformations of $\mathbb{R}^n$.

Remark 2.3 Theorem 2.1.1 extends Zhang's inequality. Indeed, if $f(x) = e^{-\|x\|_K}$ for some convex body K containing the origin, then

$$\Pi^*(f) = \frac{1}{2(n-1)!} \Pi^*(K), \qquad \|f\|_1 = n! |K|,$$

and

$$\int_{\mathbb{R}^n} \int_{\mathbb{R}^n} \min\{f(y), f(x)\} dy dx = \int_{\mathbb{R}^{2n}} e^{-\max\{\|x\|_K, \|y\|_K\}} dy dx$$

$$= \int_{\mathbb{R}^{2n}} e^{-\|(x,y)\|_{K \times K}} dy dx = (2n)! |K|^2.$$

Thus, Theorem 2.1.1 yields

$$\frac{\binom{2n}{n}}{n^n} \leq |K|^{n-1} |\Pi^*(K)|.$$

Remark 2.4 Sharp lower and upper bounds for the left hand side of the inequality in Theorem 2.1.1 in terms of the integral of f are known (see Lemmas 2.3 and 2.9 in [3], respectively).

The paper is structured as follows. In Sect. 2.2 we introduce some notation and preliminary results. A crucial role in the proof of Theorem 2.1.1 will be played by the functional form of the covariogram function, that we shall denote by g_f, associated to any $f \in \mathcal{F}(\mathbb{R}^n)$. Recall that in the geometric setting the covariogram function of a convex body K is given by $\mathbb{R}^n \ni x \to |K \cap (x + K)|$.

In this section we shall define and study the basic properties of its functional version g_f.

In Sect. 2.3 we prove the inequality in Theorem 2.1.1. The proof will rely on the following two facts: First, we will show that the polar projection body of $f \in \mathcal{F}(\mathbb{R}^n)$ can be expressed in terms of dilations of the level sets of g. This can be seen as an extension of the corresponding geometric result (see [12, Theorem 1] and [1, Propositions 4.1 and 4.3]) where the polar projection body of a convex body appears as the limit of suitable dilations of the level sets of the covariogram function. Second, we will prove a sharp relation (by inclusion) between the level sets of g and a convex body in the celebrated family of bodies introduced by Ball (cf. [5, Pg. 74]). The proof of such inclusion, that we state in full generality, follows ideas from [9, Lemmas 2.1 and 2.2].

In Sect. 2.4 we characterize the equality cases. We first show that the equality in Theorem 2.1.1 holds if and only if the function g is log-linear on every ray emanating from 0, i.e., for every $x \in \mathbb{R}^n$ and every $\lambda \in [0, 1]$, $g(\lambda x) = g(0)^{1-\lambda} g(x)^\lambda$ and then prove that such condition implies that f has to be as in the statement of the theorem.

2.2 Notation and Preliminaries

Denote by S^{n-1} the Euclidean unit sphere in $\mathbb{R}^n$ and σ denotes the uniform probability measure on S^{n-1}. If the origin is in the interior of K, the function $\rho_K : S^{n-1} \to [0, +\infty)$ given by $\rho_K(u) = \sup\{\lambda \geq 0 \mid \lambda u \in K\}$ is the radial function of K. It extends to $\mathbb{R}^n \setminus \{0\}$ via $t\rho_K(tu) = \rho_K(u)$, for any $t > 0, u \in S^{n-1}$. The volume of K is given by

$$|K| = |B_2^n| \int_{S^{n-1}} \rho_K^n(u) d\sigma(u)$$

and the boundary of K will be denoted by ∂K.

Throughout the paper, $f : \mathbb{R}^n \to [0, +\infty)$ will always denote a (non identically null) log-concave integrable function. Recall that such f is said to be log-concave if it can be written as $\frac{f}{\|f\|_\infty} = e^{-v}$ where $\|f\|_\infty$ denotes the supremum of f and $v : \mathbb{R}^n \to [0, +\infty]$ is convex or, equivalently if for every $x, y \in \mathbb{R}^n, 0 < \lambda < 1$, $f(\lambda x + (1-\lambda)y) \geq (f(x))^\lambda (f(y))^{1-\lambda}$. It is well known that f is then continuous on the interior of its support, int(supp$f$). Since Theorem 2.1.1 does not depend on the values of f on the boundary of supp$f$ we shall assume, without loss of generality, that f is continuous on its support. For any $t \in [0, \infty)$ we denote the super-level sets of v by

$$K_t(f) = \{x \in \mathbb{R}^n : f(x) \geq e^{-t} \|f\|_\infty\} = \{x \in \mathbb{R}^n : v(x) \leq t\}.$$

Since we assume that f is continuous on its support, $K_t(f)$ is a convex body for all $t > 0$. These and other basic facts on convex bodies and log-concave functions used in the paper can be found in [6].

We will use the following definition of the polar projection body of f which involves super-level sets, equivalent to the one stated in the introduction (see [2, Proposition 4.1]).

Definition 2.2.1 Let $f \in \mathcal{F}(\mathbb{R}^n)$. The polar projection body of f, denoted as $\Pi^*(f)$, is the unit ball of the norm given by

$$\|x\|_{\Pi^*(f)} := 2|x| \|f\|_\infty \int_0^\infty |P_{x^\perp} K_t(f)| \, e^{-t} dt = 2\|f\|_\infty \int_0^\infty \|x\|_{\Pi^*(K_t(f))} e^{-t} dt.$$

Remark 2.5 If $f = \chi_K$ is the characteristic function of a convex body K, then $\Pi^*(f) = \frac{1}{2}\Pi^*(K)$ and, as mentioned in Remark 2.3, if $f(x) = e^{-\|x\|_K}$, then $\Pi^*(f) = \frac{1}{2(n-1)!}\Pi^*(K)$.

We start by associating a function g_f to any function $f \in \mathcal{F}(\mathbb{R}^n)$. Such function can be regarded as the functional version of the covariogram functional. We define it by

$$g_f(x) := \int_0^\infty e^{-t} |K_t(f) \cap (x + K_t(f))| \, dt.$$

We collect its properties in the following lemma

Lemma 2.2.1 *Let* $f \in \mathcal{F}(\mathbb{R}^n)$. *Then the covariogram function* g_f *satisfies the following:*

$$g_f(x) = \int_{\mathbb{R}^n} \min\left\{ \frac{f(y)}{\|f\|_\infty}, \frac{f(y-x)}{\|f\|_\infty} \right\} dy$$

is even and log-concave, $0 \in \text{int}(\text{supp } g_f)$ *with* $\|g_f\|_\infty = g(0) = \int_0^\infty e^{-t}|$ $K_t(f)|dt = \int_{\mathbb{R}^n} \frac{f(x)}{\|f\|_\infty} dx > 0$, *and* $\int_{\mathbb{R}^n} g_f(x)dx = \int_{\mathbb{R}^n} \int_{\mathbb{R}^n} \min\left\{ \frac{f(y)}{\|f\|_\infty}, \frac{f(x)}{\|f\|_\infty} \right\} dy dx$.

Proof By Fubini's theorem, for any $x \in \mathbb{R}^n$ we have

$$\int_0^\infty e^{-t} |K_t(f) \cap (x + K_t(f))| dt = \int_{\mathbb{R}^n} \int_{-\log \min\left\{ \frac{f(y)}{\|f\|_\infty}, \frac{f(y-x)}{\|f\|_\infty} \right\}}^\infty e^{-t} dt dy$$

$$= \int_{\mathbb{R}^n} \min\left\{ \frac{f(y)}{\|f\|_\infty}, \frac{f(y-x)}{\|f\|_\infty} \right\} dy.$$

Consequently, using Fubini's theorem and a change of variables,

$$\int_{\mathbb{R}^n} g_f(x)dx = \int_{\mathbb{R}^n} \int_{\mathbb{R}^n} \min\left\{\frac{f(y)}{\|f\|_\infty}, \frac{f(x)}{\|f\|_\infty}\right\} dydx.$$

In order to prove the log-concavity of g_f, let $x_1, x_2 \in \mathbb{R}^n$, $t_1, t_2 \in [0, \infty)$, $0 \le \lambda \le 1$ and write $x = (1 - \lambda)x_1 + \lambda x_2$ and $t = (1 - \lambda)t_1 + \lambda t_2$. Clearly,

$$K_t \cap (x + K_t) \supseteq (1 - \lambda)(K_{t_1} \cap (x_1 + K_{t_1})) + \lambda(K_{t_2} \cap (x_2 + K_{t_2})).$$

By Brunn–Minkowski inequality, [6, Theorem 1.2.1],

$$|K_t \cap (x + K_t)| \ge |K_{t_1} \cap (x_1 + K_{t_1})|^{1-\lambda}|K_{t_2} \cap (x_2 + K_{t_2})|^\lambda.$$

Thus,

$$e^{-t}|K_t \cap (x + K_t)| \ge (e^{-t_1}|K_{t_1} \cap (x_1 + K_{t_1})|)^{1-\lambda}(e^{-t_2}|K_{t_2} \cap (x_2 + K_{t_2})|)^\lambda$$

and, by Prékopa–Leindler inequality, [6, Theorem 1.2.3],

$$g_f((1 - \lambda)x_1 + \lambda x_2) = \int_0^\infty e^{-t}|K_t \cap ((1 - \lambda)x_1 + \lambda x_2 + K_t)|dt \ge$$

$$\ge \left(\int_0^\infty e^{-t}|K_t \cap (x_1 + K_t)|dt\right)^{1-\lambda}$$

$$\times \left(\int_0^\infty e^{-t}|K_t \cap (x_2 + K_t)|dt\right)^\lambda$$

$$= g_f(x_1)^{1-\lambda}g_f(x_2)^\lambda.$$

Now, for any $t \in [0, \infty)$, $K_t(f) \cap (x + K_t(f)) = x + (K_t(f) \cap (-x + K_t(f)))$ and so $|K_t(f) \cap (x + K_t(f))| = |K_t(f) \cap (-x + K_t(f))|$. Therefore $g_f(x) = g_f(-x)$. Consequently, $\|g_f\|_\infty = g_f(0)$ and its value is

$$g_f(0) = \int_0^\infty e^{-t}|K_t(f)|dt = \int_{\mathbb{R}^n} \frac{f(x)}{\|f\|_\infty}dx > 0.$$

Finally, there exists $\varepsilon > 0$ such that if $|x| < \varepsilon$ then $K_1(f) \cap (x + K_1(f))$ is a non-empty convex body and has positive volume. Thus, if $|x| < \varepsilon$ then for every $t > 1$ $K_t(f) \cap (x + K_t(f))$ has positive volume and then $g_f(x) > 0$. Thus $0 \in \text{int}(\text{supp } g_f)$. $\square$

Let $g \in \mathcal{F}(\mathbb{R}^n)$ be a function such that $g(0) > 0$ and let $p > 0$. The following important family of convex bodies was introduced by K. Ball in [5, pg. 74]. We denote

$$\widetilde{K}_p(g) := \left\{ x \in \mathbb{R}^n : \int_0^\infty g(rx)r^{p-1}dr \geq \frac{g(0)}{p} \right\}.$$

It follows from the definition that the radial function of $\widetilde{K}_p(g)$ is given by

$$\rho_{\widetilde{K}_p(g)}^p(u) = \frac{1}{g(0)} \int_0^\infty p r^{p-1} g(ru)dr.$$

We will make use of the following well-known relation between the Lebesgue measure of $\widetilde{K}_n(g)$ and the integral of g.

Lemma 2.2.2 ([5]) *Let $g \in \mathcal{F}(\mathbb{R}^n)$ be such that $g(0) > 0$. Then*

$$|\widetilde{K}_n(g)| = \frac{1}{g(0)} \int_{\mathbb{R}^n} g(x)dx.$$

Proof Integrating in polar coordinates we have that

$$|\widetilde{K}_n(g)| = |B_2^n| \int_{S^{n-1}} \rho_{\widetilde{K}_n(g)}^n(u)d\sigma(u) = |B_2^n| \int_{S^{n-1}} \frac{n}{g(0)} \int_0^\infty r^{n-1} g(ru)dr d\sigma(u)$$
$$= \frac{1}{g(0)} \int_{\mathbb{R}^n} g(x)dx.$$

$\square$

2.3   Proof of the Inequality in Theorem 2.1.1

We split the main idea of the proof into two parts. We first prove Lemma 2.3.1 below, which states that $\Pi^*(f)$ equals the intersection of suitable dilations of the super-level sets $K_t(g_f)$, with g_f as defined in the previous section. We then show a sharp relation between Ball's convex body $\widetilde{K}_n(g_f)$ and the super-level set $K_t(g_f)$ in Lemma 2.3.2. Such a relation holds not only for the covariogram function g_f but for a larger class of log-concave functions.

Lemma 2.3.1 *Let $f \in \mathcal{F}(\mathbb{R}^n)$ and let $g_f : \mathbb{R}^n \to \mathbb{R}$ be the covariogram function*

$$g_f(x) = \int_0^\infty e^{-t}|K_t(f) \cap (x + K_t(f))|dt.$$

Then for every $0 < \lambda_0 < 1$

$$\bigcap_{0 < \lambda < \lambda_0} \frac{K_{-\log(1-\lambda)}(g_f)}{\lambda} = 2\|f\|_1 \, \Pi^*(f).$$

Proof For any $0 < \lambda < 1$ the convex body $\frac{K_{-\log(1-\lambda)}(g_f)}{\lambda}$ equals

$$\left\{ x \in \mathbb{R}^n \ : \ \int_0^\infty e^{-t} |K_t(f) \cap (\lambda x + K_t(f))| dt \geq (1-\lambda) \int_0^\infty e^{-t} |K_t(f)| dt \right\}$$

or equivalently

$$\left\{ x \in \mathbb{R}^n \ : \ \int_0^\infty e^{-t} \frac{|K_t(f)| - |K_t(f) \cap (\lambda|x|\frac{x}{|x|} + K_t(f))|}{\lambda} dt \right.$$
$$\left. \leq \int_0^\infty e^{-t} |K_t(f)| dt \right\}.$$

Since $|K_t(f)| - |K_t(f) \cap (\lambda|x|\frac{x}{|x|} + K_t(f))| \leq \lambda|x| |P_{x^\perp} K_t(f)|$ (see Fig. 2.1), then

$$\int_0^\infty e^{-t} \frac{|K_t(f)| - |K_t(f) \cap (\lambda|x|\frac{x}{|x|} + K_t(f))|}{\lambda} dt \leq |x| \int_0^\infty e^{-t} |P_{x^\perp} K_t(f)| dt$$
$$= \frac{\|x\|_{\Pi^*(f)}}{2\|f\|_\infty}$$

and we have that if

$$\|x\|_{\Pi^*(f)} \leq 2\|f\|_\infty \int_0^\infty e^{-t} |K_t(f)| dt = 2 \int_{\mathbb{R}^n} f(x) dx$$

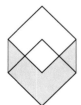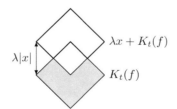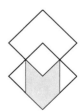

Fig. 2.1 From left to right, we represent in grey area the quantities $\lambda|x||P_{x^\perp} K_t(f)|$, $|K_t(f)| - |K_t(f) \cap (x + K_t(f))|$ and $\lambda|x||P_{x^\perp}(K_t(f) \cap (\lambda x + K_t(f)))|$

then $x \in K_{-\log(1-\lambda)}(g_f)/\lambda$. Thus

$$2\|f\|_1 \Pi^*(f) \subseteq \frac{K_{-\log(1-\lambda)}(g_f)}{\lambda}, \quad \text{for all } 0 < \lambda < 1.$$

On the other hand (see Fig. 2.1), for any $0 < \lambda < 1$ and any $x \in \mathbb{R}^n$,

$$\int_0^\infty e^{-t} \frac{|K_t(f)| - |K_t(f) \cap (\lambda|x|\frac{x}{|x|} + K_t(f))|}{\lambda} dt$$
$$\geq |x| \int_0^\infty e^{-t} |P_{x^\perp}(K_t(f) \cap (\lambda x + K_t(f)))| dt$$

and then, since $\int_0^\infty e^{-t} |P_{x^\perp}(K_t(f) \cap (\lambda x + K_t(f)))| dt$ decreases in λ and

$$\sup_{\lambda \in (0,1)} |x| \int_0^\infty e^{-t} |P_{x^\perp}(K_t(f) \cap (\lambda x + K_t(f)))| dt = |x| \int_0^\infty e^{-t} |P_{x^\perp} K_t(f)| dt$$
$$= \frac{\|x\|_{\Pi^*(f)}}{2\|f\|_\infty},$$

we have that if $\|x\|_{\Pi^*(f)} > 2\|f\|_\infty \int_0^\infty e^{-t} |K_t(f)| dt = 2\|f\|_1$, there exists $\lambda_1 > 0$ such that for every $0 < \lambda \leq \lambda_1$

$$\int_0^\infty e^{-t} \frac{|K_t(f)| - |K_t(f) \cap (\lambda|x|\frac{x}{|x|} + K_t(f))|}{\lambda} dt > \int_0^\infty e^{-t} |K_t(f)| dt$$

and then $x \notin \frac{K_{-\log(1-\lambda)}(g_f)}{\lambda}$ if $0 < \lambda \leq \lambda_1$. $\qquad\square$

In the following lemma we prove the aforementioned inclusion between the super-level sets of a function $g \in \mathcal{F}(\mathbb{R}^n)$ and the convex body $\widetilde{K}_n(g)$. We follow ideas from [9]. In that paper a similar result was stated for large values of t. Here we shall be interested in small values of t. In the second part of the Lemma we provide information on the equality case as it shall be used in the next section.

Lemma 2.3.2 Let $g \in \mathcal{F}(\mathbb{R}^n)$ be such that $0 \in \text{int}(\text{supp } g)$, $g(0) = \|g\|_\infty > 0$. Then for every $0 \leq t \leq \frac{n}{e}$,

$$\frac{t}{(n!)^{\frac{1}{n}}} \widetilde{K}_n(g) \subseteq K_t(g).$$

Moreover, for any $0 < t \leq \frac{n}{e}$ there is equality if and only if g is log-linear on every ray emanating from 0 and, furthermore, if g is not log-linear on the ray emanating *from 0 in direction $u \in S^{n-1}$, there exists $\varepsilon > 0$ such that for every $0 < t \leq \frac{n}{e}$*

$$\frac{t}{(n!)^{\frac{1}{n}}}(\rho_{\widetilde{K}_n(g)}(u) + \varepsilon) \leq \rho_{K_t(g)}(u).$$

Proof We can assume without loss of generality that $g(0) = 1$. Otherwise consider $\frac{g}{\|g\|_\infty}$. Write $g(x) = e^{-v(x)}$ for some convex function v and fix $u \in S^{n-1}$.

For any $q > 0$ the function

$$\phi(r) = v(ru) - q \log r$$

is strictly convex on $(0, \infty)$ and $\omega(r) := v(ru)$ is non-decreasing and convex on $[0, \infty)$. Since $r^q e^{-v(ru)}$ is integrable on $[0, \infty)$ and takes the value 0 at 0, we have $\lim_{r \to \infty} \phi(r) = \lim_{r \to 0^+} \phi(r) = \infty$ and consequently ϕ attains a unique minimum at some number $r_0 = r_0(q)$ and the one-sided derivatives of ϕ verify $\phi'_-(r_0) \leq 0$ and $\phi'_+(r_0) \geq 0$. This implies that the one-sided derivatives of $\omega(r)$ at r_0 satisfy $\omega'_-(r_0) \leq \frac{q}{r_0}$, and $\omega'_+(r_0) \geq \frac{q}{r_0}$. Notice that if ω is linear then necessarily $\omega(r) = \omega(r_0) + \frac{q}{r_0}(r - r_0)$.

Notice also that $r \to \omega(r_0) + \frac{q}{r_0}(r - r_0)$ is a supporting line of ω at r_0 and by convexity, $\omega(r) \geq \omega(r_0) + \frac{q}{r_0}(r - r_0)$ for every $r \in [0, \infty)$. Thus,

$$\rho^n_{\widetilde{K}_n(g)}(u) = n \int_0^\infty r^{n-1} g(ru) dr = n \int_0^\infty r^{n-1} e^{-\omega(r)} dr$$

$$\leq n e^{q - \omega(r_0)} \int_0^\infty r^{n-1} e^{-\frac{qr}{r_0}} dr$$

$$= n e^{q - \omega(r_0)} \left(\frac{r_0}{q}\right)^n \int_0^\infty r^{n-1} e^{-r} dr = g(r_0 u) \frac{e^q n!}{q^n} r_0^n.$$

Moreover, the previous inequality is equality if and only if $\omega(r) = \omega(r_0) + \frac{q}{r_0}(r - r_0)$ for every $r \in [0, \infty)$.

Consequently, for any $q > 0$

$$\frac{q}{e^{\frac{q}{n}}(n!)^{\frac{1}{n}}} \rho_{\widetilde{K}_n(g)}(u) \leq g(r_0 u)^{\frac{1}{n}} r_0,$$

with equality if and only if $\omega(r) = \omega(r_0) + \frac{q}{r_0}(r - r_0)$ for every $r \in [0, \infty)$. On the other hand, since ω is convex, $\omega'_+(r) \leq \frac{q}{r_0}$ if $r < r_0$ and then

$$\omega(r_0) = \omega(0) + \int_0^{r_0} \omega'_+(r) dr \leq v(0) + q = q,$$

thus $g(r_0 u) \geq e^{-q}$, with equality if and only if $\omega'_+(r) = \frac{q}{r_0}$ for every $r \in [0, r_0)$ and therefore $\omega(r) = \omega(r_0) + \frac{q}{r_0}(r - r_0)$ for every $r \in [0, r_0]$. Now, the definition of the log-concavity of g applied to 0 and $r_0 u$ yields

$$g\left(g(r_0 u)^{\frac{1}{n}} r_0 u\right) \geq g(r_0 u)^{g(r_0 u)^{\frac{1}{n}}} \cdot 1 = e^{g(r_0 u)^{\frac{1}{n}} \log g(r_0 u)}$$

with equality if and only if $\omega(r) = v(ru)$ is linear on $[0, r_0]$ and then $\omega(r) = \omega(r_0) + \frac{q}{r_0}(r - r_0)$ for every $r \in [0, r_0]$. Since the function $x \to x^{\frac{1}{n}} \log x$ attains its minimum at $x = e^{-n}$ and is increasing on the interval (e^{-n}, ∞), if $0 < q \leq n$ then $e^{-n} \leq e^{-q} \leq g(r_0 u)$ and hence

$$g\left(g(r_0 u)^{\frac{1}{n}} r_0 u\right) \geq e^{-q e^{-\frac{q}{n}}},$$

with equality if and only if $\omega(r)$ is linear in $[0, r_0]$ and $g(r_0 u) = e^{-q}$, which occurs if and only if $\omega(r) = \omega(r_0) + \frac{q}{r_0}(r - r_0)$ for every $r \in [0, r_0]$, that is, $g(r_0 u)^{\frac{1}{n}} r_0 u \in K_{qe^{-\frac{q}{n}}}(g)$ and $g(r_0 u)^{\frac{1}{n}} r_0 u \in \partial K_{qe^{-\frac{q}{n}}}(g)$ if and only if $\omega(r) = \omega(r_0) + \frac{q}{r_0}(r - r_0)$ for every $r \in [0, r_0]$.

Since this is valid for any $u \in S^{n-1}$,

$$\frac{qe^{-\frac{q}{n}}}{(n!)^{\frac{1}{n}}} \widetilde{K}_n(g) \subseteq K_{qe^{-\frac{q}{n}}}(g),$$

with equality if and only if for every $u \in S^{n-1}$, $\omega(r) = \omega(r_0) + \frac{q}{r_0}(r - r_0)$ for every $r \in [0, \infty)$, i.e. $v(ru)$ is linear for every $u \in S^{n-1}$. Finally, observe that when $x \in (0, n]$, the function $x \to xe^{-\frac{x}{n}}$ takes every value in $(0, \frac{n}{e}]$ thus, for every $0 \leq t \leq \frac{n}{e}$

$$\frac{t}{(n!)^{\frac{1}{n}}} \widetilde{K}_n(g) \subseteq K_t(g)$$

and for any $t \in (0, \frac{n}{e}]$ there is equality if and only if $v(ru)$ is linear for every $u \in S^{n-1}$, i.e., for every $x \in \mathbb{R}^n$ and every $\lambda \in [0, 1]$, $g(\lambda x) = g(0)^{1-\lambda} g(x)^\lambda$.

In order to establish the *furthermore* part, we are given some $u \in S^{n-1}$. We first need to prove the following

Claim The function $q \to r_0(q)$ is continuous in $(0, \infty)$ and is bounded around 0.

Indeed, we first consider a sequence $(q_k)_{k=1}^\infty$ converging to $q \in (0, \infty)$ so that there is a subsequence $(r_0(q_{k_i}))_{i=1}^\infty$ converging to some $\bar{r}$. We have

$$\omega(r_0(q_{k_i})) - q_{k_i} \log(r_0(q_{k_i})) \leq \omega(r_0(q)) - q_{k_i} \log(r_0(q))$$

and taking limits, $\omega(\bar{r}) - q\log(\bar{r}) \leq \omega(r_0(q)) - q\log(r_0(q))$. Therefore, $\bar{r} = r_0(q)$.

If, on the other hand, $q > 0$ and the subsequence $(r_0(q_{k_i}))_{i=1}^{\infty}$ tends to ∞ then, since $q_k \leq M$ for every $k \in \mathbb{N}$ and some $M > 0$, we have that

$$\omega(r_0(q_{k_i})) - M\log(r_0(q_{k_i})) + (M - q_{k_i})\log(r_0(q_{k_i})) = \omega(r_0(q_{k_i}))$$
$$- q_{k_i}\log(r_0(q_{k_i})) \leq \omega(r_0(q)) - q_{k_i}\log(r_0(q)),$$

leading to a contradiction, since the left hand side of the inequality tends to ∞. Thus, both the inferior limit and the superior limit of $r_0(q_k)$ are equal to $r_0(q)$ and we have proven continuity on $(0, \infty)$. Finally, if $(q_k)_{k=1}^{\infty}$ is a sequence converging to 0 and some subsequence $(r_0(q_{k_i}))_{i=1}^{\infty}$ tends to ∞, we would have that for every $r \in [0, \infty)$

$$\omega(r_0(q_{k_i})) + \frac{q_{k_i}}{r_0(q_{k_i})}(r - r_0(q_{k_i})) \leq \omega(r),$$

leading to a contradiction since the left hand side of this inequality tends to ∞. This finishes the proof of the claim.

As a consequence, if $(q_k)_{k=1}^{\infty}$ converges to $q \in [0, \infty)$, we have that the sequence $(\frac{q_k}{r_0(q_k)})_{k=1}^{\infty}$ is bounded, since $\frac{q_k}{r_0(q_k)} \leq \omega'_+(r_0(q_k))$.

Now, assume that there is no $\bar{\varepsilon} > 0$ such that for every $0 < q \leq n$

$$n\int_0^{\infty} r^{n-1}e^{-(\omega(r_0) + \frac{q}{r_0}(r-r_0))}\,dr - n\int_0^{\infty} r^{n-1}e^{-\omega(r)}\,dr \geq \bar{\varepsilon}.$$

Then we can find a sequence $(q_k)_{k=1}^{\infty}$ (and if necessary extract from it further subsequences which we denote in the same way) so that

$$\lim_{k\to\infty} n\int_0^{\infty} r^{n-1}\left(e^{-(\omega(r_0(q_k)) + \frac{q_k}{r_0(q_k)}(r-r_0(q_k)))} - e^{-\omega(r)}\right)\,dr = 0,$$

q_k converges to some $q \in [0, n]$, $r_0(q_k)$ converges to some $\bar{r} \in [0, \infty)$ and $\frac{q_k}{r_0(q_k)}$ converges to some $\alpha \in [0, \infty)$. Therefore, since for every $r \in [0, \infty)$

$$\omega(r_0(q_k)) + \frac{q_k}{r_0(q_k)}(r - r_0(q_k)) \leq \omega(r),$$

we have that for every $r \in [0, \infty)$, $\omega(\bar{r}) + \alpha(r - \bar{r}) \leq \omega(r)$ and, since by Fatou's lemma

$$0 \leq n\int_0^{\infty} r^{n-1}\left(e^{-(\omega(\bar{r}) + \alpha(r-\bar{r}))} - e^{-\omega(r)}\right)\,dr$$
$$\leq \lim_{k\to\infty} n\int_0^{\infty} r^{n-1}\left(e^{-(\omega(r_0(q_k)) + \frac{q_k}{r_0(q_k)}(r-r_0(q_k)))} - e^{-\omega(r)}\right)\,dr = 0,$$

we have that for every $r \in [0, \infty)$, $\omega(r) = \omega(\bar{r}) + \alpha(r - \bar{r})$ and so ω is linear.

Therefore if ω is not linear, there exists $\bar{\varepsilon} > 0$ such that for every $0 < q \leq n$,

$$n \int_0^\infty r^{n-1} g(ru)dr + \bar{\varepsilon} \leq g(r_0 u) \frac{e^q n!}{q^n} r_0^n$$

and so, for some $\varepsilon > 0$ and every $0 < q \leq n$

$$\frac{q}{e^{\frac{q}{n}}(n!)^{\frac{1}{n}}}(\rho_{\widetilde{K}_n(g)}(u) + \varepsilon) \leq \frac{q}{e^{\frac{q}{n}}(n!)^{\frac{1}{n}}}(\rho_{\widetilde{K}_n(g)}^n(u) + \bar{\varepsilon})^{\frac{1}{n}} \leq g(r_0 u)^{\frac{1}{n}} r_0.$$

Now, we continue as in the proof of the first part of the Lemma. If for some $u \in S^{n-1}$ we assume that ω is not linear, then there exists $\varepsilon > 0$ such that for every $0 < q \leq n$

$$\frac{qe^{-\frac{q}{n}}}{(n!)^{\frac{1}{n}}}(\rho_{\widetilde{K}_n(g)}(u) + \varepsilon) \leq \rho_{K_{qe^{-\frac{q}{n}}(g)}}(u).$$

and consequently, for every $0 < t \leq \frac{n}{e}$

$$\frac{t}{(n!)^{\frac{1}{n}}}(\rho_{\widetilde{K}_n(g)}(u) + \varepsilon) \leq \rho_{K_t(g)}(u)$$

$\square$

Remark 2.6 The inclusion in the lemma above cannot be extended in general to the whole range of $t \in [0, n]$. If $f = \chi_K$ is the characteristic function of a convex body, $\widetilde{K}_n(f) = K$ and we would have, taking $t = n$, $\frac{n}{(n!)^{\frac{1}{n}}}\widetilde{K}_n(f) \subseteq K_n(f)$ then, by using Stirling's formula, it would imply for large values of n that $\frac{e}{2}K \subseteq K$ which is trivially false.

On the other hand, using the same ideas as above we can obtain a more general result, namely, for every $p > 0$ and $0 \leq t \leq \frac{p}{e}$, $\frac{t}{\Gamma(1+p)^{\frac{1}{p}}}\widetilde{K}_p(g) \subseteq K_t(g)$.

Now we can prove the main result of the paper. In this section we prove the inequality.

Proof of the Inequality of Theorem 2.1.1 Let us consider the covariogram function $g_f : \mathbb{R}^n \to \mathbb{R}$ defined by

$$g_f(x) = \int_0^\infty e^{-t}|K_t(f) \cap (x + K_t(f))|dt.$$

By Lemma 2.3.1 and Lemma 2.3.2 (since $g_f(0) = \|g_f\|_\infty > 0$ by Lemma 2.2.1), for any $0 < \lambda_0 < 1 - e^{-\frac{n}{e}}$ we have

$$
2\|f\|_1 \Pi^*(f) = \bigcap_{0 < \lambda < \lambda_0} \frac{K_{-\log(1-\lambda)}(g_f)}{\lambda} \supseteq \bigcap_{0 < \lambda < \lambda_0} \frac{-\log(1-\lambda)}{(n!)^{\frac{1}{n}} \lambda} \widetilde{K}_n(g_f).
$$

Since $h(\lambda) := -(\log(1-\lambda))/\lambda$ is increasing in $\lambda \in (0, 1)$, and $\lim_{\lambda \to 0+} h(\lambda) = 1$, then

$$
2\|f\|_1 \Pi^*(f) \supseteq \frac{1}{(n!)^{\frac{1}{n}}} \widetilde{K}_n(g_f).
$$

Taking Lebesgue measure we obtain that

$$
2^n \|f\|_1^n \left|\Pi^*(f)\right| \geq \frac{1}{n!} |\widetilde{K}_n(g_f)|.
$$

One can conclude the result as a direct consequence of Lemmas 2.2.1 and 2.2.2.

$\square$

2.4 Characterization of the Equality in Theorem 2.1.1

In this section we characterize the equality case in Theorem 2.1.1. First we will show that if there is equality in the theorem for a function f attaining its maximum at the origin, then the associated covariogram function g_f has to be log-linear on every 1-dimensional ray emanating from 0. Second we will prove that such condition implies that $\frac{f}{\|f\|_\infty} = e^{-\|\cdot\|_{\Delta_n}}$ for some n-dimensional simplex Δ_n containing the origin.

Lemma 2.4.1 *Let $f \in \mathcal{F}(\mathbb{R}^n)$ and let g_f be its covariogram function*

$$
g_f(x) = \int_0^\infty e^{-t} |K_t(f) \cap (x + K_t(f))| dt.
$$

Then f attains equality in Theorem 2.1.1 if and only if for every $x \in \mathbb{R}^n$ and every $\lambda \in [0, 1]$, $g_f(\lambda x) = g_f(0)^{1-\lambda} g(x)^\lambda$.

Proof By Lemma 2.3.2, if g_f is log-linear on every 1-dimensional ray emanating from 0 then the inclusion in the proof of the inequality in Theorem 2.1.1 is an equality and then f attains equality. Assume that f attains equality in Theorem 2.1.1 and g_f is not log-linear on every 1-dimensional ray emanating from 0. Then, by Lemma 2.3.2, there exists $u \in S^{n-1}$ and $\varepsilon > 0$ such that for any $0 < \lambda_0 < 1 - e^{-\frac{n}{e}}$

and any $0 < \lambda < \lambda_0$

$$\frac{-\log(1-\lambda)}{\lambda(n!)^{\frac{1}{n}}}(\rho_{\widetilde{K}_n(g_f)}(u) + \varepsilon) \leq \rho_{\frac{K_{-\log(1-\lambda)}(g_f)}{\lambda}}(u).$$

Consequently, for such u

$$\rho_{\bigcap_{0<\lambda<\lambda_0}\frac{K_{-\log(1-\lambda)}(g_f)}{\lambda}}(u) = \inf_{0<\lambda<\lambda_0} \rho_{\frac{K_{-\log(1-\lambda)}(g_f)}{\lambda}}(u)$$

$$\geq \inf_{0<\lambda<\lambda_0} \frac{-\log(1-\lambda)}{\lambda(n!)^{\frac{1}{n}}}(\rho_{\widetilde{K}_n(g_f)}(u) + \varepsilon)$$

$$= \frac{1}{(n!)^{\frac{1}{n}}}(\rho_{\widetilde{K}_n(g_f)}(u) + \varepsilon),$$

and then

$$2\|f\|_1 \Pi^*(f) = \bigcap_{0<\lambda<\lambda_0} \frac{K_{-\log(1-\lambda)}(g_f)}{\lambda} \supsetneq \frac{1}{(n!)^{\frac{1}{n}}}\widetilde{K}_n(g_f)$$

and the volume of the left-hand side convex body is strictly greater than the volume of the right hand side convex body. $\qquad\square$

The next lemma shows that if g_f is log-linear in every ray emanating from 0 then $\frac{f}{\|f\|_\infty} = e^{-\|\cdot\|_{\Delta_n}}$ for some simplex Δ_n containing the origin.

Lemma 2.4.2 *Let $f \in \mathcal{F}(\mathbb{R}^n)$ be such that $\|f\|_\infty = f(0)$ and let g_f be its covariogram function*

$$g_f(x) = \int_0^\infty e^{-t}|K_t(f) \cap (x + K_t(f))|dt.$$

Then, for every $x \in \mathbb{R}^n$ and every $\lambda \in [0, 1]$, $g_f(\lambda x) = g_f(0)^{1-\lambda}g_f(x)^\lambda$ if and only if $\frac{f(x)}{\|f\|_\infty} = e^{-\|x\|_{\Delta_n}}$ with Δ_n an n-dimensional simplex containing the origin.

Proof The condition $g_f(\lambda x) = g_f(0)^{1-\lambda}g_f(x)^\lambda$ for every $x \in \mathbb{R}^n$ and every $\lambda \in [0, 1]$ implies that $g_f(x) \neq 0$ for every $x \in \mathbb{R}^n$, since g_f is continuous at 0, as 0 is in the interior of the support of g_f. In the following, in order to ease the notation, we will denote $K_t = K_t(f)$ for every $t \in [0, \infty)$. For $x \in \mathbb{R}^n$ and $t \in [0, \infty)$, we define

$$h_x(t) = e^{-t}|K_t \cap (x + K_t)|.$$

Notice that for any $x \in \mathbb{R}^n$, any $t_1, t_2 \in [0, \infty)$ and any $\lambda \in [0, 1]$

$$K_{(1-\lambda)t_1+\lambda t_2} \cap (x + K_{(1-\lambda)t_1+\lambda t_2}) \supseteq (1-\lambda)(K_{t_1} \cap (x + K_{t_1})) + \lambda(K_{t_2} \cap (x + K_{t_2})).$$

Therefore, by Brunn–Minkowski inequality

$$|K_{(1-\lambda)t_1+\lambda t_2} \cap (x+K_{(1-\lambda)t_1+\lambda t_2})| \geq |K_{t_1} \cap (x+K_{t_1})|^{1-\lambda} |K_{t_2} \cap (x+K_{t_2})|^{\lambda} \quad (2.1)$$

and

$$h_x((1-\lambda)t_1 + \lambda t_2) \geq \left(e^{-t_1}|K_{t_1} \cap (x+K_{t_1})|\right)^{1-\lambda} \left(e^{-t_2}|K_{t_2} \cap (x+K_{t_2})|\right)^{\lambda}$$
$$= h_x(t_1)^{1-\lambda} h_x(t_2)^{\lambda}.$$

Consequently, since h_x is log-concave and integrable in $[0, \infty)$, for any $s \in [0, \infty)$, the set $\{t \in [0, \infty) : h_x(t) \geq s\}$ is either empty or a closed interval. Let us remark that for each $x \in \mathbb{R}^n$, since $g(x) \neq 0$, the function $h_x(t)$ is not identically 0 and it attains its maximum at a unique point, since if $\|h_x\|_\infty = h_x(t_1) = h_x(t_2)$, with $t_1 < t_2$, then we have that for every $\lambda \in [0, 1]$ $h_x((1-\lambda)t_1 + \lambda t_2) \geq \|h_x\|_\infty$, so $h_x((1-\lambda)t_1 + \lambda t_2) = \|h_x\|_\infty$. Therefore, we have equality in (2.1) and by the characterization of the equality cases in Brunn–Minkowski inequality (see, for instance [4, Section 1.2]) $K_{t_2} \cap (x + K_{t_2})$ is a translation of $K_{t_1} \cap (x + K_{t_1})$ and they have the same volume. Thus $h_x(t_2) < h_x(t_1)$, which contradicts the fact that the maximum is attained both at t_1 and t_2. For any $s \in [0, \infty)$, $\lambda \in [0, 1]$, $t_1 \in \{t \in [0, \infty) : h_0(t) \geq s\|h_0\|_\infty\}$ and $t_2 \in \{t \in [0, \infty) : h_x(t) \geq s\|h_x\|_\infty\}$, let us call $t = (1-\lambda)t_1 + \lambda t_2$. Since

$$K_t \cap (\lambda x + K_t) \supseteq (1-\lambda)K_{t_1} + \lambda(K_{t_2} \cap (x + K_{t_2})),$$

by Brunn–Minkowski inequality

$$|K_t \cap (\lambda x + K_t)| \geq |K_{t_1}|^{1-\lambda} |K_{t_2} \cap (x + K_{t_2})|^{\lambda}.$$

Then

$$h_{\lambda x}(t) = e^{-t}|K_t \cap (\lambda x + K_t)| \geq (e^{-t_1}|K_{t_1}|)^{1-\lambda}(e^{-t_2}|K_{t_2} \cap (x + K_{t_2})|)^{\lambda}$$
$$= h_0(t_1)^{1-\lambda} h_x(t_2)^{\lambda} \geq s\|h_0\|_\infty^{1-\lambda}\|h_x\|_\infty^{\lambda}.$$

Consequently, for any $x \in \mathbb{R}^n$, $s \in [0, \infty)$, and $\lambda \in [0, 1]$.

$$\{t \in [0, \infty) : h_{\lambda x}(t) \geq s\|h_0\|_\infty^{1-\lambda}\|h_x\|_\infty^{\lambda}\} \supseteq$$
$$(1-\lambda)\{t \in [0, \infty) : h_0(t) \geq s\|h_0\|_\infty\} + \lambda\{t \in [0, \infty) : h_x(t) \geq s\|h_x\|_\infty\}.$$

The last sets are non-empty for every $s \in [0, 1]$. Thus, for any $x \in \mathbb{R}^n$ and $\lambda \in [0, 1]$

$$\frac{g_f(\lambda x)}{\|h_0\|_\infty^{1-\lambda}\|h_x\|_\infty^{\lambda}} = \int_0^\infty |\{t \in [0, \infty) : h_{\lambda x}(t) \geq s\|h_0\|_\infty^{1-\lambda}\|h_x\|_\infty^{\lambda}\}| ds$$

$$\geq \int_0^1 |\{t \in [0, \infty) : h_{\lambda x}(t) \geq s\|h_0\|_\infty^{1-\lambda}\|h_x\|_\infty^{\lambda}\}| ds$$

$$\geq (1-\lambda) \int_0^1 |\{t \in [0,\infty) : h_0(t) \geq s\|h_0\|_\infty\}|ds$$

$$+\lambda \int_0^1 |\{t \in [0,\infty) : h_x(t) \geq s\|h_x\|_\infty\}|ds$$

$$= (1-\lambda) \int_0^\infty \frac{h_0(t)}{\|h_0\|_\infty} dt + \lambda \int_0^\infty \frac{h_x(t)}{\|h_x\|_\infty} dt$$

$$\geq \left(\int_0^\infty \frac{h_0(t)}{\|h_0\|_\infty} dt \right)^{1-\lambda} \left(\int_0^\infty \frac{h_x(t)}{\|h_x\|_\infty} dt \right)^\lambda$$

$$= \left(\frac{g_f(0)}{\|h_0\|_\infty} \right)^{1-\lambda} \left(\frac{g_f(x)}{\|h_x\|_\infty} \right)^\lambda.$$

Since, by assumption $g_f(\lambda x) = g_f(0)^{1-\lambda} g_f(x)^\lambda$, all the inequalities in the last chain of inequalities are equalities and then

- $\|h_{\lambda x}\|_\infty = \|h_0\|_\infty^{1-\lambda} \|h_x\|_\infty^\lambda$,
- The following two sets are equal for every $s \in [0,1]$

$$\{t \in [0,\infty) : h_{\lambda x}(t) \geq s\|h_0\|_\infty^{1-\lambda}\|h_x\|_\infty^\lambda\}$$
$$= (1-\lambda)\{t \in [0,\infty) : h_0(t) \geq s\|h_0\|_\infty\} + \lambda\{t \in [0,\infty) : h_x(t) \geq s\|h_x\|_\infty\},$$

- $\int_0^\infty \frac{h_0(t)}{\|h_0\|_\infty} dt = \int_0^\infty \frac{h_x(t)}{\|h_x\|_\infty} dt$ or, equivalently $\frac{g_f(0)}{\|h_0\|_\infty} = \frac{g_f(x)}{\|h_x\|_\infty}$.

Notice that, since the sets in the second condition are intervals, if we call

- $t_1(s) = \inf\{t \in [0,\infty) : h_0(t) > s\|h_0\|_\infty\}$
- $\bar{t}_1(s) = \sup\{t \in [0,\infty) : h_0(t) > s\|h_0\|_\infty\}$
- $t_2(s,x) = \inf\{t \in [0,\infty) : h_x(t) > s\|h_x\|_\infty\}$
- $\bar{t}_2(s,x) = \sup\{t \in [0,\infty) : h_x(t) > s\|h_x\|_\infty\}$,

the second condition is equivalent to

- $h_{\lambda x}((1-\lambda)t_1(s) + \lambda t_2(s,x)) = s\|h_0\|_\infty^{1-\lambda}\|h_x\|_\infty^\lambda$
- $h_{\lambda x}((1-\lambda)\bar{t}_1(s) + \lambda\bar{t}_2(s,x)) = s\|h_0\|_\infty^{1-\lambda}\|h_x\|_\infty^\lambda$

and, since $h_0(t_1(s)) = s\|h_0\|_\infty$, $h_x(t_2(s,x)) = s\|h_x\|_\infty$, $h_0(\bar{t}_1(s)) = s\|h_0\|_\infty$, and $h_x(\bar{t}_2(s,x)) = s\|h_x\|_\infty$, the last two equalities imply that for any $s \in [0,1]$ there is equality in

$$|K_{(1-\lambda)t_1(s)+\lambda t_2(s,x)} \cap (\lambda x + K_{(1-\lambda)t_1(s)+\lambda t_2(s,x)})|$$

$$\geq |K_{t_1(s)}|^{1-\lambda} |K_{t_2(s,x)} \cap (x + K_{t_2(s,x)})|^\lambda$$

and for any $s \in (0,1]$ there is equality in

$$|K_{(1-\lambda)\bar{t}_1(s)+\lambda\bar{t}_2(s,x)} \cap (\lambda x + K_{(1-\lambda)\bar{t}_1(s)+\lambda\bar{t}_2(s)})|$$

$$\geq |K_{\bar{t}_1(s)}|^{1-\lambda} |K_{\bar{t}_2(s,x)} \cap (x + K_{\bar{t}_2(s)})|^\lambda.$$

This implies that for any $x \in \mathbb{R}^n$, any $\lambda \in [0, 1]$ and any $s \in (0, 1]$

- $K_{t_2(s,x)} \cap (x + K_{t_2(s,x)})$ is a translation of $K_{t_1(s)}$ and also $K_{(1-\lambda)t_1(s)+\lambda t_2(s,x)} \cap (\lambda x + K_{(1-\lambda)t_1(s)+\lambda t_2(s,x)})$ is a translation of $K_{t_1(s)}$,
- $K_{\bar{t}_2(s,x)} \cap (x + K_{\bar{t}_2(s,x)})$ is a translation of $K_{\bar{t}_1(s)}$ and also $K_{(1-\lambda)\bar{t}_1(s)+\lambda \bar{t}_2(s,x)} \cap (\lambda x + K_{(1-\lambda)\bar{t}_1(s)+\lambda \bar{t}_2(s,x)})$ is a translation of $K_{\bar{t}_1(s)}$.

Let us remark that for $s = 1$, since the function h_x attains its maximum at a unique point and the value of this maximum is strictly positive, $t_1(1) = \bar{t}_1(1)$ and $t_2(1, x) = \bar{t}_2(1, x)$ for every $x \in \mathbb{R}^n$. Notice also that for any $x \in \mathbb{R}^n$ and any $\lambda \in [0, 1]$, since $\|h_{\lambda x}\|_\infty = \|h_0\|_\infty^{1-\lambda} \|h_x\|_\infty^\lambda$ and $h_{\lambda x}((1-\lambda)t_1(1) + \lambda t_2(1, x)) \geq \|h_0\|_\infty^{1-\lambda} \|h_x\|_\infty^\lambda$, we have that

$$t_2(1, \lambda x) = (1 - \lambda)t_1(1) + \lambda t_2(1, x).$$

Since for every fixed $x \in \mathbb{R}^n$ and every $s \in (0, 1]$ we have that $K_{t_2(s,x)} \cap (x+K_{t_2(s,x)})$ is a translation of $K_{t_1(s)}$, $e^{-t_1(s)}|K_{t_1(s)}| = s\|h_0\|_\infty$, and $e^{-t_2(s,x)}|K_{t_2(s,x)} \cap (x + K_{t_2(s,x)})| = s\|h_x\|_\infty$, we obtain that for every $x \in \mathbb{R}^n$ and every $s \in [0, 1]$

$$\frac{s\|h_0\|_\infty}{e^{-t_1(s)}} = \frac{s\|h_x\|_\infty}{e^{-t_2(s,x)}}.$$

Equivalently

$$e^{-(t_2(s,x)-t_1(s))} = \frac{\|h_x\|_\infty}{\|h_0\|_\infty}.$$

Thus, for every $x \in \mathbb{R}^n$, the difference $t_2(s, x) - t_1(s)$ does not depend on s and $t_2(s, x) - t_1(s) = t_2(1, x) - t_1(1)$ for every $s \in [0, 1]$. Then, we have that for every $x \in \mathbb{R}^n$ and any $\lambda \in [0, 1]$

$$t_2(0, \lambda x) = t_2(0, \lambda x) - t_1(0) = t_2(1, \lambda x) - t_1(1) = \lambda(t_2(1, x) - t_1(1))$$
$$= \lambda(t_2(0, x) - t_1(0)) = \lambda t_2(0, x).$$

Consequently, since for any $x \in \mathbb{R}^n$

$$t_2(0, x) = \inf\{t \in [0, \infty) : K_t \cap (x + K_t) \neq \emptyset\}$$
$$= \inf\{t \in [0, \infty) : x \in K_t - K_t\},$$

we have that for any $x \in \mathbb{R}^n$ and any $\lambda \in [0, 1]$

$$\inf\{t \in [0, \infty) : \lambda x \in K_t - K_t\} = \lambda \inf\{t \in [0, \infty) : x \in K_t - K_t\}.$$

Fixing $t_0 \in [0, \infty)$ we have that, for every $x \in \partial(K_{t_0} - K_{t_0})$, $\inf\{t \in [0, \infty) : x \in K_t - K_t\} = t_0$. Thus, for any $\lambda \in [0, 1]$, $\lambda x \in \partial(K_{\lambda t_0} - K_{\lambda t_0})$ and then for every $t \in [0, \infty)$ and any $\lambda \in [0, 1]$

$$\lambda K_t - \lambda K_t = \lambda(K_t - K_t) = K_{\lambda t} - K_{\lambda t}.$$

Since f is log-concave and $\|f\|_\infty = f(0)$ we have that $0 \in K_t$ for every $t \in [0, \infty)$ and for every $t \in [0, \infty)$ and any $\lambda \in [0, 1]$ $\lambda K_t \subseteq K_{\lambda t}$ and then

$$\lambda K_t - \lambda K_t \subseteq K_{\lambda t} - K_{\lambda t}.$$

Since the last inclusion is an equality and $\lambda K_t \subseteq K_{\lambda t}$ we obtain that $\lambda K_t = K_{\lambda t}$. Therefore, for every $t \in [0, \infty)$, $K_t = tK_1$ and $\frac{f(x)}{\|f\|_\infty} = e^{-\|x\|_{K_1}}$. Let us call $K := K_1$. Since $t_1(0) = 0$ and for every $x \in \mathbb{R}^n$

$$\begin{aligned}
t_2(0, x) &= \inf\{t \in [0, \infty) : K_t \cap (x + K_t) \neq \emptyset\} \\
&= \inf\{t \in [0, \infty) : tK \cap (x + tK) \neq \emptyset\} \\
&= \inf\{t \in [0, \infty) : x \in tK - tK\} \\
&= \|x\|_{K-K},
\end{aligned}$$

we have that for every $x \in \mathbb{R}^n$

$$g_f(x) = g_f(0)\frac{\|h_x\|_\infty}{\|h_0\|_\infty} = g_f(0)e^{-(t_2(0,x)-t_1(0))} = g_f(0)e^{-\|x\|_{K-K}}.$$

Therefore,

$$\int_{\mathbb{R}^n} g_f(x) = (n!)^2 |K||K - K|.$$

On the other hand, if $\frac{f(x)}{\|f\|_\infty} = e^{-\|x\|_K}$, by Lemma 2.2.1,

$$\begin{aligned}
\int_{\mathbb{R}^n} g_f(x)dx &= \int_{\mathbb{R}^n} \int_{\mathbb{R}^n} \min\{e^{-\|x\|_K}, e^{-\|y\|_K}\}dydx \\
&= \int_{\mathbb{R}^{2n}} e^{-\|(x,y)\|_{K \times K}} dydx \\
&= (2n)!|K|^2.
\end{aligned}$$

Thus

$$\binom{2n}{n}|K| = |K - K|$$

and, since we have equality in Rogers–Shephard inequality [11], K is a simplex and $0 \in K = K_1$. $\square$

Lemmas 2.4.1 and 2.4.2 together characterize the equality case.

References

1. D. Alonso-Gutiérrez, C.H. Jiménez, R. Villa, Brunn–Minkowski and Zhang inequalities for convolution bodies. Adv. Math. **238**, 50–69 (2013)
2. D. Alonso-Gutiérrez, B. González Merino, C.H. Jiménez, R. Villa, John's ellipsoid and the integral ratio of a log-concave function, J. Geom. Anal. **28**(2), 1182–1201 (2018)
3. D. Alonso-Gutiérrez, S. Artstein-Avidan, B. González Merino, C.H. Jiménez, R. Villa R, Rogers–Shephard and local Loomis–Whitney type inequalities. Math. Ann. **374**(3–4), 1719–1771 (2019)
4. S. Artstein-Avidan, A. Giannopoulos, V.D. Milman, *Asymptotic Geometric Analysis, Part 1.* Mathematical Surveys and Monographs, vol. 122 (American Mathematical Society, Providence, 2015)
5. K. Ball, Logarithmically concave functions and sections of convex sets in $\mathbb{R}^n$. Stud. Math. **88**(1), 69–84 (1988)
6. S. Brazitikos, A. Giannopoulos, P. Valettas, B.-H. Vritsiou, *Geometry of Isotropic Convex Bodies.* Mathematical Surveys and Monographs, vol. 196 (American Mathematical Society, Providence, 2014)
7. A. Colesanti, *Log-concave Functions.* Convexity and Concentration. The IMA Volumes in Mathematics and its Applications, vol. 161 (Springer, New York, 2017), pp. 487–524
8. R.J. Gardner, G. Zhang, Affine inequalities and radial mean bodies. Am. J. Math. **120**(3), 505–528 (1998)
9. B. Klartag, V. Milman, Geometry of log-concave functions and measures. Geom. Dedicata **112**(1), 169–182 (2005)
10. C.M. Petty, Isoperimetric problems, in *Proceedings of the Conference on Convexity and Combinatorial Geometry* (University of Oklahoma, Norman, 1971), pp. 26–41
11. C.A. Rogers, G.C. Shephard, The difference body of a convex body. Arch. Math. **8**, 220–233 (1957)
12. M. Schmuckensläger, The distribution function of the convolution square of a convex symmetric body in $\mathbb{R}^n$. Israel J. Math. **78**, 309–334 (1992)
13. G. Zhang, Restricted chord projection and affine inequalities. Geom. Dedicata **39**(2), 213–222 (1991)
14. G. Zhang, The affine Sobolev inequality. J. Differ. Geom. **53**, 183–202 (1999)

Chapter 3
Bobkov's Inequality via Optimal Control Theory

Franck Barthe and Paata Ivanisvili

Abstract We give a Bellman proof of Bobkov's inequality using arguments of dynamic programming. As a byproduct of the method we obtain a characterization of smooth optimizers.

3.1 Bobkov's Inequality

Bobkov's inequality [4] states that

$$\int_{\mathbb{R}^n} \sqrt{I^2(f) + |\nabla f|^2}\, d\gamma^n \geq I\left(\int_{\mathbb{R}^n} f\, d\gamma^n\right) \tag{3.1.1}$$

holds for any smooth $f : \mathbb{R}^n \to [0, 1]$, where $d\gamma^n(x) = (2\pi)^{-n/2} \exp(-|x|^2/2)\, dx$ is the standard Gaussian measure on $\mathbb{R}^n$, $I(x) = \varphi(\Phi^{-1}(x))$, $\Phi(t) = \gamma^1((-\infty, t])$ and $\varphi(t) = \Phi'(t)$. We simply write γ for γ^1. This functional inequality implies the sharp isoperimetric inequality for the Gaussian measure γ^n [5, 8, 14], and has led to far-reaching extensions [2]. Bobkov's original proof of (3.1.1) relies on a

This paper is based upon work supported by the National Science Foundation under Grant No. DMS-1440140 while two of the authors were in residence at the Mathematical Sciences Research Institute in Berkeley, California, during the Fall 2017 semester. P.I. is partially supported by NSF DMS-1856486 and CAREER DMS-1945102.

F. Barthe (✉)
Institut de Mathématiques de Toulouse, UMR 5219, Université de Toulouse, CNRS, Toulouse Cedex 9, France
e-mail: franck.barthe@math.univ-toulouse.fr

P. Ivanisvili
University of California Irvine, Irvine, CA, USA
e-mail: pivanisv@uci.edu

© Springer Nature Switzerland AG 2020
B. Klartag, E. Milman (eds.), *Geometric Aspects of Functional Analysis*,
Lecture Notes in Mathematics 2256,
https://doi.org/10.1007/978-3-030-36020-7_3

delicate two-point inequality and the central limit theorem. The inequality could be reproved by interpolation along the Ornstein–Uhlenbeck semigroup [2, 10] and by stochastic calculus [3]. Actually, (3.1.1) can be deduced by applying the Gaussian isoperimetric inequality (in $\mathbb{R}^{n+1}$) to the subgraph of the function $\Phi^{-1}(f)$ (but the main interest of (3.1.1) is to give a more flexible proof of it). The calculation of the Gaussian boundary measure of a subgraph can be found in Ehrhard's paper [9].

In this note we give a new "Bellman proof" of Bobkov's inequality using the standard dynamic programming principle. The *Bellman function approach* has been extensively used by Burkholder [6] in order to obtain sharp constants for martingale transforms. One may not find the term "Bellman function" in Burkholder's work, however, one notices that a special minimal zig-zag concave function solves the problem considered in [6]. In 1997, Nazarov–Treil [11] showed that Burkholder's technique can be used to solve several harmonic analysis problems in a unified way, and coined the term "Bellman function method". The main steps of the method are similar to the ones used in solving the problems of stochastic optimal control theory [12].

Our proof of Bobkov's inequality illustrates the Bellman function approach to isoperimetric estimates, which is new, to the best of our knowledge. A closely related approach was implemented in [1, 13] for Log-Sobolev and Hardy type inequalities. We show that the existence of a special Bellman function of three variables implies the Gaussian isoperimetric inequality and moreover, we describe this function explicitly. As a byproduct of the method, we easily characterize all smooth optimizers in (3.1.1). Also, we hope that having a new proof at hand will provide new opportunities to tackle open questions about extensions of Bobkov's inequality, as its p-version, see [16].

First, in the next section we briefly describe the main ideas of the dynamic programming principle. This allows us to explain the origin of the computations of the actual proof of (3.1.1), which appears in Sect. 3.3.

3.2 Bellman Principle

We briefly sketch to the reader how the argument of optimal control theory works in general. Suppose we would like to maximize the quantity

$$\int_{\mathbb{R}} F\big(t, f(t), f'(t)\big)\, dt \tag{3.2.1}$$

in terms of $\int_{\mathbb{R}} H(t, f(t))dt$ where F and H are some given functions, f is a test function from a *sufficiently nice class* so that all the expressions involved are well defined. This means that we would like to solve the following optimization problem

$$R(y) := \sup_f \left\{ \int_{\mathbb{R}} F\big(t, f(t), f'(t)\big)\, dt \;:\; \int_{\mathbb{R}} H\big(t, f(t)\big)\, dt = y \right\}.$$

Unfortunately the function $R(y)$ may not enjoy good properties. For instance, it is unclear how to find a corresponding ODE that $R(y)$ would satisfy. Therefore, following the optimal control theory approach, we introduce some extra variables, namely, we consider the more general optimization problem below:

$$B(t, x, y) := \sup_f \left\{ \int_{-\infty}^t F\big(s, f(s), f'(s)\big)\, ds \;:\; f(t) = x, \right.$$

$$\left. \int_{-\infty}^t H\big(s, f(s)\big)\, ds = y \right\}. \tag{3.2.2}$$

Then the limit value $\sup_x \lim_{t \to \infty} B(t, x, y)$ would be a good candidate for $R(y)$. On the other hand using the standard Bellman principle (see for example [15]) one can show that

$$F(t, x, v) \le B_t(t, x, y) + B_x(t, x, y)v + B_y(t, x, y)H(t, x) \tag{3.2.3}$$

for all $v \in \mathbb{R}$. Indeed, take any (t, x, y) and assume $f^*(s)$ optimizes (assume it exists) the right hand side of (3.2.2) on the interval $(-\infty, t]$ with fixed $f^*(t) = x$ and $\int_{-\infty}^t H(s, f^*(s))ds = y$, then take a small $\varepsilon > 0$, any $v \in \mathbb{R}$, and construct a new candidate on $(-\infty, t + \varepsilon]$, namely,

$$\tilde{f}(s) = \begin{cases} f^*(s), & s \le t; \\ f^*(t) + v(s - t), & s \in [t, t + \varepsilon]. \end{cases}$$

Then

$$B\left(t + \varepsilon, x + v\varepsilon, y + \int_t^{t+\varepsilon} H\big(s, x + v(s - t)\big)ds\right)$$

$$= B\left(t + \varepsilon, \tilde{f}(t + \varepsilon), \int_{-\infty}^{t+\varepsilon} H\big(s, \tilde{f}(s)\big)ds\right)$$

$$\ge \int_{-\infty}^{t+\varepsilon} F\big(s, \tilde{f}(s), \tilde{f}'(s)\big)ds = B(t, x, y) + \int_t^{t+\varepsilon} F\big(s, x + v(s - t), v\big)ds.$$

Subtracting $B(t, x, y)$ from both sides of the latter inequality, dividing by ε and sending ε to zero we arrive at (3.2.3). Here we are omitting several details and assumptions, for example, B does not have to be differentiable.

On the other hand if one finds any function $\tilde{B}(t, x, y)$ such that (3.2.3) holds with $\tilde{B}$ instead of B, and $\tilde{B}$ has the additional property that

$$\lim_{t \to -\infty} \tilde{B}\left(t, f(t), \int_{-\infty}^{t} H(s, f(s))ds\right) = 0, \tag{3.2.4}$$

then one automatically obtains the bound $\tilde{B} \geq B$. Indeed, take $f(t)$, and notice that (3.2.3) for $\tilde{B}$ implies

$$F\left(s, f(s), f'(s)\right) \leq \frac{d}{ds}\tilde{B}\left(s, f(s), \int_{-\infty}^{s} H(u, f(u))du\right).$$

Now integrating in s on the ray $(-\infty, t]$ we obtain that

$$B \leq \tilde{B}. \tag{3.2.5}$$

So we see that the problem of solving (3.2.2) boils down to finding solutions of (3.2.3). We can optimize in the variable v in (3.2.3) for $\tilde{B}$, i.e.,

$$\sup_{v \in \mathbb{R}} \left\{ F(t, x, v) - \tilde{B}_x(t, x, y)v \right\} \leq \tilde{B}_t(t, x, y) + \tilde{B}_y(t, x, y)H(t, x). \tag{3.2.6}$$

By (3.2.5), B should be the least such possible solution. Thus it is quite natural to expect that in fact we should have equality in (3.2.6) instead of inequality. This leads us to a first order fully nonlinear PDE, the so called Hamilton–Jacobi–Bellman PDE, which can be solved by the method of characteristics.

We would like to use the above strategy to prove Bobkov's inequality (3.1.1). First we consider the case when $n = 1$. The function B that we study in Sect. 3.3.2 is the solution of the following optimization problem[1]

$$B(t, x, y) = \inf_{f \in C^1} \left\{ \int_{-\infty}^{t} \sqrt{I^2(f(s)) + (f'(s))^2}\varphi(s)ds \; : \; f(t) = x, \right.$$

$$\left. \int_{-\infty}^{t} f(s)\varphi(s)ds = y \right\}. \tag{3.2.7}$$

In solving (3.2.6)–(3.2.4) we will make a shortcut: we can *guess* an explicit expression for B and then use it as a function $\tilde{B}$ in the argument of proof. For instance, one can guess from the Euler–Lagrange equation that the optimizers in (3.2.7) should be $f(s) = \Phi(as + b)$ for two arbitrary constants $a, b \in \mathbb{R}$ (one can also argue that global extremizers f in Bobkov's inequality should be such

---

[1] Here B appears as an infimum instead of supremum, but then $-B$ is a supremum and the above reasoning applies in the same way, except that all inequalities will be reversed and sup in (3.2.6) will be replaced by inf.

that the subgraph of $\Phi^{-1} \circ f$ is a half-space, for which the Gaussian isoperimetric inequality is tight). We use this observation in order to calculate the tentative value of the function $B(t, x, y)$. First we find $a = a(t, x, y)$ and $b = b(t, x, y)$ such that $\Phi(at + b) = x$, and $\int_{-\infty}^{t} \Phi(as + b)\varphi(s)ds = y$. Then we plug $f(s) := \Phi(a(t, x, y)s + b(t, x, y))$ into the functional of the right hand side in (3.2.7) and get a function, which we call $\tilde{B}(t, x, y)$. After some technical computations we will check that $\tilde{B}$ satisfies (3.2.6) with equality, and it has a limit (3.2.4). This will imply that $\tilde{B} \leq B$. In fact $\tilde{B} = B$ but we do not have to prove it because we will see that $\tilde{B} \leq B$ implies (3.1.1), which is enough for our purpose. It will be convenient to work with $M(t, p, y) = \tilde{B}(t, \Phi(p), y)$ in the beginning, and to switch back to $\tilde{B}(t, x, y)$ later on. Also, abusing notations, we will write B instead of $\tilde{B}$ as the function (3.2.7) does not appear in the formal proof.

3.3 The Proof

3.3.1 Hamilton–Jacobi–Bellman PDE

Given any $t, p \in \mathbb{R}$, and y with $0 < y < \Phi(t)$, we claim that the following equation

$$\int_{-\infty}^{t} \Phi\big((s - t)a + p\big)\varphi(s) \, ds = y \qquad (3.3.1)$$

has a unique C^1 solution $a = a(t, p, y)$. Indeed, notice that by Fubini's theorem the left hand side of (3.3.1) represents the Gaussian measure of the "truncated halfspaces", i.e.,

$$\gamma^2 \Big(\{(s, u) \in \mathbb{R}^2 : s \leq t \text{ and } u \leq (s - t)a + p\}\Big) = y. \qquad (3.3.2)$$

Clearly the left hand side of (3.3.2) is continuously decreasing in a, when $a \to -\infty$ it tends to $\Phi(t)$, and when $a \to +\infty$ it goes to zero. Since $0 < y < \Phi(t)$ we see that there exists a unique solution $a = a(t, p, y)$. The fact that $a \in C^1$ follows from the implicit function theorem (see the computations of partial derivatives below).

Lemma 3.3.1 *Let*

$$M(t, p, y) := \varphi\left(\frac{p - a(t, p, y)\,t}{\sqrt{1 + a^2(t, p, y)}}\right)$$

$$\times \Phi\left(\frac{t + a(t, p, y)\,p}{\sqrt{1 + a^2(t, p, y)}}\right) \quad \text{for} \quad p, t \in \mathbb{R}, \ 0 < y < \Phi(t).$$

$$(3.3.3)$$

We have

$$\sqrt{\varphi^2(t)\varphi^2(p) - M_p^2} = M_t + \Phi(p)\varphi(t)M_y, \tag{3.3.4}$$

where M_t, M_p and M_y denote the partial derivatives.

Proof The derivative of the left-hand side of (3.3.1) with respect to the variable a is equal to

$$\int_{-\infty}^{t} \varphi((s-t)a + p)\varphi(s)(s-t)ds,$$

which is strictly negative. Therefore we can apply the implicit function theorem, and get a function $a = a(t, p, y)$. Next we compute the partial derivatives of a. Differentiating (3.3.1) with respect to t gives

$$\Phi(p)\varphi(t) - a\int_{-\infty}^{t} \varphi((s-t)a + p)\varphi(s)ds$$

$$+ a_t \int_{-\infty}^{t} \varphi((s-t)a + p)\varphi(s)(s-t)ds = 0. \tag{3.3.5}$$

The latter two integrals can be computed directly:

$$\int_{-\infty}^{t} \varphi((s-t)a + p)\varphi(s)ds = \frac{1}{\sqrt{1+a^2}}\varphi\left(\frac{p-at}{\sqrt{1+a^2}}\right)\Phi\left(\frac{t+ap}{\sqrt{1+a^2}}\right)$$

$$= \frac{M}{\sqrt{1+a^2}}; \quad \int_{-\infty}^{t} \varphi((s-t)a + p)\varphi(s)(s-t)ds$$

$$= -\frac{\varphi(\frac{p-at}{\sqrt{1+a^2}})\varphi(\frac{t+ap}{\sqrt{1+a^2}})}{1+a^2} - \frac{a(p-at)}{(1+a^2)^{3/2}}\varphi\left(\frac{p-at}{\sqrt{1+a^2}}\right)\Phi\left(\frac{t+ap}{\sqrt{1+a^2}}\right)$$

$$- t\frac{M}{\sqrt{1+a^2}} == -\frac{\varphi(\frac{p-at}{\sqrt{1+a^2}})\varphi(\frac{t+ap}{\sqrt{1+a^2}})}{1+a^2} - \frac{(t+ap)}{(1+a^2)^{3/2}}M.$$

These formulas suggest to introduce two auxiliary functions:

$$P := \frac{p-at}{\sqrt{1+a^2}} \quad \text{and} \quad Q := \frac{t+ap}{\sqrt{1+a^2}}. \tag{3.3.6}$$

Then $M(t, p, y) = \varphi(P)\Phi(Q)$, and the latter two integrals become

$$\int_{-\infty}^t \varphi((s-t)a + p)\varphi(s)ds = \frac{\varphi(P)\Phi(Q)}{\sqrt{1+a^2}} \tag{3.3.7}$$

$$\int_{-\infty}^t \varphi((s-t)a + p)\varphi(s)(s-t)ds = -\frac{\varphi(P)(\varphi(Q) + Q\Phi(Q))}{1+a^2}. \tag{3.3.8}$$

Thus using (3.3.5), (3.3.7), and (3.3.8) we obtain

$$a_t = \frac{\Phi(p)\varphi(t) - a\int_{-\infty}^t \varphi((s-t)a + p)\varphi(s)ds}{-\int_{-\infty}^t \varphi((s-t)a + p)\varphi(s)(s-t)ds}$$

$$= (1+a^2)\frac{\Phi(p)\varphi(t) - \frac{a}{\sqrt{1+a^2}}\varphi(P)\Phi(Q)}{\varphi(P)(\varphi(Q) + Q\Phi(Q))}.$$

In a similar way we compute

$$a_y = \frac{1}{\int_{-\infty}^t \varphi((s-t)a + p)\varphi(s)(s-t)ds} = \frac{-(1+a^2)}{\varphi(P)(\varphi(Q) + Q\Phi(Q))},$$

and

$$a_p = \frac{-\int_{-\infty}^t \varphi((s-t)a + p)\varphi(s)ds}{\int_{-\infty}^t \varphi((s-t)a + p)\varphi(s)(s-t)ds} = \frac{\Phi(Q)\sqrt{1+a^2}}{\varphi(Q) + Q\Phi(Q)}.$$

Now let us compute the partial derivatives of $M = \varphi(P)\Phi(Q)$. First we compute the partial derivatives of P and Q. We have

$$P_t = \frac{\partial}{\partial t}\left(\frac{p - at}{\sqrt{1+a^2}}\right) = -\frac{a}{\sqrt{1+a^2}} - \frac{a_t}{1+a^2}Q; \quad Q_t = \frac{1}{\sqrt{1+a^2}} + \frac{a_t}{1+a^2}P;$$

$$P_p = \frac{1}{\sqrt{1+a^2}} - \frac{a_p}{1+a^2}Q; \quad Q_p = \frac{a}{\sqrt{1+a^2}} + \frac{a_p}{1+a^2}P;$$

$$P_y = \frac{-a_y}{1+a^2}Q; \quad Q_y = \frac{a_y}{1+a^2}P.$$

Therefore we have

$$M_t = \frac{aP\varphi(P)\Phi(Q) + \varphi(P)\varphi(Q)}{\sqrt{1+a^2}} + \frac{P\varphi(P)a_t}{1+a^2}(Q\Phi(Q) + \varphi(Q))$$

$$= \frac{\varphi(P)\varphi(Q)}{\sqrt{1+a^2}} + P\varphi(t)\Phi(p);$$

$$M_p = \frac{-P\varphi(P)\Phi(Q) + \varphi(P)\varphi(Q)a}{\sqrt{1+a^2}} + \frac{a_p P\varphi(P)}{1+a^2}(Q\Phi(Q) + \varphi(Q))$$

$$= \frac{\varphi(P)\varphi(Q)a}{\sqrt{1+a^2}};$$

$$M_y = (\varphi(P)\Phi(Q))_y = -P\varphi(P)\Phi(Q)P_y + \varphi(P)\varphi(Q)Q_y$$

$$= \frac{a_y}{1+a^2}\varphi(P)P(Q\Phi(Q) + \varphi(Q)) = -P.$$

Thus

$$M_t + \Phi(p)\varphi(t)M_y = \frac{\varphi(P)\varphi(Q)}{\sqrt{1+a^2}}, \quad \varphi^2(t)\varphi^2(p) - M_p^2 = \varphi^2(P)\varphi^2(Q)\frac{1}{1+a^2},$$
$$\tag{3.3.9}$$

where in the last equality we have used that $\varphi(p)\varphi(t) = \varphi(P)\varphi(Q)$, a direct consequence of (3.3.6). The identities in (3.3.9) imply (3.3.4), and thereby the lemma is proved.

Let us point out, for further use, that the latter identity satisfied by φ gives that

$$M_p = \frac{a}{\sqrt{1+a^2}}\varphi(p)\varphi(t). \tag{3.3.10}$$

$\square$

Lemma 3.3.2 *Let M be defined as in (3.3.3), and let $f : \mathbb{R} \to (0, 1)$ be any C^1 smooth function. Then*

$$\lim_{t \to -\infty} M\left(t, \Phi^{-1}(f(t)), \int_{-\infty}^{t} f d\gamma\right) = 0; \tag{3.3.11}$$

$$\lim_{t \to \infty} M\left(t, \Phi^{-1}(f(t)), \int_{-\infty}^{t} f d\gamma\right) = I\left(\int_{\mathbb{R}} f d\gamma\right). \tag{3.3.12}$$

Proof Here we set (omitting variables) $p = p(t) := \Phi^{-1}(f(t))$, $y = y(t) := \int_{-\infty}^{t} f d\gamma$ and

$$M = M(t, p, y) = \varphi\left(\frac{p - at}{\sqrt{1+a^2}}\right)\Phi\left(\frac{t + ap}{\sqrt{1+a^2}}\right),$$

where $a = a(t, p, y)$ is defined implicitly by (3.3.1).

First we check (3.3.11). Let $\varepsilon > 0$ be an arbitrary positive number. Then there exists A such that: $|u| \geq A \implies \varphi(u) \leq \varepsilon$. If $\left|\frac{p-at}{\sqrt{1+a^2}}\right| \geq A$ then clearly $|M| \leq \varepsilon$.

On the contrary, if $\theta = \theta(t) := \frac{p - at}{\sqrt{1+a^2}}$ verifies $|\theta(t)| < A$ then

$$\frac{t + ap}{\sqrt{1 + a^2}} = t\sqrt{1 + a^2} + \theta a \leq t\sqrt{1 + a^2} + A|a| \leq (t + A)\sqrt{1 + a^2},$$

which tends to $-\infty$ when $t \to -\infty$. Therefore, for t sufficiently negative,

$$|M| \leq \Phi\left(\frac{t + ap}{\sqrt{1 + a^2}}\right) \leq \varepsilon.$$

Since $\varepsilon > 0$ was arbitrary, we have shown that

$$\lim_{t \to -\infty} M\left(t, \Phi^{-1}(f(t)), \int_{-\infty}^{t} f d\gamma\right) = 0.$$

To verify (3.3.12) we notice that (3.3.1) implies

$$y + \int_{t}^{\infty} \Phi((s - t)a + p)\varphi(s)ds = \int_{-\infty}^{\infty} \Phi((s - t)a + p)\varphi(s)ds = \Phi\left(\frac{p - at}{\sqrt{1 + a^2}}\right).$$

Therefore we obtain

$$\lim_{t \to \infty} \frac{p - at}{\sqrt{1 + a^2}} = \lim_{t \to \infty} \Phi^{-1}\left(\int_{-\infty}^{t} f d\gamma + \int_{t}^{\infty} \Phi((s - t)a + p)\varphi(s)ds\right)$$

$$= \Phi^{-1}\left(\int_{-\infty}^{\infty} f d\gamma\right)$$

regardless of the values of the function a. Since f takes values in $(0, 1)$, we have proved that the function $\theta(t) = \frac{p - at}{\sqrt{1+a^2}}$ has a (finite) limit when t tends to $+\infty$ and therefore, $|\theta|$ is bounded on $[0, +\infty)$ by a constant Θ. Thus

$$\frac{t + ap}{\sqrt{1 + a^2}} = t\sqrt{1 + a^2} + \theta a \geq t\sqrt{1 + a^2} - \Theta|a| \geq (t - \Theta)\sqrt{1 + a^2},$$

tends to $+\infty$ when $t \to +\infty$ (recall that Θ is a constant). Thus

$$\lim_{t \to \infty} M\left(t, \Phi^{-1}(f(t)), \int_{-\infty}^{t} f d\gamma\right) = \lim_{t \to \infty} \varphi\left(\frac{p - at}{\sqrt{1 + a^2}}\right) \Phi\left(\frac{t + ap}{\sqrt{1 + a^2}}\right)$$

$$= \varphi\left(\Phi^{-1}\left(\int_{-\infty}^{\infty} f d\gamma\right)\right).$$

$\square$

3.3.2 Deriving Bobkov's Inequality

Let $B(t, x, y) := M(t, \Phi^{-1}(x), y)$ for $t \in \mathbb{R}$, $x \in (0, 1)$ and $0 < y < \Phi(t)$. Lemma 3.3.1 implies that

$$I(x)\sqrt{\varphi^2(t) - B_x^2} = B_t + x\varphi(t)B_y. \tag{3.3.13}$$

One can easily check by studying the derivative in v that

$$\min_{v \in \mathbb{R}} \left\{ \varphi(t)\sqrt{I^2(x) + v^2} - vB_x \right\} = I(x)\sqrt{\varphi^2(t) - B_x^2}, \tag{3.3.14}$$

and that the minimum is attained only when[2] $v = \dfrac{I(x)B_x}{\sqrt{\varphi^2(t) - B_x^2}}$. Therefore (3.3.13) and (3.3.14) imply that for any $v \in \mathbb{R}$ we have

$$\varphi(t)\sqrt{I^2(x) + v^2} \geq B_t(t, x, y) + B_x(t, x, y)v + B_y(t, x, y)x\varphi(t), \tag{3.3.15}$$

where the inequality is strict when $v \neq \dfrac{I(x)B_x}{\sqrt{\varphi^2(t) - B_x^2}}$.

Now take any $f \in C^1(\mathbb{R})$ with values in $(0, 1)$ such that $\int_{\mathbb{R}} \sqrt{I(f)^2 + (f')^2}d\gamma < \infty$ (otherwise there is nothing to prove). Applying (3.3.15) for $x = f(t)$, $v = f'(t)$ and $y = \int_{-\infty}^{t} f\varphi$, we get:

$$\Psi(t) := \sqrt{I^2(f(t)) + (f'(t))^2}\varphi(t)$$
$$- \frac{d}{dt}B\left(t, f(t), \int_{-\infty}^{t} f\varphi\right) \geq 0 \quad \text{for all} \quad t \in \mathbb{R}.$$

Therefore

$$\int_{-T}^{T} \sqrt{I^2(f(t)) + (f'(t))^2}\varphi(t)dt - M\left(T, \Phi^{-1}(f(T)), \int_{-\infty}^{T} fd\gamma\right)$$
$$+ M\left(-T, \Phi^{-1}(f(-T)), \int_{-\infty}^{-T} fd\gamma\right)$$
$$= \int_{-T}^{T} \left[\sqrt{I^2(f(t)) + (f'(t))^2}\varphi(t) - \frac{d}{dt}B\left(t, f(t), \int_{-\infty}^{t} f\varphi\right)\right]dt$$
$$= \int_{-T}^{T} \Psi(t)dt \geq 0.$$

---

[2] Notice that $\varphi^2(t) - B_x^2 \neq 0$ follows from the second equation in (3.3.9)

Finally sending $T \to \infty$ and using Lemma 3.3.2 we obtain

$$\int_{\mathbb{R}} \sqrt{I^2(f(t)) + (f'(t))^2}\varphi(t)dt - I\left(\int_{\mathbb{R}} f\varphi\right) = \lim_{T \to \infty} \int_{-T}^{T} \Psi(t)dt \geq 0.$$
(3.3.16)

Using standard approximation arguments we can extend (3.3.16) to any $C^1(\mathbb{R})$ smooth f with values in $[0, 1]$. This proves Bobkov's inequality (3.1.1) in dimension $n = 1$. To obtain (3.1.1) in higher dimensions, we use the standard tenzorization argument [3]. Let us illustrate the argument for $n = 2$. Take any $C^1(\mathbb{R}^2)$ smooth $g(x, y)$ with values in $[0, 1]$. We have

$$I\left(\int_{\mathbb{R}}\int_{\mathbb{R}} g(x, y)d\gamma(x)d\gamma(y)\right)$$

$$\overset{(3.3.16)}{\leq} \int_{\mathbb{R}} \sqrt{I^2\left(\int_{\mathbb{R}} g(x, y)d\gamma(y)\right) + \left(\int_{\mathbb{R}} g_x(x, y)d\gamma(y)\right)^2}\, d\gamma(x)$$

$$\overset{(3.3.16)}{\leq} \int_{\mathbb{R}} \sqrt{\left(\int_{\mathbb{R}} \sqrt{I^2(g) + g_y^2}d\gamma(y)\right)^2 + \left(\int_{\mathbb{R}} g_x(x, y)d\gamma(y)\right)^2}\, d\gamma(x)$$

$$\overset{\text{minkowski}}{\leq} \int_{\mathbb{R}}\int_{\mathbb{R}} \sqrt{I^2(g) + g_x^2 + g_y^2}d\gamma(x)d\gamma(y) = \int_{\mathbb{R}^2} \sqrt{I^2(g) + |\nabla g|^2}d\gamma^2.$$
(3.3.17)

This finishes the proof of Bobkov's inequality.

3.3.3 Optimizers

Assume that a C^1 function $f : \mathbb{R} \to (0, 1)$ is such that Bobkov's inequality (3.1.1) is an equality. Then the left hand side of (3.3.16) is zero. Since Ψ is a non-negative continuous function, it follows that $\Psi(t) = 0$ for all $t \in \mathbb{R}$. This means that (3.3.15) was an equality when we applied it to prove that $\Psi \geq 0$, therefore $v = \frac{I(x)B_x}{\sqrt{\varphi^2(t) - B_x^2}}$ where $x = f(t)$, $v = f'(t)$, and B_x stands for $B_x\left(t, f(t), \int_{-\infty}^{t} f\varphi\right)$. Hence for all $t \in \mathbb{R}$,

$$\frac{f'(t)}{I(f(t))} = \frac{B_x}{\sqrt{\varphi^2(t) - B_x^2}}.$$

Let us rewrite this equation, by setting $h(t) := \Phi^{-1}(f(t))$ and using as before $M(t, p, y) := B(t, \Phi(p), y)$. Since $h'(t) = \frac{f'(t)}{I(f(t))}$ and $M_p(t, p, y) = $

$\varphi(p)B_x(t, \Phi(p), y)$ we get after simplification

$$h'(t) = \frac{M_p\left(t, h(t), \int_{-\infty}^t \Phi(h)\varphi\right)}{\sqrt{\varphi(t)^2\varphi(h(t))^2 - M_p^2\left(t, h(t), \int_{-\infty}^t \Phi(h)\varphi\right)}}$$

$$\overset{(3.3.10)}{=} a\left(t, h(t), \int_{-\infty}^t \Phi(h)\varphi\right).$$

Since a is C^1, and so is h by hypothesis, this equation shows that h is C^2. Using (3.3.1) we obtain

$$\int_{-\infty}^t \Phi((s-t)h'(t) + h(t))\varphi(s)ds = \int_{-\infty}^t \Phi(h(s))\varphi(s)ds. \qquad (3.3.18)$$

After differentiation of (3.3.18) in t and some simplifications we obtain

$$h''(t) \int_{-\infty}^t \varphi((s-t)h'(t) + h(t))\varphi(s)(s-t)ds = 0.$$

The latter equality can hold if and only if $h'' = 0$, and thereby $f(t) = \Phi(ut + v)$ for some constants $u, v \in \mathbb{R}$.

One can extend this result to higher dimensions by showing that all C^1 functions $f : \mathbb{R}^n \to (0, 1)$ which reach equality in Bobkov's inequality are of the form $f = \Phi \circ \ell$ for some linear form ℓ. Indeed, for this we need to carefully examine the equality cases in the tensorization argument. Let us again illustrate the argument for $n = 2$. Take any $g \in C^1(\mathbb{R}^2)$ which takes values in $(0, 1)$, and which achieves the equality in Bobkov's inequality. Equality on the second step in the chain of inequalities (3.3.17) implies that $g(x, y) = \Phi(yu(x) + v(x))$ for some functions $u(x), v(x)$. Since $g \in C^1$ and Φ is a smooth diffeomorphism we see that $u, v \in C^1(\mathbb{R})$. On the other hand equality in the part of Minkowski inequality (3.3.17) implies that

$$\sqrt{I^2(g) + g_y^2} = k(x)g_x(x, y)$$

for a nonvanishing function $k(x)$. Simplifying the latter equality we obtain

$$\sqrt{1 + u(x)^2} = k(x)(yu'(x) + v'(x)) \quad \text{for all} \quad x, y \in \mathbb{R}.$$

It follows that $u(x) = C_1$ is a constant, i.e., $g(x, y) = \Phi(yC_1 + v(x))$. Repeating the same reasonings in a different order for the variables x, y one obtains that $g(x, y) = \Phi(xC_2 + \tilde{v}(y))$, and thereby $yC_1 + v(x) = xC_2 + \tilde{v}(y)$ for all $x, y \in \mathbb{R}$. Then it easily follows that $g(x, y) = \Phi(xC_1 + yC_2 + C_3)$ for some constants C_1, C_2 and C_3.

Clearly, these functions, $f = \Phi \circ \ell$ for some linear ℓ, do give equality cases (the subgraph of $\Phi^{-1} \circ f = \ell$ is a half-space, which gives equality in the Gaussian isoperimetric inequality). Hence we have characterized all C^1 equality cases in Bobkov's inequality. However, at this stage our approach is not sufficiently developed and cannot handle functions of lower regularity. Carlen and Kerce [7] have studied equality cases in the natural larger class of functions with bounded variations, where additional equality cases are given by indicator functions of half-spaces.

Acknowledgement We would like to thank the anonymous referee for his/her helpful suggestions and comments.

References

1. R.A. Adams, F.H. Clarke, Gross's Logarithmic Sobolev inequality: a simple proof. Am. J. Math. **101**(6), 1265–1269 (1979)
2. D. Bakry, M. Ledoux, Lévy–Gromov's isoperimetric inequality for an infinite dimensional diffusion generator. Invent. Math. **123**, 259–281 (1996)
3. F. Barthe, B. Maurey, Some remarks on isoperimetry of Gaussian type. Ann. Inst. H. Poincaré Probab. Statist. **36**(4), 419–434 (2000)
4. S.G. Bobkov, An isoperimetric inequality on the discrete cube, and an elementary proof of the isoperimetric inequality in Gauss space. Ann. Probab. **25**(1), 206–214 (1997)
5. C. Borell, The Brunn–Minkowski inequality in Gauss space. Invent. Math. **30**, 207–216 (1975)
6. D.L. Burkholder, Boundary value problems and sharp inequalities in for martingale transforms. Ann. Prob. **12**(3), 647–702 (1984)
7. E.A. Carlen, C. Kerce, On the cases of equality in Bobkov's inequality and Gaussian rearrangement. Calc. Var. Partial Differ. Equ. **13**(1), 1–18 (2001)
8. A. Ehrhard, Symétrisation dans l'espace de Gauss. Math. Scand. **53**, 281–301 (1983)
9. A. Ehrhard, Inégalités isopérimétriques et intégrales de Dirichlet gaussiennes. Ann. Sci. École Norm. Sup. (4) **17**(2), 317–332 (1984)
10. M. Ledoux, *A Short Proof of the Gaussian Isoperimetric Inequality. High Dimensional Probability*. High Dimensional Probability (Oberwolfach, 1996). Progress in Probability, vol. 43 (Birkhäuser, Basel, 1998), pp. 229–232
11. F. Nazarov, S. Treil, Hunting the Bellman function: application to estimates of singular integrals and other classical problems of harmonic analysis. Algebra i Analiz **8**(5), 32–162 (1996) (in Russian); translated in St.-Petersburg Math. J. **8**(5), 721–824 (1997)
12. F. Nazarov, S. Treil, A. Volberg, Bellman function in stochastic optimal control and harmonic analysis (how our Bellman function got its name). Oper. Th.: Adv. Appl. **129**, 393–424 (2001)
13. A. Osekowski, A new approach to Hardy-type inequalities. Arch. Math. **104**(2), 165–176 (2015)
14. V.N. Sudakov, B.S. Tsirel'son, Extremal properties of half-spaces for spherically invariant measures. J. Soviet Math. **9**, 9–18 (1978)
15. L.C. Young, *Lectures on the Calculus of Variations and Optimal Control Theory* (Saunders, Philadelphia, 1969)
16. B. Zegarlinski, Isoperimetry for Gibbs measures. Ann. Probab. **29**(2), 802–819 (2001)

Chapter 4
Arithmetic Progressions in the Trace of Brownian Motion in Space

Itai Benjamini and Gady Kozma

Abstract It is shown that the trace of three dimensional Brownian motion contains arithmetic progressions of length 5 and no arithmetic progressions of length 6 a.s.

4.1 Introduction

In this note we comment that a.s. the trace of a three dimensional Brownian motion contains arithmetic progressions of length 5, and no arithmetic progressions of length 6.

Similarly, the maximal arithmetic progression in the trace of Brownian motion in $\mathbb{R}^d$ is 3 for $d = 4, 5$ and 2 above that (we will only prove the three dimensional result here, the higher dimensional results are proved similarly). On the other hand, the trace of a two dimensional Brownian motion a.s. contains arbitrarily long arithmetic progressions starting at the origin and having a fixed difference.

Consider n steps simple random walk on the d dimensional square grid $\mathbb{Z}^d$, look at the number of arithmetic progressions of length 3 in the range, study the distribution and large deviations?

Question In the large deviations regime, is there a deterministic limiting shape?

4.2 Proofs

We start with the two dimensional case.

I. Benjamini · G. Kozma (✉)
Weizmann Institute of Science, Rehovot, Israel
e-mail: gady.kozma@weizmann.ac.il

© Springer Nature Switzerland AG 2020
B. Klartag, E. Milman (eds.), *Geometric Aspects of Functional Analysis*,
Lecture Notes in Mathematics 2256,
https://doi.org/10.1007/978-3-030-36020-7_4

Proposition 4.1 *The trace of two dimensional Brownian motion a.s. contains arbitrarily long arithmetic progressions starting at the origin and having a fixed difference.*

Proof Given a set S of Hausdorff dimension 1 in the Euclidean plane, two dimensional Brownian motion W running for unit time will intersect S in a set of Hausdorff dimension 1 as well, with positive probability, see e.g. [1, 10]. Examine the unit circle. With positive probability Brownian motion run for unit time intersects the unit circle in a set S_1 of dimension 1. To each point in S_1 add it to itself to get S_2 a set of dimension 1. Let $\tau_1 \geq 1$ be the first time (after 1) our Brownian motion hits the circle with radius $3/2$. Examine it now in the time interval $[\tau_1, \tau_1 + 1]$. By the Harnack principle [1, Theorem 3.42], the probability that Brownian motion started from $W(\tau_1)$ to intersect S_2 in a set of dimension 1 is comparable to that of Brownian motion starting from 0 which, as already stated, is bounded away from 0. Hence $W[\tau_1, \tau_1 + 1]$ will again intersect S_2 in a set of dimension 1. To each point in the intersection of the form $2x, x \in S_1$ add x and call the resulting set S_3, again of dimension 1. Continue in the same manner to get arbitrarily long arithmetic progressions. Scale invariance implies that we get arbitrarily long arithmetic progression with probability 1.   □

The argument above shows that with positive probability the trace of a unit time two dimensional Brownian motion admits uncountably many arithmetic progressions of arbitrary length and difference 1.

We now prove the high dimensional result.

Lemma 4.2 *A three-dimensional Brownian motion contains no arithmetic progressions of length 6, a.s.*

Proof By scaling invariant we may restrict our attention to arithmetic progressions contained in the unit ball B, and to spacings at least δ for some $\delta > 0$. Denote the Brownian motion by W. If it contains an arithmetic progression then for every $\varepsilon > 0$ one may find $x_1, \ldots, x_6 \in B \cap \frac{1}{3}\varepsilon\mathbb{Z}^d$ such that $W \cap B(x_i, \varepsilon) \neq \emptyset$ and such that the x_i form an ε-approximate arithmetic progressions, by which we mean that $|x_{i-1} + x_{i+1} - 2x_i| \leq 4\varepsilon$ for $i = 2, 3, 4, 5$. Further, the x_i are δ-separated in the sense that $|x_i - x_{i+1}| \geq \delta - 2\varepsilon$ for $i = 1, 2, 3, 4, 5$. Denote the set of such x_i by $\mathscr{X}$ and define

$$H_x = \{W \cap B(x_i, \varepsilon) \neq \emptyset \; \forall i \in \{1, \ldots, 6\}\} \qquad x = (x_1, \ldots, x_6)$$

$$X = X(\varepsilon) = \sum_{x \in \mathscr{X}} \mathbb{1}\{H_x\}.$$

We now claim that

$$\mathbb{E}(X) \leq C \qquad \mathbb{E}(X^2) \geq c|\log \varepsilon| \tag{4.1}$$

where the constants c and C may depend on δ. Both calculations are standard: the first (that of $\mathbb{E}(X)$), is an immediate corollary of the fact that $3d$ Brownian motion starting from 0 hits the ball $B(v, \varepsilon)$ with probability $\approx \varepsilon/(|v| + \varepsilon)$, see e.g. [1, corollary 3.19]. Here and below, $\approx$ means that the ratio of the two quantities is bounded above and below by constants that depend only on δ. This gives

$$\mathbb{P}(H_x) \approx \frac{\varepsilon^6}{d(0, x) + \varepsilon} \tag{4.2}$$

where $d(0, x) := \min\{d(0, x_i) : i = 1, \ldots, 6\}$. Denote by $\mathscr{X}_n$ the set of $x \in \mathscr{X}$ such that $\varepsilon 2^n < d(0, x) \le \varepsilon 2^{n+1}$, with $\mathscr{X}_0$ having the lower bound removed. We can now write

$$\mathbb{E}(X) = \sum_{n=0}^{\log 1/\varepsilon} \sum_{x \in \mathscr{X}_n} \mathbb{P}(H_x) \overset{(4.2)}{\le} C \sum_{n=0}^{\log 1/\varepsilon} 2^{3n} \cdot 2^{-n} \cdot \varepsilon^{-3} \cdot \varepsilon^5 \le C$$

where 2^{3n} is the number of possibilities for the x_i closest to 0 for $x \in \mathscr{X}_n$ (this, and all other quantities in this explanation are up to constants); where 2^{-n} is $\mathbb{P}(W \cap B(x_i, \varepsilon) \ne \emptyset)$; where ε^{-3} is the number of possibilities for $x_2 - x_1$ (we use here that the determination of x_1 and $x_2 - x_1$ leave only a constant number of possibilities for $x_3, \ldots, x_4$); and where ε^5 is the probability to hit all of $B(x_1, \varepsilon), \ldots, B(x_6, \varepsilon)$ except $B(x_i, \varepsilon)$ given that you have hit $B(x_i, \varepsilon)$.

The calculation of $\mathbb{E}(X^2)$ is similar, we write $\mathbb{E}(X^2) = \sum_{x, y \in \mathscr{X}} \mathbb{P}(H_x \cap H_y)$ and estimate the probability directly. We get about constant contribution from each set $\{x, y : |x_i - y_i| \approx 2^{-n} \, \forall i\}$ for every n, hence the $|\log \varepsilon|$ term.

We now make a somewhat stronger claim on the interaction between different x. We claim that there exists $\lambda > 0$ such that, for any x,

$$\mathbb{P}(H_x \cap \{X \le \lambda |\log \varepsilon|\}) \le \frac{C}{|\log \varepsilon|} \mathbb{P}(H_x). \tag{4.3}$$

To see this fix x and let, for each scale $k \in \{1, \ldots, \lfloor |\log \varepsilon| \rfloor\}$,

$$X_k := \sum_{y \in \mathscr{Y}_k} \mathbb{1}\{H_y\}$$

$$\mathscr{Y}_k := \{y \in \mathscr{X} : 2^k \varepsilon \le |y_i - x_i| < 2^{k+1} \varepsilon \quad \forall i \in \{1, \ldots, 6\}\}$$

(X_k depends on x, of course, but we omit this dependency from the notation). A calculation identical to the above shows that $\mathbb{E}(X_k \mid H_x) \ge c$ and $\mathbb{E}(X_k^2 \mid H_x) \le C$ so

$$\mathbb{P}(X_k > 0 \mid H_x) \ge c. \tag{4.4}$$

Further, the events $X_k > 0$ (still conditioned on H_x) are approximately independent in the following sense:

Lemma 4.3 *For each $x \in \mathscr{X}$ and $k \in \{1, \ldots, \lfloor \log(\delta/4\varepsilon) \rfloor\}$,*

$$\operatorname{cov}(X_k > 0, X_l > 0 \mid H_x) \leq 2e^{-c|k-l|}. \tag{4.5}$$

Proof Assume for concreteness that $k < l$ and that $l - k$ is sufficiently large (otherwise the claim holds trivially, if only the c in the exponent is taken sufficiently small). Define two radii $r < s$ between $2^k \varepsilon$ and $2^l \varepsilon$ as follows:

$$r := 2^{(2/3)k+(1/3)l} \varepsilon \qquad s := 2^{(1/3)k+(2/3)l} \varepsilon.$$

Next, define a sequence of stopping times: the even ones for exiting balls of radius s and the odd ones for entering balls of radius r. In a formula, let $\tau_0 = 0$ and

$$\tau_{2m+1} := \inf \left\{ t \geq \tau_{2m} : W(t) \in \bigcup_{i=1}^{6} B(x_i, r) \right\}$$

$$\tau_{2m} := \inf \left\{ t \geq \tau_{2m-1} : W(t) \notin \bigcup_{i=1}^{6} B(x_i, s) \right\}$$

Let M be the first number such that $\tau_{2M+1} = \infty$. With probability 1 M is finite. We now claim that

$$\mathbb{P}(H_x \cap \{M \geq 6 + \lambda\}) \leq \frac{\varepsilon^6}{d(0, x) + \varepsilon} \left(Cr/s \right)^{\lambda} \qquad \forall \lambda = 1, 2, \ldots \tag{4.6}$$

To see (4.6) assume $d(0, x) > c$ for simplicity. Then every visit to $B(x_i, \varepsilon)$ from $\partial B(x_i, r)$ "costs" ε/r in the probability, while every visit of $B(x_j, r)$ from $\partial B(x_i, s)$ costs r/s if $i = j$ and r if $i \neq j$. Since H_x requires a visit to all of $x_1, \ldots, x_6$ we have to pay the costs ε/r and r at least 6 times, and the costs of r/s (or r, which is smaller) at least λ times. Counting over the order in which these visits happen adds no more than a C^{λ}. This shows (4.6) in the case that $d(0, x) > c$. The other case is identical and we skip the details.

Since $\mathbb{P}(H_x) \approx \varepsilon^6/(d(0, x) + \varepsilon)$ (recall (4.2)) this shows that the case $M > 6$ is irrelevant. Indeed, if we define $\mathscr{K} = \{X_k > 0\} \cap \{M = 6\}$ and $\mathscr{L} = \{X_l > 0\} \cap \{M = 6\}$ then

$$|\operatorname{cov}(\mathscr{K}, \mathscr{L} \mid H_x) - \operatorname{cov}(X_k > 0, X_l > 0 \mid H_x)| \leq \frac{Cr}{s} \tag{4.7}$$

(for $l - k$ sufficiently large) and we may concentrate on $\operatorname{cov}(\mathscr{K}, \mathscr{L} \mid H_x)$.

Let μ be the measure on $\mathbb{R}^{36}$ giving the distribution of $W(\tau_1), \ldots, W(\tau_{12})$ (we will not distinguish between $(\mathbb{R}^3)^{12}$ and $\mathbb{R}^{36}$). For an event E we will use $\mathbb{P}(E \mid W = u)$ as a short for

$$\mathbb{P}(E \mid W(\tau_i) = u_i \; \forall i \in \{1, \ldots, 12\}, M = 6)$$

(which is of course a μ-almost everywhere defined function). We next observe that for E equal to any of $\mathscr{L}$, H_x and $\mathscr{K} \cap H_x$ the function $\mathbb{P}(E \mid W = u)$ is nearly constant i.e.

$$\frac{\operatorname{ess\,max} \mathbb{P}(E \mid W = u)}{\operatorname{ess\,min} \mathbb{P}(E \mid W = u)} \le 1 + 2e^{-c|k-l|} \tag{4.8}$$

This is because $\mathscr{K}$ and H_x depend only on the behaviour inside the balls $B(x_i, 2^k \varepsilon)$ while u_{2m+1} are on $\partial B(x_i, r)$. This follows from the well-known fact that the distribution of W on the first hitting times (after τ_{2m+1}) of $B(x_i, 2^k \varepsilon)$ is independent of u_{2m+1}, up to an error of $(2^k \varepsilon)/r$; and similarly, the conditioning on exiting $B(x_i, s)$ at u_{2m+2} only adds an error of $(2^k \varepsilon)/s$. For the convenience of the reader we recall briefly how this is shown: consider Brownian motion started from a $y_1 \in \partial B(x_i, r)$ and let $y_2 \in \partial B(x_i, 2^{k+1} \varepsilon)$ be the first point visited in $B(x_i, 2^{k+1} \varepsilon)$, let y_3 be the last, and let y_4 be the first point visited in $B(x_i, s)$. Then the joint distribution of y_2, y_3 and y_4 can be written easily using the Poisson kernel (see [1, Theorem 3.44] for its formula). For example, the density of y_2 is $(r - \varepsilon 2^{k+1}) |y_2 - y_1|^{-3}$ times the uniform density on the sphere of radius $\varepsilon 2^{k+1}$ (the density in $\mathbb{R}^3$) from which we need to subtract the density after exiting $B(x_i, s)$, which is given by an integral of similar expressions. The exact form does not matter, only the fact that the y_1 dependency comes from the term $|y_2 - y_1|$ is nearly constant in y_1 in the sense above. The same holds for the density of the transition from y_3 to y_4 and the density between y_2 and y_3 is of course completely independent of y_1 and y_4. Conditioning on exiting in a given y_4 is merely restricting to a subspace and normalising, conserving the near independence. This justifies (4.8) in this case.

We have ignored here the case that $0 \in B(x_i, r)$ for some i, in which case u_1 is inside $B(x_i, r)$ rather than on its boundary, but in this case u_1 is constant and certainly does not affect anything. This shows (4.8) for $E = H_x$ and $\mathscr{K} \cap H_x$.

The argument for the other case is similar, because $\mathscr{L}$ depends only on what happens outside $B(x_i, 2^l \varepsilon)$ and u_{2m} is on $\partial B(x_i, s)$ (this time without exceptions). Hence we have only an error of $s/(2^l \varepsilon)$. All these errors are exponential in $l - k$. This shows (4.8) is all 3 cases. In particular we get, for all three cases for which (4.8) holds, that

$$\mathbb{P}(E \mid W = u) = \mathbb{P}(E)(1 + O(e^{-c|k-l|})) \tag{4.9}$$

which holds for μ-almost every u.

The last point to note is that, conditioning on $W = u$ makes $\mathscr{L}$ independent of H_x and of $\mathscr{K}$ as the first depends only on what happens in the odd time intervals, i.e. between τ_{2m} and τ_{2m+1}, $m = 0, \ldots, 6$ while the other two depend on what happens in the even time intervals, between τ_{2m-1} and τ_{2m}, $m = 1, \ldots, 6$. Hence

$$\mathbb{P}(\mathscr{L} \cap H_x) = \int \mathbb{P}(\mathscr{L} \cap H_x \mid W = u)\, d\mu(u)$$

$$\text{by independence} \quad = \int \mathbb{P}(\mathscr{L} \mid W = u)\mathbb{P}(H_x \mid W = u)\, d\mu(u)$$

$$\text{by (4.9)} \quad = \mathbb{P}(\mathscr{L})\mathbb{P}(H_x)(1 + O(e^{-c|k-l|})).$$

A similar argument gives

$$\mathbb{P}(\mathscr{L} \cap \mathscr{K} \cap H_x) = \mathbb{P}(\mathscr{L})\mathbb{P}(\mathscr{K} \cap H_x)(1 + O(e^{-c|k-l|})).$$

Together these two inequalities bound $\mathrm{cov}(\mathscr{K}, \mathscr{L} \mid H_x)$. With (4.7) the lemma is proved.   □

With (4.5) established we can easily see (4.3), by using Chebyshev's inequality for the variable $\#\{k : X_k > 0\}$, with (4.4) giving the first moment and (4.5) the covariance. (In fact, it is not difficult to get a much better estimate than $C/|\log \varepsilon|$, an ε^c is also possible. But we will not need it).

Summing (4.3) over all x and using (4.1) gives

$$\mathbb{P}(X \in (0, \lambda|\log \varepsilon|)) \leq \frac{C}{|\log \varepsilon|}$$

This, with $\mathbb{E}(X) \leq C$ shows that $\mathbb{P}(X > 0) \leq C/|\log \varepsilon|$, proving Lemma 4.2.   □

Lemma 4.4 *A three-dimensional Brownian motion contains arithmetic progressions of length 5, a.s.*

Proof Let ε and $X = X(\varepsilon)$ be as in the proof of the previous lemma (except we now fix δ to be, say, $\frac{1}{10}$). It is straightforward to calculate

$$\mathbb{E}(X(\varepsilon)) \geq \frac{c}{\varepsilon} \qquad \mathbb{E}(X(\varepsilon)^2) \leq \frac{C}{\varepsilon^2}$$

which show that $\mathbb{P}(X(\varepsilon) > 0) \geq c$. A simple calculation shows that for some $\lambda > 0$ we have that $X(\lambda \varepsilon) > 0 \implies X(\varepsilon) > 0$. Hence $\{X(\lambda^k) > 0\}$ is a sequence of decreasing events with probabilities bounded below. This implies that

$$\mathbb{P}\left(\bigcap_k \{X(\lambda^k) > 0\} \right) > 0.$$

The event of the intersection can be described in words as follows: for every k there exists $x_1^{(k)}, \ldots, x_5^{(k)} \in B$ which are $\frac{1}{10}$-separated and λ^k-approximate arithmetic progression such that $W \cap B(x_i^{(k)}, \lambda^k) \neq \emptyset$ for $i \in \{1, \ldots, 5\}$. Taking a subsequential limit we get $x_i^{(k_n)} \to x_i$ and these x_i will be $\frac{1}{10}$-separated, will form an arithmetic progression, and will be on the path of W. So we conclude

$$\mathbb{P}(W \text{ contains a 5-term arithmetic progression in } B) > 0.$$

Scaling invariance now shows that the probability is in fact 1. □

Reference

1. P. Mörters, Y. Peres, *Brownian Motion*. Cambridge Series in Statistical and Probabilistic Mathematics, vol. 30. (Cambridge University Press, Cambridge, 2010), pp. xii+403

Chapter 5
Edgeworth Corrections in Randomized Central Limit Theorems

Sergey G. Bobkov

Abstract We consider rates of approximation of distributions of weighted sums of independent, identically distributed random variables by the Edgeworth correction of the 4-th order.

5.1 Introduction

Given independent, identically distributed random variables $X_1, \ldots, X_n$ (for short - i.i.d.), we consider weighted sums

$$S_\theta = \theta_1 X_1 + \cdots + \theta_n X_n, \qquad \theta = (\theta_1, \ldots, \theta_n),$$

with $\theta_1^2 + \cdots + \theta_n^2 = 1$, thus indexed by the points from the unit sphere S^{n-1} in $\mathbb{R}^n$ ($n \geq 2$). Throughout it is assumed that $\mathbb{E}X_1 = 0$, $\mathbb{E}X_1^2 = 1$, so that $\mathbb{E}S_\theta = 0$, $\mathbb{E}S_\theta^2 = 1$. According to the central limit theorem, if all the coefficients θ_k's are small, the distribution function

$$F_\theta(x) = \mathbb{P}\{S_\theta \leq x\}, \qquad x \in \mathbb{R},$$

Research was partially supported by the Simons Fellowship and NSF grant DMS-1855575. The book chapter was prepared within the framework of the HSE University Basic Research Program and funded by the Russian Academic Excellence Project '5-100.

S. G. Bobkov (✉)
School of Mathematics, University of Minnesota, Minneapolis, MN, USA
e-mail: bobko001@umn.edu

HSE University, Moscow, Russia

© Springer Nature Switzerland AG 2020
B. Klartag, E. Milman (eds.), *Geometric Aspects of Functional Analysis*,
Lecture Notes in Mathematics 2256,
https://doi.org/10.1007/978-3-030-36020-7_5

is close to the normal distribution function $\Phi(x) = \int_{-\infty}^{x} \varphi(y)\,dy$ with density $\varphi(y) = \frac{1}{\sqrt{2\pi}}\,e^{-y^2/2}$. This property can be quantified in terms of the Kolmogorov distance

$$\rho(F_\theta, \Phi) = \sup_x |F_\theta(x) - \Phi(x)|,$$

by involving absolute moments $\beta_s = \mathbb{E}\,|X_1|^s$. In particular, if the 3-rd absolute moment β_3 is finite, then

$$\rho(F_\theta, \Phi) \leq c\beta_3 \sum_{k=1}^{n} |\theta_k|^3 \tag{5.1.1}$$

up to some absolute constant c (cf. [11]). As best, here the right-hand side is of order $1/\sqrt{n}$ which is optimal in general, including the case of equal coefficients; moreover, this rate may not be improved under higher order moment assumptions.

Nevertheless, the situation is different when one is concerned about the typical behavior of these distances for most of θ in the sense of the normalized Lebesgue measure $\mathfrak{s}_{n-1}$ on S^{n-1}. In particular, Klartag and Sodin [7] have showed that, under the 4-th moment condition, the value $\rho(F_\theta, \Phi)$ is actually at most of order $1/n$ on average. More precisely, with some absolute constants $c, r_0 > 0$, for any $r \geq r_0$, we have

$$\rho(F_\theta, \Phi) \leq \frac{cr}{n}\,\beta_4 \tag{5.1.2}$$

for all $\theta \in S^{n-1}$ except for a set of $\mathfrak{s}_{n-1}$-measure $\leq 2\exp\{-r^{1/2}\}$. This cannot be obtained on the basis of (5.1.1), since the average of $\sum_{k=1}^{n} |\theta_k|^3$ is proportional to $1/\sqrt{n}$.

As it turns out, under rather general conditions, the relation (5.1.2) admits a further refinement, by replacing Φ with the corrected normal "distribution" function

$$G(x) = \Phi(x) - \frac{\alpha}{n}\,(x^3 - 3x)\,\varphi(x), \qquad \alpha = \frac{\beta_4 - 3}{8}. \tag{5.1.3}$$

We will use $\mathbb{E}_\theta$ to denote integrals with respect to the measure $\mathfrak{s}_{n-1}$. Put $\alpha_3 = \mathbb{E}X_1^3$.

Theorem 5.1.1 *If $\alpha_3 = 0$, $\beta_5 < \infty$, then with some positive absolute constant c*

$$\mathbb{E}_\theta\,\rho(F_\theta, G) \leq \frac{c}{n^{3/2}}\,\beta_5. \tag{5.1.4}$$

Moreover, there exists an absolute constant $r_0 > 0$ such that for all $r \geq r_0$,

$$\mathfrak{s}_{n-1}\left\{\rho(F_\theta, G) \geq \frac{cr}{n^{3/2}}\,\beta_5\right\} \leq 2\exp\left\{-r^{1/2}\right\}. \tag{5.1.5}$$

Theorem 5.1.1 involves all symmetric probability distributions with finite 5-th absolute moment in which case $G = \Phi$. Moreover, this bound is optimal in the sense that it can be reversed in a typical situation, where the 4-th moment of X_1 is different than the 4-th moment of the standard normal law. The same is also true about (5.1.2) on average, when the 3-rd moment of X_1 is not zero. Denote by $\mathcal{G}$ the collection of all functions G of bounded variation on the real line such that $G(-\infty) = 0$ and $G(\infty) = 1$.

Theorem 5.1.2 *If $\alpha_3 \neq 0$, $\beta_4 < \infty$, then the inequality*

$$\inf_{G \in \mathcal{G}} \mathbb{E}_\theta \, \rho(F_\theta, G) \geq \frac{c}{n} \qquad (5.1.6)$$

holds for all n with a constant $c > 0$ depending on α_3 and β_4 only. Moreover, if $\alpha_3 = 0$, $\beta_4 \neq 3$, $\beta_5 < \infty$, then

$$\inf_{G \in \mathcal{G}} \mathbb{E}_\theta \, \rho(F_\theta, G) \geq \frac{c}{n^{3/2}}, \qquad (5.1.7)$$

where the constant $c > 0$ depends on β_4 and β_5.

The paper is organized as follows. First, we recall a general scheme of Edgeworth corrections. Being specialized to the weighted sums, the corresponding asymptotic expansions contain as parameters special functions on the sphere, which we discuss in Sect. 5.3. The behavior of characteristic functions of the weighted sums on large intervals is analyzed separately in Sect. 5.4. These preparations are sufficient for the proof of Theorem 5.1.1, cf. Sect. 5.5 (where we also give a slight refinement of Klartag–Sodin's theorem in the i.i.d. situation). Sections 5.6 and 5.7 deal with lower bounds on the Kolmogorov distance, which are used to prove Theorem 5.1.2 (Sect. 5.8).

In the sequel, we use c, C to denote positive absolute constants, in general different in different places; similarly, c_q, C_q denote constants depending on a parameter q.

5.2 Construction of Asymptotic Expansions

Let $\xi_1, \ldots, \xi_n$ be independent, not necessarily identically distributed random variables such that $\mathbb{E}\xi_k = 0$ and $\sum_{k=1}^n \mathbb{E}\xi_k^2 = 1$. Consider the sum $S_n = \xi_1 + \cdots + \xi_n$, which thus has mean zero and variance one. An asymptotic behaviour of the distribution of S_n in a weak sense is usually analyzed in terms of its characteristic function $f_n(t) = \mathbb{E} \, e^{it S_n}$. In turn, the behaviour of $f_n(t)$ on large t-intervals is controlled by the Lyapunov coefficients

$$L_s = \sum_{k=1}^n \mathbb{E} \, |\xi_k|^s, \qquad s \geq 2.$$

Note that $L_s \geq n^{-\frac{s-2}{2}}$. In fact, these quantities are often of the order $n^{-\frac{s-2}{2}}$. For example,

$$L_s = n^{-\frac{s-2}{2}} \, \mathbb{E} \, |X_1|^s \quad \text{in case} \quad \xi_k = \frac{1}{\sqrt{n}} X_k$$

with identically distributed X_k. Since the function $s \to L_s^{\frac{1}{s-2}}$ is non-decreasing on the half-axis $s > 2$ (due to $L_2 = 1$), we have $L_3 \leq L_4^{1/2} \leq L_5^{1/3}$.

If L_s is finite for a fixed integer $s \geq 2$, the cumulants

$$\gamma_p(\xi_k) = \frac{d^p}{i^p \, dt^p} \, \log \mathbb{E} \, e^{it\xi_k} \Big|_{t=0}$$

are well-defined and finite for all $p = 1, \ldots, s$. Every cumulant $\gamma_p(\xi_k)$ is determined by the first p moments $\alpha_{r,k} = \mathbb{E} \, \xi_k^r$, $r = 1, \ldots, p$. The first cumulants are

$$\gamma_1(\xi_k) = \alpha_{1,k} = 0, \quad \gamma_2(\xi_k) = \alpha_{2,k}^2, \quad \gamma_3(\xi_k) = \alpha_{3,k}, \quad \gamma_4(\xi_k) = \alpha_{4,k} - 3\alpha_{2,k}^2.$$

A result of Bikjalis asserts that $|\gamma_p(\xi_k)| \leq (p-1)! \, \mathbb{E} \, |\xi_k|^p$ (cf. [1, 3]). The cumulants of S_n exist for the same values of p and have an additive structure:

$$\gamma_p(S_n) = \frac{d^p}{i^p \, dt^p} \, \log \mathbb{E} \, e^{it S_n} \Big|_{t=0} = \sum_{k=1}^{n} \gamma_p(\xi_k).$$

Hence, they admit a similar upper estimate

$$|\gamma_p(S_n)| \leq (p-1)! \, L_p. \tag{5.2.1}$$

The Lyapunov coefficients may also be used to bound absolute moments of S_n. The well-known Rosenthal inequality indicates that $\mathbb{E} \, |S_n|^p \leq C_p \, (1 + L_p)$ for $p \geq 2$.

We refer an interested reader to [2, 3, 11] for more references and here only mention a few definitions and basic results.

Definition 5.2.1 Let L_s be finite for an integer $s \geq 3$. An Edgeworth approximation of order $s - 1$ for the characteristic function $f_n(t) = \mathbb{E} \, e^{it S_n}$ is given by

$$g_{s-1}(t) = e^{-t^2/2} + e^{-t^2/2} \sum \frac{1}{k_1! \ldots k_{s-3}!} \left(\frac{\gamma_3}{3!} \right)^{k_1} \ldots \left(\frac{\gamma_{s-1}}{(s-1)!} \right)^{k_{s-3}} (it)^k, \quad t \in \mathbb{R}.$$

Here $\gamma_p = \gamma_p(S_n)$, $k = 3k_1 + \cdots + (s-1)k_{s-3}$, and the summation is performed over all tuples $(k_1, \ldots, k_{s-3})$ of non-negative integers, not all zero, such that $k_1 + 2k_2 + \cdots + (s-3)k_{s-3} \leq s - 3$.

The function g_{s-1} is also called the corrected normal characteristic function (although it is not a characteristic function in the usual sense). The index $s - 1$ indicates that the cumulants up to γ_{s-1} participate in the constructions. Note that the above sum represents a polynomial of degree at most $3(s - 3)$ in variable t.

When $s = 3$, we have $g_2(t) = e^{-t^2/2}$ which is the standard normal characteristic function. The next Edgeworth correction is given by

$$g_3(t) = e^{-t^2/2} \left(1 + \gamma_3 \frac{(it)^3}{3!}\right), \qquad \gamma_3 = \sum_{k=1}^{n} \mathbb{E}\xi_k^3. \tag{5.2.2}$$

For $s = 5$, if $\gamma_3 = 0$, we have

$$g_4(t) = \left(1 + \gamma_4 \frac{(it)^4}{4!}\right) e^{-t^2/2}, \qquad \gamma_4 = \sum_{k=1}^{n} \left(\mathbb{E}\xi_k^4 - 3\,(\mathbb{E}\xi_k^2)^2\right). \tag{5.2.3}$$

We will need the following general statement about the Edgeworth approximations.

Proposition 5.2.2 *Let $L_s < \infty$ for an integer $s \geq 3$. Then in the interval $|t| \leq \frac{1}{L_3}$,*

$$\left|f_n(t) - g_{s-1}(t)\right| \leq C_s L_s \min\left\{1, |t|^s\right\} e^{-t^2/8}. \tag{5.2.4}$$

When $s = 3$, (5.2.4) leads to the popular inequality

$$\left|f_n(t) - e^{-t^2/2}\right| \leq C L_3 \min\{1, |t|^3\} e^{-t^2/8}.$$

By (5.2.1), the cumulants of S_n satisfy $|\gamma_p| \leq (p - 1)!\, L_p \leq (p - 1)!\, L_s^{\frac{p-2}{s-2}}$ implying that

$$\left|\left(\frac{\gamma_3}{3!}\right)^{k_1} \cdots \left(\frac{\gamma_{s-1}}{(s-1)!}\right)^{k_{s-3}}\right| \leq \frac{L_s^{k/(s-2)}}{3^{k_1} \cdots (s-1)^{k_{s-3}}} \tag{5.2.5}$$

with some $1 \leq k \leq s - 3$. Applying this bound in Definition 5.2.1, it readily follows that

$$\left|g_{s-1}(t) - e^{-t^2/2}\right| \leq C_s \max\{|t|^3, |t|^{3(s-3)}\} e^{-t^2/2} \max\left\{L_s^{\frac{1}{s-2}}, L_s^{\frac{s-3}{s-2}}\right\}.$$

In particular,

$$\int_{-\infty}^{\infty} \left|g_{s-1}(t) - e^{-t^2/2}\right| dt \leq C_s \max\{1, L_s\}.$$

Being integrable, the function g_{s-1} appears as the Fourier–Stieltjes transform of a certain signed Borel measure μ_{s-1} on the real line with density

$$\varphi_{s-1}(x) = \varphi(x) + \varphi(x) \sum \frac{1}{k_1! \dots k_{s-3}!} \left(\frac{\gamma_3}{3!}\right)^{k_1} \dots \left(\frac{\gamma_{s-1}}{(s-1)!}\right)^{k_{s-3}} H_k(x), \quad x \in \mathbb{R}.$$

Here, the summation is as before, and H_k are the Chebyshev-Hermite polynomials of degrees $k = 3k_1 + \dots + (s-1)k_{s-3}$ with leading coefficient 1. By the construction, $g_{s-1}(0) = 1$, that is, μ_{s-1} has total mass 1. Moreover, g_{s-1} and f_n have equal derivatives at zero up to order $s - 1$, which is equivalent to

$$\mathbb{E} \, S_n^p = \int_{-\infty}^{\infty} x^p \, d\mu_{s-1}(x) = \int_{-\infty}^{\infty} x^p \, \varphi_{s-1}(x) \, dx, \qquad p = 1, \dots, s - 1.$$

Using (5.2.5), we can also see that

$$\sup_x |\varphi_{s-1}(x)| \leq C_s \max\{1, L_s\}, \qquad \int_{-\infty}^{\infty} |\varphi_{s-1}(x)| \, dx \leq C_s \max\{1, L_s\}. \tag{5.2.6}$$

The associated "distribution" function

$$\Phi_{s-1}(x) = \mu_{s-1}\big((-\infty, x]\big) = \int_{-\infty}^{x} \varphi_{s-1}(y) \, dy, \qquad x \in \mathbb{R},$$

has a similar description

$$\Phi_{s-1}(x) = \Phi(x) - \varphi(x) \sum \frac{1}{k_1! \dots k_{s-3}!} \left(\frac{\gamma_3}{3!}\right)^{k_1} \dots \left(\frac{\gamma_{s-1}}{(s-1)!}\right)^{k_{s-3}} H_{k-1}(x).$$

This function has bounded total variation and satisfies $\Phi_{s-1}(-\infty) = 0$ and $\Phi_{s-1}(\infty) = 1$.

The measure μ_2 is just the standard Gaussian measure with distribution function $\Phi_2 = \Phi$. The Edgeworth correction g_3 corresponds to the signed measure with "distribution" function

$$\Phi_3(x) = \Phi(x) - \frac{\gamma_3}{3!} (x^2 - 1)\varphi(x). \tag{5.2.7}$$

If $\gamma_3 = 0$, the next Edgeworth correction g_4 corresponds to the "distribution" function

$$\Phi_4(x) = \Phi(x) - \frac{\gamma_4}{4!} (x^3 - 3x) \, \varphi(x). \tag{5.2.8}$$

Since Proposition 5.2.2 quantifies closeness of $f_n(t)$ to $g_{s-1}(t)$ on large t-intervals (when L_s is small), one may hope that, under some additional assumptions, the distribution function F_n will be properly approximated by Φ_{s-1} in Kolmogorov distance. This may be achieved by applying Berry–Esseen-type theorems such as the following:

Proposition 5.2.3 *If L_s is finite for an integer $s \geq 3$, then*

$$c_s \, \rho(F_n, \Phi_{s-1}) \leq L_s + 1_{\{L_s \leq L_3 \leq 1\}} \int_{1/L_3}^{1/L_s} \frac{|f_n(t)|}{t} \, dt. \qquad (5.2.9)$$

Proof A classical theorem due to Esseen asserts the following: Let F be a non-decreasing bounded function, and G be a differentiable function of bounded variation such that $F(-\infty) = G(-\infty) = 0$. If $|G'(x)| \leq M$ for all x, then for any $T > 0$,

$$c \, \rho(F, G) \leq \int_0^T \left| \frac{f(t) - g(t)}{t} \right| dt + \frac{M}{T}. \qquad (5.2.10)$$

Here,

$$f(t) = \int_{-\infty}^{\infty} e^{itx} \, dF(x), \qquad g(t) = \int_{-\infty}^{\infty} e^{itx} \, dG(x)$$

denote the Fourier–Stieltjes transforms of F and G, respectively.

First, assume that $L_s \leq 1$. Necessarily $L_3 \leq L_s^{\frac{1}{s-2}} \leq 1$. Assuming moreover that $L_s \leq L_3$, we choose $T = 1/L_s$ and apply the bound (5.2.9) with $F = F_n$ and $G = \Phi_{s-1}$, in which case we have $|G'(x)| = |\varphi_{s-1}(x)| \leq C_s$, by (5.2.6). Then, applying (5.2.4) in (5.2.10), we get

$$c_s \, \rho(F_n, \Phi_{s-1}) \leq L_s + \int_{1/L_3}^{1/L_s} \frac{|f_n(t)|}{t} \, dt + \int_{1/L_3}^{1/L_s} \frac{|g_{s-1}(t)|}{t} \, dt. \qquad (5.2.11)$$

To estimate the last integral, one may use the bound (cf. [3], Proposition 17.1)

$$|g_{s-1}(t)| \leq C_s L_s \, e^{-t^2/8}, \quad \text{if } |t| \max \left\{ L_s^{\frac{1}{s-2}}, L_s^{\frac{1}{3(s-2)}} \right\} \geq \frac{1}{8}.$$

Since $L_3 \leq L_s^{\frac{1}{s-2}}$, it holds for $t \geq 1/(8L_3)$, and (5.2.11) thus yields (5.2.9).

Now, suppose that $L_3 \leq L_s \leq 1$. Then we choose $T = 1/L_3$ in (5.2.10) and apply (5.2.4) again, which leads to $c_s \, \rho(F_n, \Phi_{s-1}) \leq L_3$. Finally, if $L_s > 1$, one may use the second bound (5.2.6) which immediately implies that

$$\rho(F_n, \Phi_{s-1}) \leq \rho(F_n, \Phi) + \rho(\Phi, \Phi_{s-1}) \leq C_s L_s.$$

$\square$

5.3   Moments and Deviations of Lyapunov Coefficients

Let us return to the scheme of the weighted sums. In the rest of the paper, we assume that

$$S_\theta = \theta_1 X_1 + \cdots + \theta_n X_n, \qquad \theta = (\theta_1, \ldots, \theta_n) \in S^{n-1},$$

where X_k's are i.i.d. random variables such that $\mathbb{E}X_1 = 0$, $\mathbb{E}X_1^2 = 1$, and $\beta_s = \mathbb{E}|X_1|^s < \infty$ for an integer $s \geq 3$. First, we will be focusing on the application of Proposition 5.2.3 to the approximation of the distribution functions F_θ of S_θ for most of θ's by the corresponding Edgeworth corrections $\Phi_{s-1} = \Phi_{s-1,\theta}$, especially with $s = 4$ and $s = 5$.

According to (5.2.9), in order to control the Kolmogorov distance from F_θ to $\Phi_{s-1,\theta}$, one should estimate the Lyapunov coefficients $L_s = L_s(\theta)$; we also need information about the magnitude of the characteristic functions $f_\theta(t) = \mathbb{E}\,e^{it\,S_\theta}$ on large t-intervals such as $|t| \leq 1/L_s(\theta)$. Note that the Lyapunov coefficients take the form

$$L_p(\theta) = \beta_p\, l_p(\theta) \quad \text{where } l_p(\theta) = \sum_{k=1}^{n} |\theta_k|^p \quad (2 \leq p \leq s).$$

On the other hand, according to Definition 5.2.1, the construction of the functions $\Phi_{s-1,\theta}$ is based on the cumulants $\gamma_p(\theta) = \gamma_p(S_\theta)$ for $p \leq s - 1$, which are given in terms of the cumulants $\gamma_p = \gamma_p(X_1)$ of the underlying distribution by

$$\gamma_p(\theta) = \gamma_p\, \alpha_p(\theta), \quad \alpha_p(\theta) = \sum_{k=1}^{n} \theta_k^p.$$

In particular, $\gamma_1(\theta) = 0$, $\gamma_2(\theta) = 1$, and

$$\gamma_3(\theta) = \gamma_3\, \alpha_3(\theta) = \alpha_3 \sum_{k=1}^{n} \theta_k^3 \qquad (\alpha_3 = \mathbb{E}X_1^3),$$

$$\gamma_4(\theta) = \gamma_4\, l_4(\theta) = (\beta_4 - 3) \sum_{k=1}^{n} \theta_k^4 \qquad (\beta_4 = \mathbb{E}X_1^4).$$

Thus, in order to study the typical behaviour of distances $\rho(F_\theta, \Phi_{s-1,\theta})$, we have to explore the distribution of the functionals l_p and α_p under the measure $\mathfrak{s}_{n-1}$ (note that $\alpha_p = l_p$ for even p). The behaviour of distributions of l_p for large n is mainly described by their means and variances. Since the distribution of the first coordinate

θ_1 under $\mathfrak{s}_{n-1}$ has density

$$c_n (1 - x^2)^{\frac{n-3}{2}} \quad (|x| \leq 1), \qquad c_n = \frac{\Gamma(\frac{n}{2})}{\sqrt{\pi}\,\Gamma(\frac{n-1}{2})},$$

we get

$$\mathbb{E}_\theta |\theta_1|^p = 2c_n \int_0^1 x^p (1 - x^2)^{\frac{n-3}{2}}\, dx = \frac{\Gamma(\frac{p+1}{2})\,\Gamma(\frac{n}{2})}{\sqrt{\pi}\,\Gamma(\frac{p+n}{2})}.$$

In particular, since $\mathbb{E}_\theta\, l_p(\theta) = n\, \mathbb{E}_\theta |\theta_1|^p$, we have

$$\mathbb{E}_\theta\, l_{2k}(\theta) = \mathbb{E}_\theta\, \alpha_{2k}(\theta) = \frac{(2k-1)\,!!}{(n+2)\ldots(n+2k-2)} \tag{5.3.1}$$

for even powers $p = 2k$, $k = 2, 3, \ldots$ Hence,

$$\mathbb{E}_\theta\, l_{2k} < 2^k k!\, n^{-(k-1)} < p^{p/2}\, n^{-\frac{p-2}{2}}.$$

Here, the resulting bound also holds for $p = 2k - 1$. Indeed, using the property that the function $p \to l_p^{\frac{1}{p-2}}$ is non-decreasing in $p > 2$, we have $l_{2k-1} \leq l_{2k}^{\frac{2k-3}{2k-2}}$ and $\mathbb{E}_\theta\, l_{2k-1} \leq (\mathbb{E}_\theta\, l_{2k})^{\frac{2k-3}{2k-2}}$. Therefore

$$\begin{aligned}
\mathbb{E}_\theta |\theta_1|^{2k-1} &\leq \left(\mathbb{E}_\theta\, \theta_1^{2k}\right)^{\frac{2k-3}{2k-2}} \\
&< \left(2^k k!\, n^{-(k-1)}\right)^{\frac{2k-3}{2k-2}} = (2^k k!)^{\frac{2k-3}{2k-2}}\, n^{-\frac{p-2}{2}} \\
&< (2k-1)^{\frac{k(2k-3)}{2k-2}}\, n^{-\frac{p-2}{2}} < (2k-1)^{\frac{2k-1}{2}}\, n^{-\frac{p-2}{2}},
\end{aligned}$$

where we used a simple inequality $2^k k! < (2k-1)^k$. That is, we have:

Lemma 5.3.1 *For all integers $p \geq 3$, we have $\mathbb{E}_\theta\, l_p < p^{p/2}\, n^{-\frac{p-2}{2}}$.*

For the first Lyapunov coefficients, the p-dependent constant can slightly be improved. For example,

$$\mathbb{E}_\theta\, l_3 \leq \mathbb{E}_\theta\, l_4^{1/2} \leq \left(\mathbb{E}_\theta\, l_4\right)^{1/2} = \left(\frac{3}{n+2}\right)^{1/2} < \frac{2}{n^{1/2}}.$$

Similarly, since $l_5^{1/3} \leq l_6^{1/4}$,

$$\mathbb{E}_\theta\, l_5 \leq \mathbb{E}_\theta\, l_6^{3/4} \leq \left(\mathbb{E}_\theta\, l_6\right)^{3/4} = \left(\frac{15}{(n+2)(n+4)}\right)^{3/4} < \frac{8}{n^{3/2}}.$$

Using (5.3.1) together with a similar formula

$$\mathbb{E}\,|Z|^{2k} = 2^k\,\frac{\Gamma(\frac{2k+n}{2})}{\Gamma(\frac{n}{2})} = n(n+2)\dots(n+2k-2), \qquad k=1,2,\dots,$$

where Z is a standard normal random vector in $\mathbb{R}^n$ (needed with $k=4$), we also find that

$$\mathrm{Var}_\theta(\alpha_3) = \frac{15}{(n+2)(n+4)} < \frac{15}{n^2},$$

$$\mathrm{Var}_\theta(l_4) = \frac{24\,(n-1)}{(n+2)^2(n+4)(n+6)} < \frac{24}{n^3}.$$

This means that the deviations of α_3 are of order $1/n$, while the deviations of l_4 from its mean are of order $n^{-3/2}$.

With worse numerical constants these bounds can also be obtained by applying the spherical Poincaré inequality. However, by virtue of the (stronger) logarithmic Sobolev inequality on the unit sphere with an optimal constant [8, 10], namely

$$\int u^2 \log u^2 \, d\mathfrak{s}_{n-1} - \int u^2 \, d\mathfrak{s}_{n-1} \log \int u^2 \, d\mathfrak{s}_{n-1} \le \frac{2}{n-1} \int |\nabla u|^2 \, d\mathfrak{s}_{n-1},$$

one can get more information, such as the bound on the growth of moments

$$\|u - \mathbb{E}_\theta\, u\|_p \le \frac{\sqrt{p-1}}{\sqrt{n-1}} \|\nabla u\|_p, \qquad p \ge 2. \qquad (5.3.2)$$

Both inequalities hold true for any smooth function u on $\mathbb{R}^n$ with gradient ∇u, and with L^p-norms being understood with respect to the measure $\mathfrak{s}_{n-1}$ (cf. e.g. [4], Theorem 4.1).

Generalizing $\alpha_3(\theta)$ and $l_4(\theta)$, now consider the functions

$$Q_3(\theta) = \sum_{k=1}^n a_k \theta_k^3, \qquad Q_4(\theta) = \sum_{k=1}^n a_k \theta_k^4.$$

Lemma 5.3.2 *Assume that $\frac{1}{n}\sum_{k=1}^n a_k^2 = 1$ and put $\bar{a} = \frac{1}{n}\sum_{k=1}^n a_k$. For all $r > 0$,*

$$\mathfrak{s}_{n-1}\{n\,|Q_3| \ge r\} \le 2\,\exp\left\{-\frac{1}{23}\,r^{2/3}\right\},$$

$$\mathfrak{s}_{n-1}\left\{n^{3/2}\left|Q_4 - \frac{3\,\bar{a}}{n+2}\right| \ge r\right\} \le 2\,\exp\left\{-\frac{1}{38}\,r^{1/2}\right\}.$$

Proof We apply (5.3.2) to the function $u = n Q_3$. Using $\frac{p-1}{n-1} \leq \frac{2p}{n}$, by Jensen's inequality, for any $p \geq 2$,

$$\|u\|_p^p \leq n^p \left(\frac{2p}{n}\right)^{p/2} \|\nabla Q_3\|_p^p$$

$$= n^{p/2} (2p)^{p/2} 3^p \int \left(\sum_{k=1}^n a_k^2 \theta_k^4\right)^{p/2} d\mathfrak{s}_{n-1}(\theta)$$

$$\leq n^p (2p)^{p/2} 3^p \cdot \frac{1}{n} \sum_{k=1}^n a_k^2 \int |\theta_k|^{2p} d\mathfrak{s}_{n-1}(\theta) = n^p (18\,p)^{p/2} \mathbb{E}_\theta |\theta_1|^{2p}.$$

If $p = m$ is integer, applying the relation (5.3.1), we get

$$\|u\|_m^m \leq n^m (18\,m)^{m/2} \mathbb{E}_\theta |\theta_1|^{2m}$$

$$\leq (18\,m)^{m/2} (2m-1)!! \leq 18^{m/2} 2^m m^{3m/2},$$

where we used the bound $(2m-1)!! < (2m)^m$. Thus, $\|u\|_m \leq 6\sqrt{2}\,m^{3/2}$. At the expense of a larger absolute factor, this inequality can be extended to all real $p \geq 2$ in place of m. Indeed, pick up an integer m such that $m \leq p < m+1$. Then

$$\|u\|_p \leq \|u\|_{m+1} \leq 6\sqrt{2}\,(m+1)^{3/2} \leq 6\sqrt{2}\,(p+1)^{3/2} \leq 9\sqrt{3}\,p^{3/2},$$

i.e. $\|u\|_p^p \leq (bp)^{3p/2}$ with $b = (9\sqrt{3})^{2/3}$. By Markov's inequality, choosing $p = \frac{1}{2^{1/3} b} r^{2/3}$ $(r > 0)$, we get

$$\mathfrak{s}_{n-1}\{|u| \geq r\} \leq \frac{(bp)^{3p/2}}{r^p} = \exp\left\{-\frac{p}{2} \log 2\right\},$$

provided that $p \geq 2$. But, in the case $0 < p < 2$, the above right-hand side is greater than $1/2$, so that we have

$$\mathfrak{s}_{n-1}\{|u| \geq r\} \leq 2 \exp\left\{-\frac{p}{2} \log 2\right\}$$

for all $p > 0$. It remains to note that $\frac{p}{2} \log 2 = \frac{\log 2}{2^{4/3}\,3^{5/3}} r^{2/3} > \frac{1}{23} r^{2/3}$ thus proving the first inequality of the lemma.

To derive the second one, let us apply (5.3.2) to the function $u = n^{3/2} (Q_4 - \frac{3\bar{a}}{n+2})$. Similarly, for any $p \geq 2$,

$$\|u\|_p^p \leq n^{3p/2} \left(\frac{2p}{n}\right)^{p/2} \|\nabla Q_4\|_p^p$$

$$= n^{3p/2} \left(\frac{2p}{n}\right)^{p/2} 4^p \int \left(\sum_{k=1}^{n} a_k^2 \theta_k^6\right)^{p/2} d\mathfrak{s}_{n-1}(\theta)$$

$$\leq n^{2p} \left(\frac{2p}{n}\right)^{p/2} 4^p \cdot \frac{1}{n} \sum_{k=1}^{n} a_k^2 \int |\theta_k|^{3p} d\mathfrak{s}_{n-1}(\theta) = n^{3p/2} (32\, p)^{p/2} \, \mathbb{E}_\theta |\theta_1|^{3p}.$$

Let us replace p with $2m$ assuming that $m \geq 1$ is integer. By (5.3.1), we get

$$\|u\|_{2m}^{2m} \leq n^{3m} (64\, m)^m \, \mathbb{E}_\theta |\theta_1|^{6m}$$

$$\leq (64\, m)^m (6m-1)!! \leq (48\sqrt{6}\, m^2)^{2m}.$$

Hence $\|u\|_{2m} \leq 48\sqrt{6}\, m^2$. To extend this inequality to real $p \geq 2$, pick up an integer m such that $2m \leq p < 2(m+1)$. Then

$$\|u\|_p \leq \|u\|_{2(m+1)} \leq 48\sqrt{6}\, (m+1)^2 \leq 48\sqrt{6}\, p^2 = (bp)^{2p}, \quad b = (48\sqrt{6})^{1/2}.$$

By Markov's inequality, given $r > 0$ and choosing $p = \frac{1}{2^{1/4} b} \sqrt{r}$, we get

$$\mathfrak{s}_{n-1}\{|u| \geq r\} \leq \frac{(bp)^{2p}}{r^p} = \exp\left\{-\frac{p}{2} \log 2\right\}$$

provided that $p \geq 2$. In the case $0 < p < 2$, the right-hand side is greater than $1/2$, so that

$$\mathfrak{s}_{n-1}\{|u| \geq r\} \leq 2 \exp\left\{-\frac{p}{2} \log 2\right\}$$

for all $p > 0$. It remains to note that $\frac{p}{2} \log 2 > \frac{1}{38} r^{1/2}$. $\square$

Let us now consider deviations of l_p above their means.

Lemma 5.3.3 *For all real $r \geq 1$ and integer $p > 2$,*

$$\mathfrak{s}_{n-1}\left\{n^{\frac{p-2}{2}} l_p \geq c_p\, r\right\} \leq \exp\left\{-(rn)^{2/p}\right\}, \tag{5.3.3}$$

where one may take $c_3 = 33$, $c_4 = 121$, and $c_p = (\sqrt{p}+2)^p$ in general.

Proof If u is a function on S^{n-1} with Lipschitz semi-norm $\|u\|_{\mathrm{Lip}} \leq 1$ with respect to the Euclidean distance, then (cf. e.g. [9])

$$\mathfrak{s}_{n-1}\{u \geq \mathbb{E}_\theta\, u + t\} \leq e^{-nt^2/4}, \qquad t \geq 0.$$

As a partial case, one may consider the ℓ_p^n-norms $u(\theta) = l_p(\theta)^{1/p}$ on $\mathbb{R}^n$ with $p \geq 2$, for which we thus have that

$$\mathfrak{s}_{n-1}\{l_p^{1/p} \geq (\mathbb{E}_\theta\, l_p)^{1/p} + t\} \leq e^{-nt^2/4}.$$

Using the bound $\mathbb{E}_\theta\, l_p \leq A_p\, n^{-\frac{p-2}{2}}$ with $A_p = p^{p/2}$ as in Lemma 5.3.1, the choice $t = 2r^{1/p}\, n^{-\frac{p-2}{2p}}$ leads to

$$\mathfrak{s}_{n-1}\{n^{\frac{p-2}{2p}}\, l_p^{1/p} \geq A_p^{1/p} + 2r\} \leq \exp\{-(rn)^{2/p}\}.$$

Hence, we obtain (3.3.4) with $c_p = (A_p^{1/p} + 2)^p \leq (\sqrt{p} + 2)^p$. Using $A_3 = \sqrt{3}$ and $A_4 = 3$, one may take $c_3 = (A_3^{1/3} + 2)^3 < 33$ and $(A_4^{1/4} + 2)^4 < 121$. $\qquad\square$

5.4 Upper Bounds on Characteristic Functions

The property that the values of the characteristic functions $f_\theta(t) = \mathbb{E}\, e^{itS_\theta}$ are small in absolute value for most of $\theta \in S^{n-1}$ with large t may be seen from the following:

Lemma 5.4.1 *For all $t \in \mathbb{R}$,*

$$\mathbb{E}_\theta\, |f_\theta(t)|^2 \leq 5\, e^{-t^2/2} + 5\, e^{-n/(12\,\beta_4)}. \tag{5.4.1}$$

Proof Using an independent copy $Y = (Y_1 \ldots, Y_n)$ of the random vector $X = (X_1, \ldots, X_n)$ in $\mathbb{R}^n$, write

$$|f_\theta(t)|^2 = \mathbb{E}\, e^{it\langle X-Y,\theta\rangle}, \qquad \theta \in S^{n-1},$$

and integrate over the sphere, which gives

$$\mathbb{E}_\theta\, |f_\theta(t)|^2 = \mathbb{E}\, J_n(t\, |X-Y|),$$

where by $J_n(t) = \mathbb{E}_\theta\, e^{it\theta_1}$ we denote the characteristic function of the first coordinate on the sphere under $\mathfrak{s}_{n-1}$. One may split the last expectation to the event $|X - Y| \leq \sqrt{n}$ and to the opposite one, which implies

$$\mathbb{E}_\theta\, |f_\theta(t)|^2 \leq \sup_{u \geq t\sqrt{n}} |J_n(u)| + \mathbb{P}\{|X - Y|^2 \leq n\}.$$

To proceed, we employ the bound $|J_n(u)| \leq 5\,e^{-u^2/2n} + 4\,e^{-n/12}$ derived in [5], cf. Proposition 3.3. Consequently, since $\beta_4 \geq 1$, the inequality (5.4.1) would follow from

$$\mathbb{P}\{|X - Y|^2 \leq n\} \leq e^{-n/(16\,\beta_4)} \tag{5.4.2}$$

But, this bound is a particular case of the following well-known observation: Given i.i.d. random variables $\xi_k \geq 0$ such that $\mathbb{E}\xi_1 = 1$, the sum $U_n = \xi_1 + \cdots + \xi_n$ satisfies

$$\mathbb{P}\{U_n \leq \lambda n\} \leq \exp\left\{-\frac{(1-\lambda)^2}{2\,\mathbb{E}\xi_1^2}\,n\right\}, \qquad 0 < \lambda < 1. \tag{5.4.3}$$

To recall a standard argument, note that

$$\mathbb{E}\,e^{-rU_n} \geq e^{-\lambda rn}\,\mathbb{P}\{U_n \leq \lambda n\}, \qquad r \geq 0. \tag{5.4.4}$$

The function $\psi(r) = \mathbb{E}\,e^{-r\xi_1}$ is positive and admits Taylor's expansion near zero up to the quadratic form, which implies that

$$\psi(r) \leq 1 - r\,\mathbb{E}\xi_1 + \frac{r^2}{2}\,\mathbb{E}\xi_1^2 \leq \exp\left\{-r\,\mathbb{E}\xi_1 + \frac{r^2}{2}\,\mathbb{E}\xi_1^2\right\}.$$

Hence

$$\mathbb{E}\,e^{-rU_n} = \psi(r)^n \leq \exp\left\{-rn + \frac{nr^2}{2}\,\mathbb{E}\xi_1^2\right\}.$$

In view of (5.4.4), this bound yields

$$\mathbb{P}\{U_n \leq \lambda n\} \leq \exp\left\{-(1-\lambda)nr + \frac{nr^2}{2}\,\mathbb{E}\xi_1^2\right\},$$

and after optimization over r we arrive at (5.4.3).

In the case $\xi_k = \frac{1}{2}(X_k - Y_k)^2$ with i.i.d. X_k such that $\mathbb{E}X_1 = 0$, $\mathbb{E}X_1^2 = 1$, $\mathbb{E}X_1^4 = \beta_4$, we have

$$\mathbb{E}\xi_1^2 = \frac{1}{2}\,\mathbb{E}\xi_1^4 + \frac{3}{2}(\mathbb{E}\xi_1^2)^2 \leq 2\beta_4,$$

and (5.4.3) yields

$$\mathbb{P}\{|X - Y|^2 \leq 2\lambda n\} \leq \exp\left\{-\frac{(1-\lambda)^2}{4\beta_4}\,n\right\}.$$

To obtain (5.4.2), it remains to put here $\lambda = 1/2$. □

Let us now turn to the integrals

$$I_s(\theta) = 1_{\Omega_s} \int_{1/L_3}^{1/L_s} \frac{|f_\theta(t)|}{t} dt, \qquad \Omega_s = \{\theta \in S^{n-1} : L_s \leq L_3 \leq 1\}, \qquad (5.4.5)$$

appearing in the Berry–Esseen-type bound (5.2.9) for the scheme of the weighted sums with $L_s = L_s(\theta)$. Since in general $L_s \geq \beta_s\, n^{-\frac{s-2}{2}}$, necessarily $I_s(\theta) = 0$, if $\beta_s > n^{\frac{s-2}{2}}$.

Lemma 5.4.2 *Given an integer $s \geq 4$, we have*

$$c_s\, \mathfrak{s}_{n-1}\{I_s(\theta) \geq \beta_s\, n^{-\frac{s-2}{2}}\} \leq \exp\{-n^{2/3}\} + \exp\{-\frac{cn}{\beta_4}\}. \qquad (5.4.6)$$

In particular,

$$\mathbb{E}_\theta\, I_s(\theta) \leq C_s \beta_s\, n^{-\frac{s-2}{2}}. \qquad (5.4.7)$$

Proof Introduce the sets on the unit sphere $\Omega_0 = \{L_3 < 2cb_n\}$, $\Omega_1 = \{L_3 \geq 2cb_n\}$, where $b_n = \beta_3/\sqrt{n}$ and $c = 33$. By Lemma 5.3.3 with $p = 3$,

$$\mathfrak{s}_{n-1}(\Omega_1) \leq \exp\{-(2n)^{2/3}\}.$$

Since $L_s \geq n^{-\frac{s-2}{2}}$, while $|f_\theta(t)| \leq 1$, we get, for all $\theta \in S^{n-1}$,

$$I_s(\theta) = 1_{\Omega_s} \int_{1/L_3}^{1/L_s} \frac{|f_\theta(t)|}{t} dt \leq \int_1^{n^{\frac{s-2}{2}}} \frac{1}{t} dt = \frac{s-2}{2} \log n,$$

and conclude that

$$\mathbb{E}_\theta\left[I_s(\theta)\, 1_{\Omega_1}\right] \leq \frac{s-2}{2} \log n\; \mathfrak{s}_{n-1}(\Omega_1) \leq \frac{s-2}{2} \log n\, \exp\{-(2n)^{2/3}\}$$

$$\leq Cs\, \exp\{-n^{2/3}\}. \qquad (5.4.8)$$

Given $\theta \in \Omega_0 \cap \Omega_s$, let us extend the integration in (5.4.5) to the interval $[T_0, T]$ with endpoints $T_0 = \max\{1, (2cb_n)^{-1}\}$ and $T = \frac{1}{\beta_s}\, n^{\frac{s-2}{2}}$, and with the requirement that $T_0 \leq T$. Since, by Lemma 5.4.1,

$$\mathbb{E}_\theta\, |f_\theta(t)| \leq 3\, e^{-t^2/4} + 3\, e^{-n/(32\beta_4)},$$

we obtain that

$$\mathbb{E}_\theta \, I_s(\theta) \, 1_{\Omega_0} \leq 3 \, \mathbb{E}_\theta \int_{T_0}^{T} \left(e^{-t^2/4} + e^{-n/(32\beta_4)} \right) \frac{dt}{t}$$

$$\leq 6 \, e^{-T_0^2/4} + 3 \log \frac{T}{T_0} e^{-n/(32\beta_4)}$$

$$\leq 6 \exp\left\{ -\frac{1}{(4cb_n)^2} \right\} + 3 \log \frac{n^{\frac{s-2}{2}}}{\beta_s} e^{-n/(32\beta_4)}. \qquad (5.4.9)$$

Due to the assumption $\mathbb{E}X_1^2 = 1$, the function $p \to \beta_p^{1/(p-2)}$ is non-decreasing in p, so, $\beta_4 \leq \beta_s^{2/(s-2)}$ and

$$\log \frac{n^{\frac{s-2}{2}}}{\beta_s} e^{-n/(64\beta_4)} \leq \log \frac{n^{\frac{s-2}{2}}}{\beta_s} \exp\left\{ -\frac{1}{64} \beta_s^{-\frac{2}{s-2}} n \right\} \leq C_s.$$

This simplifies (5.4.9) to

$$c_s \, \mathbb{E}_\theta \, I_s(\theta) \, 1_{\Omega_0} \leq \exp\left\{ -\frac{n}{(4c\beta_3)^2} \right\} + e^{-n/(64\beta_4)} \leq 2e^{-c'n/\beta_4},$$

where we used $\beta_3^2 \leq \beta_4$. Together with (5.4.8), we thus arrive at

$$c_s \, \mathbb{E}_\theta \, I_s(\theta) \leq \exp\left\{ -n^{2/3} \right\} + \exp\left\{ -\frac{cn}{\beta_4} \right\},$$

which yields (5.4.6)–(5.4.7), by applying Markov's inequality and using $\beta_4 \leq \beta_s^{2/(s-2)}$.   □

5.5   Proof of Theorem 5.1.1

We continue to keep our standard notations in the scheme of the weighted sums

$$S_\theta = \theta_1 X_1 + \cdots + \theta_n X_n, \qquad \theta = (\theta_1, \ldots, \theta_n) \in S^{n-1},$$

with i.i.d. random variables $X_1, \ldots, X_n$ such that $\mathbb{E}X_1 = 0$, $\mathbb{E}X_1^2 = 1$, $\beta_s = \mathbb{E}|X_1|^s < \infty$ for an integer $s \geq 3$. Let us write down the bound of Proposition 5.2.3 for this scheme:

$$c_s \, \rho(F_\theta, \Phi_{s-1,\theta}) \leq L_s(\theta) + I_s(\theta).$$

Here $L_s = L_s(\theta) = \beta_s l_s(\theta) = \beta_s \sum_{k=1}^n |\theta_k|^s$ and

$$I_s(\theta) = 1_{\Omega_s} \int_{1/L_3}^{1/L_s} \frac{|f_\theta(t)|}{t} \, dt, \qquad \Omega_s = \{\theta \in S^{n-1} : L_s \le L_3 \le 1\}.$$

As we know from Lemma 5.3.1, $\mathbb{E}_\theta L_s(\theta) \le c_s \beta_s \, n^{-\frac{s-2}{2}}$, which is sharpened in Lemma 5.3.3 to

$$\mathfrak{s}_{n-1}\{L_s(\theta) \ge c_s \beta_s r \, n^{-\frac{s-2}{2}}\} \le \exp\{-(rn)^{2/s}\}, \qquad r \ge 1,$$

with $c_s = (\sqrt{s} + 2)^s$. Since Lemma 5.4.2 provides similar bounds for $I_s(\theta)$, we obtain:

Theorem 5.5.1 *Assuming that $\beta_s < \infty$, let $\Phi_{s-1,\theta}$ be the Edgeworth correction for F_θ of an integer order $s \ge 4$. Then*

$$\mathbb{E}_\theta \, \rho(F_\theta, \Phi_{s-1,\theta}) \le c_s \beta_s \, n^{-\frac{s-2}{2}}. \tag{5.5.1}$$

Moreover, for all $r \ge 1$,

$$\mathfrak{s}_{n-1}\left\{\rho(F_\theta, \Phi_{s-1,\theta}) \ge c_s r \, \beta_s \, n^{-\frac{s-2}{2}}\right\} \le C_s \varepsilon_s(n, r), \tag{5.5.2}$$

where

$$\varepsilon_s(n, r) = \exp\{-\min((rn)^{2/s}, n^{2/3}, cn/\beta_4)\}.$$

The upper bound on the right-hand side of (5.5.2) has almost an exponential decay with respect to n. For example, when $s = 4$ and with $r = 1$ in (5.5.2), we get

$$\mathfrak{s}_{n-1}\left\{\rho(F_\theta, \Phi_{3,\theta}) \ge \frac{C\beta_4}{n}\right\} \le C \, e^{-c\sqrt{n}}, \qquad n \ge \beta_4^2. \tag{5.5.3}$$

However, F_θ is still approximated by a function depending on θ. According to (5.2.7),

$$\Phi_{3,\theta}(x) = \Phi(x) - \frac{\gamma_3(\theta)}{3!} \, (x^2 - 1)\varphi(x)$$

$$= \Phi(x) - \frac{\alpha_3}{6} \, (x^2 - 1)\varphi(x) \, \alpha_3(\theta), \qquad \alpha_3 = \mathbb{E}X_1^3, \quad \alpha_3(\theta) = \sum_{k=1}^n \theta_k^3.$$

To eliminate the correction term in the case $\alpha_3 \neq 0$, note that $|x^2 - 1| \varphi(x) \leq 1$, leading to

$$\rho(\Phi_{3,\theta}, \Phi) \leq \beta_3 |\alpha_3(\theta)|.$$

But, $\alpha_3(\theta)$ is of order $1/n$, as indicated in Lemma 5.3.2. Using (5.5.1)–(5.5.2), this gives

$$\mathbb{E}_\theta \, \rho(F_\theta, \Phi) \leq \frac{C}{n} \beta_4$$

and

$$\mathfrak{s}_{n-1}\left\{\rho(F_\theta, \Phi) \geq \frac{Cr}{n} \beta_4\right\} \leq C \exp\left\{-c \min\left((rn)^{1/2}, n^{2/3}, n/\beta_4, r^{2/3}\right)\right\}$$

with arbitrary $r \geq 1$, which may be assumed to satisfy $r \leq n/(C\beta_4)$. But, in this case, within a universal factor the quantities $(rn)^{1/2}$, $n^{2/3}$ and n/β_4 dominate $r^{2/3}$. We thus arrive at the Klartag-Sodin theorem for the i.i.d. situation with a slight improvement of the power of r (which was actually mentioned in [7]). In addition, one may emphasize a concentration threshold phenomenon as in (5.5.3) for the case where $\alpha_3 = 0$.

Corollary 5.5.2 *If β_4 is finite, then for all $r \geq 1$,*

$$\mathfrak{s}_{n-1}\left\{\rho(F_\theta, \Phi) \geq \frac{Cr}{n} \beta_4\right\} \leq C \exp\left\{-cr^{2/3}\right\}.$$

Moreover, if $\alpha_3 = 0$ and $n \geq \beta_4^2$, then

$$\mathfrak{s}_{n-1}\left\{\rho(F_\theta, \Phi) \geq \frac{C\beta_4}{n}\right\} \leq C \exp\left\{-c\sqrt{n}\right\}.$$

On the other hand, if $\alpha_3 = 0$ and β_5 is finite, one may turn to the next Edgeworth correction which is given according to (5.2.8) by

$$\Phi_{4,\theta}(x) = \Phi(x) - \frac{\gamma_4(\theta)}{4!} H_3(x) \varphi(x)$$

$$= \Phi(x) - \frac{\beta_4 - 3}{24} H_3(x) \varphi(x) l_4(\theta), \qquad l_4(\theta) = \sum_{k=1}^{n} \theta_k^4, \quad (5.5.4)$$

with $H_3(x) = x^3 - 3x$. This approximation also depends on θ, but the correction term does not have mean zero.

Proof of Theorem 5.1.1. Using $\mathbb{E}_\theta \, l_4(\theta) = \frac{3}{n+2}$, let us rewrite the above as

$$\Phi_{4,\theta}(x) = G(x) + \frac{\beta_4 - 3}{4n(n+2)} H_3(x)\,\varphi(x) - \frac{\beta_4 - 3}{24} H_3(x)\,\varphi(x)\left(l_4(\theta) - \frac{3}{n+2}\right),$$

where

$$G(x) = \Phi(x) - \frac{\beta_4 - 3}{8n} H_3(x)\,\varphi(x)$$

does not contain θ anymore. Since $|\beta_4 - 3| \leq 2\beta_4$ and $|H_3(x)|\,\varphi(x) \leq 1$, it follows that

$$\rho(\Phi_{4,\theta}, G) \leq \frac{\beta_4}{2n^2} + \frac{\beta_4}{12}\left|l_4(\theta) - \frac{3}{n+2}\right|, \qquad (5.5.5)$$

which in turn, recalling the bound $\mathrm{Var}_\theta(l_4) < 24\,n^{-3}$, yields

$$\mathbb{E}_\theta \, \rho(\Phi_{4,\theta}, G) \leq 2\beta_4 \, n^{-3/2}.$$

But, according to Theorem 5.5.1 with $s = 5$,

$$\mathbb{E}_\theta \, \rho(F_\theta, \Phi_{4,\theta}) \leq C\beta_5 \, n^{-3/2}.$$

These two bounds yield the inequality (5.1.4) of Theorem 5.1.1, by applying the triangle inequality for the distance ρ.

Moreover, applying Lemma 5.3.2, from (5.5.5) it also follows that

$$\mathfrak{s}_{n-1}\left\{\rho(\Phi_{4,\theta}, G) \geq Cr\beta_4 \, n^{-3/2}\right\} \leq C \exp\left\{-c r^{1/2}\right\}, \qquad r \geq 1. \qquad (5.5.6)$$

Combining this with the inequality (5.5.2) and recalling that $\beta_4 \leq \beta_5^{2/3} \leq \beta_5$, we get

$$\mathfrak{s}_{n-1}\left\{\rho(F_\theta, G) \geq Cr\beta_5 \, n^{-3/2}\right\} \leq C \exp\left\{-c \min\left((rn)^{2/5}, n^{2/3}, n\beta_5^{-2/3}, r^{1/2}\right)\right\}$$

with an arbitrary value $r \geq 1$. Note that $|G(x)| \leq C\beta_4 \leq C\beta_5$, so that we may restrict ourselves to the region

$$1 \leq r \leq n^{3/2}/(C\beta_5)$$

(since otherwise the left probability is zero). But in this case, necessarily $\beta_5 \leq n^{3/2}/C$, and both $(rn)^{1/2}$ and $n\beta_5^{-2/3}$ dominate $r^{1/2}$. Hence, the above bound is simplified to

$$\mathfrak{s}_{n-1}\left\{\rho(F_\theta, G) \geq Cr\beta_5 \, n^{-3/2}\right\} \leq C \exp\left\{-c \min\left(n^{2/3}, r^{1/2}\right)\right\}.$$

Here, $r^{1/2}$ is dominated by $n^{2/3}$ in the region $r \leq n^{4/3}$, in which case we arrive at the desired inequality

$$\mathfrak{s}_{n-1}\{\rho(F_\theta, G) \geq Cr\beta_5\, n^{-3/2}\} \leq C \exp\{-cr^{1/2}\}. \tag{5.5.7}$$

As for larger values of r, the usual Berry–Esseen inequality (5.1.1) with a purely Gaussian approximation is more accurate. Indeed, together with Lemma 5.3.3 for $p = 3$, the bound (5.1.1), which is known to hold with $c = 1$, gives, for all $r \geq n$,

$$\mathfrak{s}_{n-1}\{\rho(F_\theta, \Phi) \geq Cr\beta_3\, n^{-3/2}\} \leq \mathfrak{s}_{n-1}\{l_3(\theta) \geq Crn^{-3/2}\}$$
$$\leq \exp\{-r^{2/3}\}, \tag{5.5.8}$$

which sharpens (5.5.7). At the expense of a worse rate, one may replace here Φ with $\Phi_{4,\theta}$. From (5.5.4), $\rho(\Phi_{4,\theta}, \Phi) \leq \beta_4\, l_4(\theta)$. Hence, applying Lemma 5.3.3 with $p = 4$ and with $r/\sqrt{n}$ in place of r (which is justified as long as $Cr \geq 121\sqrt{n}$), we get

$$\mathfrak{s}_{n-1}\{\rho(\Phi_{4,\theta}, \Phi) \geq Cr\beta_4\, n^{-3/2}\} \leq \mathfrak{s}_{n-1}\{n\, l_4(\theta) \geq Crn^{-1/2}\}$$
$$\leq \exp\{-r^{1/2}\, n^{1/4}\} \leq \exp\{-r^{1/2}\}.$$

Combining this with (5.5.8), we get

$$\mathfrak{s}_{n-1}\{\rho(F_\theta, \Phi_{4,\theta}) \geq Cr\beta_4\, n^{-3/2}\} \leq 2 \exp\{-r^{1/2}\}, \qquad r \geq n.$$

Finally, by (5.5.6),

$$\mathfrak{s}_{n-1}\{\rho(F_\theta, G) \geq Cr\beta_4\, n^{-3/2}\} \leq (C+2) \exp\{-cr^{1/2}\}.$$

This means that we have obtained the required bound (5.5.7) for all values $r \geq 1$. It remains to rescale the parameter r to arrive at the inequality (5.1.5) of Theorem 5.1.1. $\square$

5.6   General Lower Bounds

Let U be a function of bounded variation on the real line with $U(-\infty) = U(\infty) = 0$. By analogue with Berry–Esseen-type theorems, a standard approach to the problem of lower bounds for the L^∞-norm $\|U\| = \sup_x |U(x)|$ may be based on the study of the associated Fourier–Stieltjes transform

$$u(t) = \int_{-\infty}^{\infty} e^{itx}\, dU(x), \qquad t \in \mathbb{R}.$$

For example, we have the following estimate derived in [2], cf. Theorem 19.2.

Lemma 5.6.1 *For any $T > 0$,*

$$\|U\| \geq \frac{1}{3T} \left| \int_0^T u(t) \left(1 - \frac{t}{T}\right) dt \right|. \tag{5.6.1}$$

In the scheme of the weighted sums, introduce the characteristic function $f(t) = \mathbb{E}_\theta f_\theta(t)$ of the average distribution function $F(x) = \mathbb{E}_\theta F_\theta(x) = \mathbb{E}_\theta \mathbb{P}\{S_\theta \leq x\}$. Lemma 5.6.1 may be used to derive:

Lemma 5.6.2 *Given a function G of bounded variation such that $G(-\infty) = 0$ and $G(\infty) = 1$, for any $T > 0$,*

$$\mathbb{E}_\theta \, \rho(F_\theta, G) \geq \frac{1}{6\sqrt{2}\, T} \, \mathbb{E}_\theta \left| \int_0^T (f_\theta(t) - f(t)) \left(1 - \frac{t}{T}\right) dt \right|. \tag{5.6.2}$$

Proof Given a complex-valued random variable ξ with finite first absolute moment, for any complex number b,

$$\mathbb{E} \, |\xi - b| \geq \frac{1}{2\sqrt{2}} \, \mathbb{E} \, |\xi - \mathbb{E}\xi|. \tag{5.6.3}$$

For the proof of this claim, first note that, by the triangle inequality,

$$\mathbb{E}\sqrt{\eta_0^2 + \eta_1^2} = \mathbb{E} \, |\eta| \geq |\mathbb{E}\eta| = \sqrt{(\mathbb{E}\eta_0)^2 + (\mathbb{E}\eta_1)^2}$$

for any complex-valued random variable η with $\eta_0 = \mathrm{Re}(\eta)$, $\eta_1 = \mathrm{Im}(\eta)$. Replacing η_0 with $|\eta_0|$ and η_1 with $|\eta_1|$, the above can be formally sharpened to

$$\mathbb{E}\sqrt{\eta_0^2 + \eta_1^2} \geq \sqrt{(\mathbb{E} \, |\eta_0|)^2 + (\mathbb{E} \, |\eta_1|)^2}. \tag{5.6.4}$$

Now, write $\xi = \xi_0 + i\xi_1$. Since the inequality (5.6.3) is shift invariant, we may assume that both ξ_0 and ξ_1 have median at zero. In that case, for any $b = b_0 + ib_1$, $b_0, b_1 \in \mathbb{R}$,

$$\mathbb{E} \, |\xi| \leq \mathbb{E} \, |\xi_0| + \mathbb{E} \, |\xi_1| \leq \mathbb{E} \, |\xi - b_0| + \mathbb{E} \, |\xi - b_1|.$$

so,

$$\mathbb{E} \, |\xi - \mathbb{E}\xi| \leq 2 \, \mathbb{E} \, |\xi| \leq 2 \, \mathbb{E} \, |\xi - b_0| + 2 \, \mathbb{E} \, |\xi - b_1|.$$

Using (5.6.4) with $\eta_0 = \xi - b_0$ and $\eta_1 = \xi - b_1$, this gives

$$\mathbb{E}\,|\xi - b| \geq \sqrt{(\mathbb{E}\,|\xi - b_0|)^2 + (\mathbb{E}\,|\xi - b_1|)^2}$$

$$\geq \frac{1}{\sqrt{2}}\,\mathbb{E}\,|\xi - b_0| + \frac{1}{\sqrt{2}}\,\mathbb{E}\,|\xi - b_1| \;\geq\; \frac{1}{2\sqrt{2}}\,\mathbb{E}\,|\xi - \mathbb{E}\xi|,$$

and we arrive at (5.6.3).

Finally, denote by g the Fourier–Stieltjes transform of G. We apply (5.6.3) on the probability space $(S^{n-1}, \mathfrak{s}_{n-1})$ with

$$\xi = \int_0^T f_\theta(t)\left(1 - \frac{t}{T}\right)dt, \qquad b = \int_0^T g(t)\left(1 - \frac{t}{T}\right)dt.$$

In view of (5.6.1) applied to $U = F_\theta - G$, we then get

$$\mathbb{E}_\theta\,\rho(F_\theta, G) \;=\; \mathbb{E}_\theta\,\|F_\theta - G\| \geq \frac{1}{3T}\,\mathbb{E}_\theta\left|\int_0^T (f_\theta(t) - g(t))\left(1 - \frac{t}{T}\right)dt\right|$$

$$\geq \frac{1}{6\sqrt{2}\,T}\,\mathbb{E}_\theta\left|\int_0^T (f_\theta(t) - f(t))\left(1 - \frac{t}{T}\right)dt\right|.$$

$$\square$$

5.7 Approximation by Mean Characteristic Functions

To apply the lower bound (5.6.2), we need to look once more at the asymptotic behaviour of characteristic functions $f_\theta(t)$, at least near zero. To this aim, Proposition 5.2.2 may still be used. In the scheme of the weighted sums, it gives the next two assertions for Edgeworth approximations of orders 4 and 5. Put $\alpha_3 = \mathbb{E}X_1^3$ and $f(t) = \mathbb{E}_\theta\,f_\theta(t)$.

Lemma 5.7.1 *If β_4 is finite, then for all $\theta \in S^{n-1}$ except for a set on the sphere of measure at most $C\beta_3\,e^{-\sqrt{n}}$, in the interval $|t| \leq T_n = \sqrt{n}/(33\,\beta_3)$, we have*

$$f_\theta(t) - f(t) \;=\; \alpha_3\,\alpha_3(\theta)\,\frac{(it)^3}{3!}\,e^{-t^2/2} + \varepsilon \tag{5.7.1}$$

with

$$|\varepsilon| \;\leq\; C\beta_4\,n^{-1}\,t^4\,e^{-t^2/8} + C\beta_3\,\exp\{-\sqrt{n}\}.$$

Lemma 5.7.2 *If β_5 is finite and $\alpha_3 = 0$, then for all $\theta \in S^{n-1}$ except for a set of measure at most $C\beta_4 \exp\{-n^{2/5}\}$, in the interval $|t| \leq T_n$, we have*

$$f_\theta(t) - f(t) = (\beta_4 - 3)\left(l_4(\theta) - \frac{3}{n+2}\right)\frac{t^4}{4!}e^{-t^2/2} + \varepsilon, \tag{5.7.2}$$

where

$$|\varepsilon| \leq C\beta_5\, n^{-3/2}\,|t|^5\, e^{-t^2/8} + C\beta_4 \exp\left\{-n^{2/5}\right\}.$$

Proof According to Definition 5.2.1 with $s = 4$ and $s = 5$, $f_\theta(t)$ is approximated by the functions of the form (5.2.2)–(5.2.3), that is, by

$$g_{3,\theta}(t) = e^{-t^2/2}\left(1 + \gamma_3(\theta)\frac{(it)^3}{3!}\right), \qquad \gamma_3(\theta) = \alpha_3\,\alpha_3(\theta),$$

$$g_{4,\theta}(t) = e^{-t^2/2}\left(1 + \gamma_4(\theta)\frac{t^4}{4!}\right), \qquad \gamma_4(\theta) = (\beta_4 - 3)\,l_4(\theta),$$

where we assume that $\alpha_3 = 0$ in the second case. More precisely, if $\beta_4 < \infty$, then for any $\theta \in S^{n-1}$, in the interval $|t| \leq 1/L_3(\theta)$, we have that

$$|f_\theta(t) - g_{3,\theta}(t)| \leq C\beta_4\, l_4(\theta)\, t^4\, e^{-t^2/8}. \tag{5.7.3}$$

Moreover, if $\beta_5 < \infty$ and $\alpha_3 = 0$, then in the same interval,

$$|f_\theta(t) - g_{4,\theta}(t)| \leq C\beta_5\, l_5(\theta)\,|t|^5\, e^{-t^2/8}. \tag{5.7.4}$$

Using the results from Sect. 5.3, one can simplify these relations for a majority of the coefficients. As was already stressed (as a consequence of Lemma 5.3.3 with $p = 3$),

$$L_3(\theta) \leq 33\frac{\beta_3}{\sqrt{n}} = \frac{1}{T_n}$$

for all θ from a set Ω on the sphere of measure at least $1 - \exp\{-n^{2/3}\}$. Therefore, the bounds (5.7.3)–(5.7.4) are fulfilled for all $|t| \leq T_n$ and for all $\theta \in \Omega$.

Moreover, by Lemma 5.3.3,

$$\mathfrak{s}_{n-1}\left\{l_4(\theta) \geq Cn^{-1}\right\} \leq \exp\left\{-\sqrt{n}\right\}.$$

Therefore, (5.7.3) leads to a simpler version

$$\left|f_\theta(t) - g_{\theta,3}(t)\right| \leq C\beta_4\, n^{-1}\, t^4\, e^{-t^2/8}, \qquad |t| \leq T_n, \tag{5.7.5}$$

which holds for all θ except for a set $\mathcal{F}$ on the sphere of measure at most $2\exp\{-\sqrt{n}\}$.

Clearly, one may replace f_θ and $g_{3,\theta}$ in (5.7.5) by their mean values $f(t) = \mathbb{E}_\theta\, f_\theta(t)$ and $g(t) = \mathbb{E}_\theta\, g_{3,\theta}(t) = e^{-t^2/2}$ at the expense of an error not exceeding

$$\mathfrak{s}_{n-1}(\mathcal{F})\,\sup_t\,|f_\theta(t) - g_{3,\theta}(t)| \leq C\beta_3\,\exp\{-\sqrt{n}\}.$$

Averaging over θ in (5.7.5), it thus yields in the same interval

$$\left|f(t) - g(t)\right| \leq C\beta_4\, n^{-1}\, t^4\, e^{-t^2/8} + C\beta_3\,\exp\{-\sqrt{n}\}. \tag{5.7.6}$$

Finally, combining the latter with (5.7.5), one may bound the expression

$$(f_\theta(t) - g_{3,\theta}(t)) - (f(t) - g(t)) = (f_\theta(t) - f(t)) - \alpha_3\,\alpha_3(\theta)\,\frac{(it)^3}{3!}\,e^{-t^2/2}$$

by the same quantity as on the right-hand side of (5.7.6). This proves Lemma 5.7.1.

Now, turning to (5.7.4), we apply Lemma 5.3.3 with $p = 5$, when it gives

$$\mathfrak{s}_{n-1}\{l_5(\theta) \geq Cn^{-3/2}\} \leq \exp\{-n^{2/5}\}.$$

Hence, we get a simpler version

$$\left|f_\theta(t) - g_{4,\theta}(t)\right| \leq C\beta_5\, n^{-3/2}\, |t|^5\, e^{-t^2/8}, \qquad |t| \leq T_n, \tag{5.7.7}$$

which holds for all θ except for a set $\mathcal{F}$ on the sphere of measure at most $2\exp\{-n^{2/5}\}$. Again, at the expense of an error not exceeding

$$\mathfrak{s}_{n-1}(\mathcal{F})\,\sup_t\,|f_\theta(t) - g_{4,\theta}(t)| \leq C\beta_4\,\exp\{-n^{2/5}\},$$

one may replace f_θ and $g_{4,\theta}$ in (5.7.7) by their mean values $f(t)$ and $g(t)$, where now

$$g(t) = \mathbb{E}_\theta\, g_\theta(t) = e^{-t^2/2}\left(1 + \frac{\alpha}{n+2}\,t^4\right), \qquad \alpha = \frac{\beta_4 - 3}{8}.$$

Averaging over θ in (5.7.7), it thus yields

$$\left|f(t) - g(t)\right| \leq C\beta_5\, n^{-3/2}\, |t|^5\, e^{-t^2/8} + C\beta_4\,\exp\{-n^{2/5}\}. \tag{5.7.8}$$

Finally, combining the latter with (5.7.5), one may bound the expression

$$(f_\theta(t) - g_{4,\theta}(t)) - (f(t) - g(t)) = (f_\theta(t) - f(t)) - (\beta_4 - 3)\left(l_4(\theta) - \frac{3}{n+2}\right)\frac{t^4}{4!}\,e^{-t^2/2}$$

by the same quantity as on the right-hand side of (5.7.8). This proves Lemma 5.7.2.

□

5.8 Proof of Theorem 5.1.2

First, let us apply Lemma 5.6.2 by virtue of the representation (5.7.1) from Lemma 5.7.1, which holds for all θ in $\mathcal{F} \subset S^{n-1}$ of measure at least $1 - C\beta_3 \exp\{-n^{1/2}\}$ in the interval $|t| \leq \sqrt{n}/(33\,\beta_3)$. Given $0 < T \leq 1$, we have

$$\int_0^T t^3 e^{-t^2/2} \left(1 - \frac{t}{T}\right) dt \geq \frac{1}{4} \int_0^{T/2} t^3 \, dt = \frac{1}{256} T^4.$$

On the other hand,

$$\int_0^T t^4 e^{-t^2/8} \left(1 - \frac{t}{T}\right) dt \leq \frac{1}{5} T^5.$$

Therefore, for all $\theta \in \mathcal{F}$ and $n \geq (33\,\beta_3)^2$,

$$\left| \int_0^T (f_\theta(t) - f(t)) \left(1 - \frac{t}{T}\right) dt \right| \geq c\,|\alpha_3|\,|\alpha_3(\theta)|\,T^4$$
$$- \frac{C}{n}\,\beta_4 T^5 - C\beta_3 \exp\left\{-n^{1/2}\right\} T.$$

Integrating this inequality over the set $\mathcal{F}$ and using

$$\mathbb{E}_\theta\,|\alpha_3(\theta)|\,1_{\mathcal{F}} \geq \frac{c}{n},$$

we arrive at

$$\frac{1}{T}\,\mathbb{E}_\theta \left| \int_0^T (f_\theta(t) - f(t)) \left(1 - \frac{t}{T}\right) dt \right| \geq \frac{T^3}{n} \left(c\,|\alpha_3| - C\beta_4 T\right) - C\beta_4 \exp\left\{-n^{1/2}\right\}.$$

Choosing an appropriate value of $T \sim |\alpha_3|/\beta_4$ and applying Lemma 5.6.2, we get

$$\mathbb{E}_\theta\,\rho(F_\theta, G) \geq c\,\frac{|\alpha_3|^4}{\beta_4^3\,n} - C\beta_4 \exp\left\{-n^{1/2}\right\}$$

with an arbitrary function G of bounded variation such that $G(-\infty) = 0$, $G(\infty) = 1$. The latter immediately yields the required relation (5.1.6) for the range $n \geq n_0$ with constant $c = c_0\,|\alpha_3|^4/\beta_4^3$ and for a sufficiently large n_0 depending α_3 and β_4.

To involve the remaining values $2 \leq n < n_0$, let us to note that the infimum in (5.1.6) is positive. Indeed, assuming the opposite for a fixed n, there would exist $G \in \mathcal{G}$ such that $\mathbb{E}_\theta \, \rho(F_\theta, G) = 0$ and hence $F_\theta(x) = G(x)$ for all $\theta \in S^{n-1}$ and all points x. In particular, all the weighted sums S_θ would be equidistributed. But this is only possible when all the random variables X_k have a standard normal distribution, according to the Pólya characterization theorem [12], cf. also [6]. And this contradicts to the assumption $\alpha_3 \neq 0$.

The second assertion, where $\alpha_3 = 0$, but $\beta_4 \neq 0$, is similar. We now apply Lemma 5.6.2 by virtue of the representation (5.7.2) of Lemma 5.7.2, which holds for all θ in a set $\mathcal{F} \subset S^{n-1}$ of measure at least $1 - C\beta_4 \exp\{-n^{2/5}\}$ in the same interval $|t| \leq \sqrt{n}/(33 \, \beta_3)$. Given $0 < T \leq 1$, we have

$$\int_0^T t^4 \, e^{-t^2/2} \left(1 - \frac{t}{T}\right) dt \geq \frac{1}{4} \int_0^{T/2} t^4 \, dt = \frac{1}{640} \, T^5.$$

On the other hand,

$$\int_0^T t^5 \, e^{-t^2/8} \left(1 - \frac{t}{T}\right) dt \leq \frac{1}{6} \, T^6.$$

Therefore, for all $\theta \in \mathcal{F}$ and $n \geq (33 \, \beta_3)^2$,

$$\left| \int_0^T (f_\theta(t) - f(t)) \left(1 - \frac{t}{T}\right) dt \right| \geq c \, |\beta_4 - 3| \left| l_4(\theta) - \frac{3}{n+2} \right| T^5$$
$$- \frac{C}{n^{3/2}} \, \beta_5 \, T^6 - C\beta_4 \exp\left\{ -n^{2/5} \right\}.$$

Integrating this inequality over the set $\mathcal{F}$ and using

$$\mathbb{E}_\theta \left| l_4(\theta) - \frac{3}{n+2} \right| 1_\mathcal{F} \geq \frac{c}{n^{3/2}},$$

we arrive at

$$\frac{1}{T} \, \mathbb{E}_\theta \left| \int_0^T (f_\theta(t) - f(t)) \left(1 - \frac{t}{T}\right) dt \right| \geq \frac{T^4}{n^{3/2}} \left(c \, |\beta_4 - 3| - C\beta_5 \, T\right)$$
$$- C\beta_5 \exp\left\{ -n^{2/5} \right\}.$$

Choosing an appropriate value of $T \sim |\beta_4 - 3|/\beta_5$ and applying Lemma 5.6.2, we get

$$\mathbb{E}_\theta \, \rho(F_\theta, G) \geq c \, \frac{|\beta_4 - 3|^5}{\beta_5^4 \, n} - C\beta_5 \exp\left\{ -n^{2/5} \right\}.$$

The latter yields the required relation (5.1.7) for the range $n \geq n_0$ with constant

$$c = c_0 \, |\beta_4 - 3|^5 / \beta_5^4$$

and with a sufficiently large n_0 depending β_4 and β_5. A similar argument as before allows us to involve the remaining values $2 \leq n < n_0$ as well.   □

References

1. A. Bikjalis, Remainder terms in asymptotic expansions for characteristic functions and their derivatives. (Russian) Litovsk. Mat. Sb. **7**, 571–582 (1967). Selected Transl. Math. Stat. Probab. **11**, 149–162 (1973)
2. S.G. Bobkov, Closeness of probability distributions in terms of Fourier–Stieltjes transforms. (Russian) Uspekhi Mat. Nauk **71**(6), 37–98 (2016); translation in Russian Math. Surv. **71**(6), 1021–1079 (2016)
3. S.G. Bobkov, Asymptotic expansions for products of characteristic functions under moment assumptions of non-integer orders, in *Convexity and Concentration*. The IMA Volumes in Mathematics and its Applications, vol. 161 (2017), pp. 297–357
4. S. Bobkov, M. Ledoux, Weighted Poincaré-type inequalities for Cauchy and other convex measures. Ann. Probab. **37**(2), 403–427 (2009)
5. S.G. Bobkov, G.P. Chistyakov, Götze, F. Berry–Esseen bounds for typical weighted sums. Electron. J. Probab. **23**(92), 22 (2018)
6. A.M. Kagan, Yu.V. Linnik, C.R. Rao, *Characterization Problems in Mathematical Statistics*. Translated from the Russian by B. Ramachandran. Wiley Series in Probability and Mathematical Statistics. (Wiley, London 1973), pp. xii+499
7. B. Klartag, S. Sodin, Variations on the Berry–Esseen theorem. (Russian summary) Teor. Veroyatn. Primen. **56**(3), 514–533 (2011); reprinted in Theory Probab. Appl. **56**(3), 403–419 (2012)
8. M. Ledoux, Concentration of measure and logarithmic Sobolev inequalities, in *Séminaire de Probabilités XXXIII*. Lecture Notes in Mathematics, vol. 1709 (Springer, Berlin, 1999), pp. 120–216
9. M. Ledoux, *The Concentration of Measure Phenomenon*. Mathematical Surveys and Monographs, vol. 89 (American Mathematical Society, Providence, 2001), pp. x+181
10. C.E. Mueller, F.B. Weissler, Hypercontractivity for the heat semigroup for ultraspherical polynomials and on the n-sphere. J. Funct. Anal. **48**(2), 252–283 (1982)
11. V.V. Petrov, *Sums of Independent Random Variables* (Springer, Berlin, 1975), pp. x + 345
12. G. Pólya, Herleitung des Gaußschen Fehlergesetzes aus einer Funktionalgleichung (German). Math. Z. **18**(1), 96–108 (1923)

Chapter 6
Three Applications of the Siegel Mass Formula

Jean Bourgain and Ciprian Demeter

Abstract We present three applications of the Siegel mass formula, using the explicit upper bounds for densities derived in Bourgain and Demeter (Int Math Res Not 2015(11):3150–3184, 2014).

6.1 Background on the Siegel Mass Formula

Let $m \geq n + 1$ and let $\gamma \in M_{m,m}(\mathbb{Z})$ and $\Lambda \in M_{n,n}(\mathbb{Z})$ be two positive definite matrices with integer entries. Denote by $A(\gamma, \Lambda)$ the number of solutions $\mathcal{L} \in M_{m,n}(\mathbb{Z})$ for

$$\mathcal{L}^* \gamma \mathcal{L} = \Lambda. \tag{6.1}$$

Then Siegel's mass formula [6] asserts that

$$A(\gamma, \Lambda) \lesssim_{n,m,\gamma} (\det(\Lambda))^{\frac{m-n-1}{2}} \prod_{p \text{ prime}} \nu_p(\gamma, \Lambda). \tag{6.2}$$

In our forthcoming applications $m = n + 1$ and γ will always be the identity matrix I_{n+1}. In this case, the factor $(\det(\Lambda))^{\frac{m-n-1}{2}}$ is 1.

The authors are partially supported by the Collaborative Research NSF grant DMS-1800305.

AMS Subject Classification Primary 11L03

J. Bourgain
School of Mathematics, Institute for Advanced Study, Princeton, NJ, USA
e-mail: bourgain@math.ias.edu

C. Demeter (✉)
Department of Mathematics, Indiana University, Bloomington, IN, USA
e-mail: demeterc@indiana.edu

© Springer Nature Switzerland AG 2020   99
B. Klartag, E. Milman (eds.), *Geometric Aspects of Functional Analysis*,
Lecture Notes in Mathematics 2256,
https://doi.org/10.1007/978-3-030-36020-7_6

In evaluating the densities $v_p(I_{n+1}, \Lambda)$ we distinguish two separate cases: $p \nmid \det(\Lambda)$ and $p|\det(\Lambda)$. We recall the following estimate (Proposition 5.6.2. (ii) in [5]), see also Proposition 4.2 in [1].

Proposition 1.1 *We have*

$$\prod_{p \nmid \det(\Lambda)} v_p(I_{n+1}, \Lambda) \lesssim 1, \qquad (6.3)$$

with some universal implicit constant.

Let us next consider the primes p which divide $\det(\Lambda)$. Recall that the number of such primes is

$$O\left(\frac{\log \det(\Lambda)}{\log \log \det(\Lambda)}\right). \qquad (6.4)$$

We will denote by $o_p(T)$ the largest α such that $p^\alpha \mid T$. We denote by $d(T)$ the number of divisors on T, and by gcd the greatest common divisor. If T has factorization

$$T = \prod_i p_i^{\alpha_i}$$

then

$$d(T) = \prod (1 + \alpha_i).$$

Recall that we have the bound

$$d(T) \lesssim_\epsilon T^\epsilon.$$

For an $n \times n$ matrix Λ and for $A, B \subset \{1, \ldots, n\}$ with $|A| = |B|$ we define

$$\mu_{A,B} = \det((\Lambda_{i,j})_{i \in A, j \in B}).$$

We recall the following result from [1].

Proposition 1.2 *Let $\Lambda \in M_{n,n}(\mathbb{Z})$ be a positive definite matrix and let $p|\det(\Lambda)$. Then*

$$v_p(I_{n+1}, \Lambda) \lesssim \sum_{\substack{0 \le l_i : 1 \le i \le n \\ l_1 + l_2 + \ldots + l_n \le o_p(\det(\Lambda))}} p^{\beta_2(l_1, \ldots, l_n) + \ldots + \beta_n(l_1, \ldots, l_n)},$$

where $\beta_i = \beta_i(l_1, \ldots, l_n)$ satisfies

$$\beta_i = \min\{(i-1)l_i, \ (i-2)l_i + \min_{|A|=1} o_p(\mu_{\{1\},A}) - l_1, \ (i-3)l_i + \min_{|A|=2} o_p(\mu_{\{1,2\},A})$$

$$- l_1 - l_2, \ldots, \ \min_{|A|=i-1} o_p(\mu_{\{1,2,\ldots,i-1\},A}) - l_1 - l_2 - \ldots - l_{i-1}\}$$

Let us list two consequences that will be used in the next sections.

Corollary 1.3 $(n = 2)$ *Let $\Lambda \in M_{2,2}(\mathbb{Z})$ be a positive definite symmetric matrix. Then*

$$A(I_3, \Lambda) \lesssim_\epsilon (\det(\Lambda))^\epsilon \gcd(\Lambda_{1,1}, \Lambda_{1,2}, \Lambda_{2,2}).$$

Proof Let $p \mid \det(\Lambda)$. Proposition 1.2 and its symmetric version implies that

$$\nu_p(I_3, \Lambda) \lesssim [o_p(\det(\Lambda))]^2 p^{\min\{o_p(\Lambda_{1,1}), o_p(\Lambda_{1,2})\}}$$

and

$$\nu_p(I_3, \Lambda) \lesssim [o_p(\det(\Lambda))]^2 p^{\min\{o_p(\Lambda_{2,1}), o_p(\Lambda_{2,2})\}}.$$

Combining them leads to

$$\nu_p(I_3, \Lambda) \leq C[o_p(\det(\Lambda))]^2 p^{\min\{o_p(\Lambda_{1,1}), o_p(\Lambda_{1,2}), o_p(\Lambda_{2,2})\}},$$

with C independent of p. Thus

$$\prod_{p \mid \det(\Lambda)} \nu_p(I_3, \Lambda) \lesssim C^{O\left(\frac{\log \det(\Lambda)}{\log \log \det(\Lambda)}\right)} d(\det(\Lambda))^2 \gcd(\Lambda_{1,1}, \Lambda_{1,2}, \Lambda_{2,2})$$

$$\lesssim_\epsilon (\det(\Lambda))^\epsilon \gcd(\Lambda_{1,1}, \Lambda_{1,2}, \Lambda_{2,2}).$$

The result now follows by combining this inequality with (6.2) and (6.3).   ∎

Corollary 1.4 $(n = 3)$ *Let $\Lambda \in M_{3,3}(\mathbb{Z})$ be a positive definite matrix. Then*

$$A(I_4, \Lambda) \lesssim_\epsilon (\det(\Lambda))^\epsilon \gcd(\Lambda_{A,B} : A, B \subset \{1, 2, 3\}, |A| = |B| = 2).$$

Proof Use the bound

$$\beta_2(l_1, l_2, l_3) \leq l_2$$

and

$$\beta_3(l_1, l_2, l_3) \le \min_{|A|=2} o_p(\mu_{\{1,2\},A}) - l_1 - l_2$$

(and its symmetric versions). The rest is the same as in Corollary 1.3. ∎

In the next sections we present three applications of these corollaries.

6.2   Uneven Parsell–Vinogradov Sums

Our first application concerns an essentially sharp estimate for the eighth moment of the quadratic Parsell–Vinogradov sums, in the general case when N and M are unrelated. The special case $N \sim M$ was proved in [2], as a consequence of the $l^8(L^8)$ decoupling for the surface

$$(t, s, ts, t^2, s^2).$$

Theorem 2.1 *For each $N, M \ge 1$*

$$\| \sum_{n=1}^{N} \sum_{m=1}^{M} e(nx_1 + mx_2 + nmx_3 + n^2 x_4 + m^2 x_5) \|_{L^8([0,1]^5)}^8$$

$$\lesssim (NM)^{1+\epsilon} (N^3 M^3 + N^4 + M^4).$$

Proof The left hand side represents the number of integral solutions $n_1, \ldots, n_8 \in [1, N]$, $m_1, \ldots, m_8 \in [1, M]$ of the system

$$\begin{cases} n_1 + \ldots + n_4 = n_5 + \ldots + n_8 \\ m_1 + \ldots + m_4 = m_5 + \ldots + m_8 \\ n_1^2 + \ldots + n_4^2 = n_5^2 + \ldots + n_8^2 \\ m_1^2 + \ldots + m_4^2 = m_5^2 + \ldots + m_8^2 \\ n_1 m_1 + \ldots + n_4 m_4 = n_5 m_5 + \ldots + n_8 m_8 \end{cases} \tag{6.5}$$

Let us start by proving that, apart from the small multiplicative term $(NM)^\epsilon$, the proposed upper bound is sharp. First, the system (6.5) is easily seen to have $\sim (NM)^4$ trivial solutions, those satisfying $n_i = n_{i+4}$ and $m_i = m_{i+4}$ for each $1 \le i \le 4$. Second, for each $1 \le m \le M$ the number of solutions with all m_i equal to m

is given by

$$\left\| \sum_{n=1}^{N} e(nx + n^2 y) \right\|_{L^8([0,1]^2)}^{8} \sim N^5.$$

Thus, (6.5) has at least $MN^5 + NM^5$ solutions.

Let us next prove the upper bound. For positive integers $A \lesssim N, B \lesssim M, C \lesssim N^2, D \lesssim M^2, E \lesssim NM$, we note that the number of integral solutions of the system

$$\begin{cases} n_1 + \ldots + n_4 = A \\ m_1 + \ldots + m_4 = B \\ n_1^2 + \ldots + n_4^2 = C \\ m_1^2 + \ldots + m_4^2 = D \\ n_1 m_1 + \ldots + n_4 m_4 = E \end{cases}$$

is smaller than the number $N_{A,B,C,D,E}$ of solutions of the following system

$$\begin{cases} n_1 + \ldots + n_4 = 4A \\ m_1 + \ldots + m_4 = 4B \\ n_1^2 + \ldots + n_4^2 = 16C \\ m_1^2 + \ldots + m_4^2 = 16D \\ n_1 m_1 + \ldots + n_4 m_4 = 16E \end{cases}.$$

The number of solutions of (6.5) is clearly dominated by

$$\sum_{A \lesssim N} \sum_{B \lesssim M} \sum_{C \lesssim N^2} \sum_{D \lesssim M^2} \sum_{E \lesssim NM} N_{A,B,C,D,E}^2. \tag{6.6}$$

We also note that $N_{A,B,C,D,E}$ is equal to the number N_{C_1,D_1,E_1} of solutions for the system

$$\begin{cases} n_1 + \ldots + n_4 = 0 \\ m_1 + \ldots + m_4 = 0 \\ n_1^2 + \ldots + n_4^2 = C_1 \\ m_1^2 + \ldots + m_4^2 = D_1 \\ n_1 m_1 + \ldots + n_4 m_4 = E_1 \end{cases} \tag{6.7}$$

where

$$\begin{cases} C_1 = 16C - 4A^2 \\ D_1 = 16D - 4B^2 \\ E_1 = 16E - 4AB \end{cases}.$$

So

$$\sum_{A,B,C,D,E} N^2_{A,B,C,D,E}$$

$$\lesssim \sum_{\substack{C_1=O(N^2) \\ D_1=O(M^2) \\ E_1=O(NM)}} N^2_{C_1,D_1,E_1} |\{(A, B, C, D, E) : C_1 = 16C - 4A^2,$$

$$D_1 = 16D - 4B^2, \quad E_1 = 16E - 4AB\}|.$$

Note that for each such C_1, D_1, E_1 we have

$$|\{(A, B, C, D, E) : A = O(N), \quad B = O(M), \quad C_1 = 16C - 4A^2,$$

$$D_1 = 16D - 4B^2, \quad E_1 = 16E - 4AB\}|$$

$$\lesssim NM.$$

It follows that

$$\sum_{A,B,C,D,E} N^2_{A,B,C,D,E} \lesssim NM \sum_{C_1,D_1,E_1} N^2_{C_1,D_1,E_1}.$$

The system (6.7) can be rewritten as follows

$$\begin{cases} n_1^2 + n_2^2 + n_3^2 + (n_1 + n_2 + n_3)^2 = C_1 \\ m_1^2 + m_2^2 + m_3^2 + (m_1 + m_2 + m_3)^2 = D_1 \\ n_1m_1 + n_2m_2 + n_3m_3 + (n_1 + n_2 + n_3)(m_1 + m_2 + m_3) = E_1 \end{cases}.$$

This has at most as many solutions as the system

$$\begin{cases} n_1^2 + n_2^2 + n_3^2 = C_1 \\ m_1^2 + m_2^2 + m_3^2 = D_1 \\ n_1m_1 + n_2m_2 + n_3m_3 = E_1 \end{cases}. \tag{6.8}$$

Indeed, this can be seen by changing variables

$$\begin{cases} (n_1 + n_2, n_2 + n_3, n_3 + n_1) \mapsto (n_1, n_2, n_3) \\ (m_1 + m_2, m_2 + m_3, m_3 + m_1) \mapsto (m_1, m_2, m_3) \end{cases}.$$

On the other hand, (6.8) is equivalent with

$$\begin{bmatrix} n_1 & n_2 & n_3 \\ m_1 & m_2 & m_3 \end{bmatrix} I_3 \begin{bmatrix} n_1 & m_1 \\ n_2 & m_2 \\ n_3 & m_3 \end{bmatrix} = \begin{bmatrix} C_1 & E_1 \\ E_1 & D_1 \end{bmatrix}.$$

Let us analyze the number $\tilde{N}_{C_1, D_1, E_1}$ of solutions of this system. We start with the singular case, when $C_1 D_1 = E_1^2$. In this case the number of solutions can be estimated directly. We split the discussion into two subcases.

First, if both $C_1, D_1 \neq 0$, we have equality in the Cauchy–Schwarz inequality. Thus, $(n_1, n_2, n_3) = \lambda(m_1, m_2, m_3)$ for some nonzero rational λ. Thus, $(n_1, n_2, n_3) = \lambda_1(a, b, c)$ and $(m_1, m_2, m_3) = \lambda_2(a, b, c)$ for some integers $\lambda_1, \lambda_2, a, b, c$ satisfying $\gcd(a, b, c) = 1$. We estimate, adding over all admissible lines

$$\sum_{0 < |C_1| \lesssim N^2} \sum_{0 < |D_1| \lesssim M^2} \tilde{N}^2_{C_1, D_1, (C_1 D_1)^{1/2}}$$

$$\leq \Big(\sum_{0 < |C_1| \lesssim N^2} \sum_{0 < |D_1| \lesssim M^2} \tilde{N}_{C_1, D_1, (C_1 D_1)^{1/2}} \Big)^2$$

$$\leq \Big(\sum_{\substack{|a|, |b|, |c| \lesssim \min(N, M) \\ \gcd(a,b,c)=1}} \frac{N}{\max(a, b, c)} \frac{M}{\max(a, b, c)} \Big)^2$$

$$\leq \Big(\sum_{|a|, |b|, |c| \lesssim \min(N, M)} \frac{NM}{\max(a, b, c)^2} \Big)^2$$

$$\lesssim (NM \min(N, M))^2.$$

When $D_1 = 0$, (6.8) boils down to one equation $n_1^2 + n_2^2 + n_3^2 = C_1$. We estimate

$$\sum_{0 < |C_1| \lesssim N^2} \tilde{N}^2_{C_1, 0, 0} \leq \sum_{0 < |C_1| \lesssim N^2} (C_1)^{1+\epsilon} \lesssim N^{4+\epsilon}$$

This proves that the contribution from the singular case satisfies

$$NM \sum_{C_1 \lesssim N^2} \sum_{D_1 \lesssim M^2} \tilde{N}^2_{C_1, D_1, (C_1 D_1)^{1/2}} \lesssim_\epsilon (NM)^{1+\epsilon} (N^4 + M^4 + (NM \min(N, M))^2).$$

$$(6.9)$$

Let us assume that $C_1 D_1 \neq E_1^2$. Corollary 1.3 shows that

$$\tilde{N}_{C_1, D_1, E_1} \lesssim_\epsilon (NM)^\epsilon \gcd(C_1, D_1, E_1).$$

Fix $\lambda \lesssim \min(N^2, M^2)$. The number of triples (C_1, D_1, E_1) with $C_1 \lesssim N^2$, $D_1 \lesssim M^2$, $E_1 \lesssim NM$, such that $\lambda = \gcd(C_1, D_1, E_1)$ is trivially dominated by

$$\frac{N^2}{\lambda} \frac{M^2}{\lambda} \frac{NM}{\lambda} = \frac{N^3 M^3}{\lambda^3}.$$

Using these, the contribution from the nonsingular case can be dominated by

$$NM \sum_{\lambda \lesssim N^2} \sum_{\substack{C_1, D_1, E_1 \\ \gcd(C_1, D_1, E_1) = \lambda}} \tilde{N}_{C_1, D_1, E_1}^2 \lesssim (NM)^{1+\epsilon} \sum_{\lambda \lesssim \min(N^2, M^2)} \sum_{\substack{C_1, D_1, E_1 \\ \gcd(C_1, D_1, E_1) = \lambda}} \lambda^2$$

$$\lesssim (NM)^{4+\epsilon} \sum_{\lambda \lesssim \min(N^2, M^2)} \frac{1}{\lambda}$$

$$\lesssim_\epsilon (NM)^{4+\epsilon}.$$

This together with (6.9) completes the proof of our result, as $\min(NM)^2 \leq NM$. ∎

6.3 Non-congruent Tetrahedra in the Truncated Lattice $[0, q]^3 \cap \mathbb{Z}^3$

Our goal here is to answer the following question asked in [4].

Question 3.1 Let $T_3([0, q]^3 \cap \mathbb{Z}^3)$ denote the collection of all equivalence classes of congruent tetrahedra with vertices in $[0, q]^3 \cap \mathbb{Z}^3$. Is there a $\delta > 0$ and some $C > 0$, both independent of q such that

$$\#T_3([0, q]^3 \cap \mathbb{Z}^3) \leq Cq^{9-\delta}$$

for each $q > 1$?

A positive answer to this question would have implications on producing lower bounds for the Falconer distance-type problem for tetrahedra. We refer to [4] for details on this interesting problem.

Here we give a negative answer to this question. We hope that our approach to answering this question will inspire further constructions which might eventually improve the lower bound for the Falconer-type problem.

Theorem 3.2 *We have for each $\epsilon > 0$ and each $q > 1$*

$$\#T_3([0, q]^3 \cap \mathbb{Z}^3) \gtrsim_\epsilon q^{9-\epsilon}.$$

Note the following trivial upper bound, which shows the essential tightness of our result

$$\#T_3([0, q]^3 \cap \mathbb{Z}^3) \le Cq^9.$$

Indeed, by translation invariance it suffices to fix one vertex at the origin. The upper bound follows since there are $(q + 1)^3$ possibilities for each of the remaining three vertices.

Proof of Theorem 3.2 As mentioned before, we fix one vertex to be the origin $\mathbf{0} = (0, 0, 0)$. A class of congruent tetrahedra in $T_3([0, q]^3 \cap \mathbb{Z}^3)$ can be identified with a matrix $\Lambda \in M_{3,3}(\mathbb{Z})$. Namely, the congruence class of the tetrahedron with vertices $\mathbf{0}, \mathbf{x}, \mathbf{y}, \mathbf{z} \in [0, q]^3 \cap \mathbb{Z}^3$ is represented by the matrix

$$\Lambda = \begin{bmatrix} \langle \mathbf{x}, \mathbf{x} \rangle & \langle \mathbf{x}, \mathbf{y} \rangle & \langle \mathbf{x}, \mathbf{z} \rangle \\ \langle \mathbf{y}, \mathbf{x} \rangle & \langle \mathbf{y}, \mathbf{y} \rangle & \langle \mathbf{y}, \mathbf{z} \rangle \\ \langle \mathbf{z}, \mathbf{x} \rangle & \langle \mathbf{z}, \mathbf{y} \rangle & \langle \mathbf{z}, \mathbf{z} \rangle \end{bmatrix}.$$

A tetrahedron is called non-degenerate if $\mathbf{x}, \mathbf{y}, \mathbf{z}$ are linearly independent. We will implicitly assume the congruence classes correspond to non-degenerate tetrahedra.

We seek for an upper bound on the number N_Λ of integral solutions $\mathcal{L} = (\mathbf{x}, \mathbf{y}, \mathbf{z}) \in (\mathbb{Z}^3)^3$ to the equation

$$\mathcal{L}^*\mathcal{L} = \Lambda.$$

This will represent the number of congruent tetrahedra with side lengths specified by Λ.

In the numerology from Sect. 6.1, this corresponds to $n = m = 3$. To make the theorems in that section applicable we reduce the counting problem to the $m = 3, n = 2$ case as follows. One can certainly bound N_Λ by $q^\epsilon N'_\Lambda$, where N'_Λ is the number of integral solutions $\mathcal{L}' = (\mathbf{x}, \mathbf{y}) \in (\mathbb{Z}^3)^2$ satisfying

$$(\mathcal{L}')^*\mathcal{L}' = \Lambda'$$

and Λ' is the 2×2 minor of Λ obtained from the first two rows and columns of Λ. Indeed, if $\mathbf{x}, \mathbf{y}$ are fixed, the matrix Λ forces $\mathbf{z}$ to lie on the intersection of the sphere of radius $\Lambda_{3,3}^{1/2}$ centered at the origin with, say, a sphere centered at $\mathbf{x}$ whose radius is determined only by the entries of Λ. These radii are $O(q)$, so the resulting circle can only have $O(q^\epsilon)$ points.

Note also that we only care about those Λ' for which there exist $\mathbf{x}, \mathbf{y} \in [0, q]^3 \cap \mathbb{Z}^3$ linearly independent, such that

$$\Lambda' = \begin{bmatrix} \langle \mathbf{x}, \mathbf{x} \rangle & \langle \mathbf{x}, \mathbf{y} \rangle \\ \langle \mathbf{y}, \mathbf{x} \rangle & \langle \mathbf{y}, \mathbf{y} \rangle \end{bmatrix}.$$

This in particular forces Λ' to be positive definite.

Apply now Corollary 1.3. This will bound N'_Λ by

$$q^\epsilon \gcd(\Lambda_{i,j} : i, j \neq 3) \leq q^\epsilon \gcd(\Lambda_{1,1}, \Lambda_{2,2}).$$

Thus

$$N_\Lambda \lesssim_\epsilon q^\epsilon \gcd(\Lambda_{1,1}, \Lambda_{2,2}), \qquad (6.10)$$

for each Λ corresponding to a non-degenerate tetrahedron.

Denote by M_r the number of lattice points on the sphere or radius $r^{1/2}$ centered at the origin. In our case $r \leq q^2$ so we know that $M_r \lesssim_\epsilon q^{1+\epsilon}$. We need to work with spheres that contain many points. Let

$$A := \{r \leq q^2 : M_r \geq q/2\}.$$

Since for each $\epsilon > 0$ we have $M_r \leq C_\epsilon q^{1+\epsilon}$, a double counting argument shows that $q^3 \leq C_\epsilon \#A q^{1+\epsilon} + \frac{1}{2}q^2 q$. Thus $\#A \gtrsim_\epsilon q^{2-\epsilon}$.

Note that for each $r_i \in A$ there are $\sim M_{r_1} M_{r_2} M_{r_3}$ non-degenerate tetrahedrons with vertices $\mathbf{x}, \mathbf{y}, \mathbf{z}$ on the spheres centered at the origin and with radii $r_1^{1/2}, r_2^{1/2}, r_3^{1/2}$ respectively. The congruence class of such a tetrahedron contains

$$\lesssim_\epsilon q^\epsilon \gcd(r_1, r_2)$$

elements, according to (6.10).

We conclude that there are at least

$$\frac{M_{r_1} M_{r_2} M_{r_3}}{q^\epsilon \gcd(r_1, r_2)}$$

congruence classes generated by such non-degenerate tetrahedra. As distinct radii necessarily give rise to distinct congruence classes, we obtain the lower bound

$$\#T_3([0, q]^3 \cap \mathbb{Z}^3) \gtrsim_\epsilon \sum_{r_i \in A} \frac{M_{r_1} M_{r_2} M_{r_3}}{q^\epsilon \gcd(r_1, r_2)} \gtrsim_\epsilon q^{3-\epsilon} \sum_{r_1, r_2, r_3 \in A} \frac{1}{\gcd(r_1, r_2)}.$$

It is immediate that for each integer d there can be at most $\frac{q^6}{d^2}$ tuples $(r_1, r_2, r_3) \in [0, q^2]^3$, hence also in A^3, with $\gcd(r_1, r_2) = d$. Using this observation and the bound $\#(A^3) \geq C_\epsilon q^{6-\epsilon}$, it follows that for each $\epsilon > 0$ at least $\frac{1}{2}\#(A^3)$ among the triples $(r_1, r_2, r_3) \in A^3$ will have $\gcd(r_1, r_2) \leq \frac{10q^\epsilon}{C_\epsilon}$.

This implies that

$$\sum_{r_1, r_2, r_3 \in A} \frac{1}{\gcd(r_1, r_2)} \gtrsim_\epsilon q^{6-\epsilon},$$

which finishes the proof of the theorem. ∎

6.4 Distribution of Lattice Points on Caps

Let $n \geq 2$ and $\lambda \geq 1$ be two integers. Define the lattice points of the sphere

$$\mathcal{F}_{n,\lambda} = \{\xi = (\xi_1, \ldots, \xi_n) \in \mathbb{Z}^n : |\xi_1|^2 + \ldots |\xi_n|^2 = \lambda\}.$$

It was proved in [3] that

$$|\{(\mathbf{x}, \mathbf{y}) \in \mathbb{Z}^3 \times \mathbb{Z}^3 : |\mathbf{x}|^2 = |\mathbf{y}|^2 = \lambda, \ |\mathbf{x} - \mathbf{y}| < \lambda^{1/4}\}| \lesssim_\epsilon \lambda^{\frac{1}{2}+\epsilon}.$$

This is a statement on the average distribution of the lattice points on $\mathcal{F}_{3,\lambda}$ in caps of size $\lambda^{1/4}$. Roughly speaking it states that most such caps contain at most $O(\lambda^\epsilon)$ lattice points. It seems reasonable to conjecture that for $n \geq 3$

$$|\{(\mathbf{x}^1, \ldots, \mathbf{x}^{n-1}) \in (\mathbb{Z}^n)^{n-1} : |\mathbf{x}^i|^2 = \lambda, \ |\mathbf{x}^i - \mathbf{x}^j| < \lambda^{\frac{1}{2(n-1)}} \text{ for } i \neq j\}| \lesssim_\epsilon \lambda^{\frac{n-2}{2}+\epsilon}. \tag{6.11}$$

We next prove this conjecture for $n = 4$.

Note that if $|\mathbf{x}|^2 = |\mathbf{y}|^2 = \lambda$ then $\mathbf{x} \cdot \mathbf{y} = \lambda - \frac{1}{2}|\mathbf{x} - \mathbf{y}|^2$. Denote by X the $(n-1) \times n$ matrix with entries $x_{ij} = x^i_j$, the latter being the jth entry of $\mathbf{x}^i$. We can thus bound the left hand side from (6.11) by

$$\sum_\Lambda |\{X \in \mathbb{Z}^{(n-1) \times n} : XX^T = \Lambda\}|, \tag{6.12}$$

with the sum extending over all symmetric $(n-1) \times (n-1)$ matrices with integer entries of the form

$$\begin{cases} \Lambda_{i,i} = \lambda \text{ for } 1 \leq i \leq n-1 \\ |\lambda - \Lambda_{i,j}| \leq \rho^2 \text{ for } 1 \leq i \neq j \leq n. \end{cases} \tag{6.13}$$

We use $\rho = \lambda^{\frac{1}{2(n-1)}}$.

Replacing $(\mathbf{x}^1, \ldots, \mathbf{x}^{n-1})$ by $(\mathbf{x}^1, \mathbf{y}^2, \ldots, \mathbf{y}^{n-1})$ with $\mathbf{y}^i = \mathbf{x}^i - \mathbf{x}^1$ for $2 \leq i \leq n - 1$, an alternative expression for (6.12) is

$$\sum_{\Lambda'} |\{X' \in \mathbb{Z}^{(n-1) \times n} : X'(X')^T = \Lambda'\}|, \qquad (6.14)$$

with the sum over symmetric $(n - 1) \times (n - 1)$ matrices Λ' with integer entries of the form

$$\begin{cases} \Lambda'_{1,1} = \lambda \\ |\Lambda'_{i,j}| \leq \rho^2 \text{ for } i, j \neq 1 \\ \Lambda'_{i,1} = \Lambda'_{1,i} = -\frac{1}{2}\Lambda'_{i,i} \text{ for } 2 \leq i \leq n - 1. \end{cases} \qquad (6.15)$$

We will now estimate (6.14) when $n = 4$, using the bounds on local densities from Sect. 6.1. We assume $\mathbf{x}^1 \wedge \mathbf{x}^2 \wedge \mathbf{x}^3 \neq 0$ and leave the other more immediate case to the reader. Note that a typical Λ' in our summation has the form

$$\Lambda' = \begin{bmatrix} \lambda & -a & -b \\ -a & 2a & c \\ -b & c & 2b \end{bmatrix},$$

with $\det(\Lambda') \neq 0$.

Using Corollary 1.4 we bound (6.14) by

$$\lesssim_\epsilon \lambda^\epsilon \sum_{|a|,|b|,|c| \leq \rho^2} \gcd(\Lambda'_{A,B} : A, B \subset \{1, 2, 3\}, |A| = |B| = 2)$$

$$\lesssim_\epsilon \lambda^\epsilon \sum_{|a|,|b|,|c| \leq \rho^2} \gcd(a(2\lambda - a), b(2\lambda - b), 4ab - c^2)$$

$$\lesssim_\epsilon \lambda^\epsilon \sum_{d \in \mathcal{D}} d|\{(a, b, c) : |a|, |b|, |c| \leq \rho^2, \ d|a(2\lambda - a), \ d|b(2\lambda - b), \ d|4ab - c^2\}|,$$

where $|\mathcal{D}| \lesssim_\epsilon \rho^{2+\epsilon}$ and $\mathcal{D}$ has all elements $O(\rho^4)$.

Write $d = d_1 d_2$ where $\prod_{p|d_1} p$ divides λ and $(d_2, \lambda) = (d_1, d_2) = 1$. Let also d_1^* be the smallest number such that $d_1^*|d_1$ and $d_1|(d_1^*)^2$. The Chinese Remainder Theorem shows that $d|a(2\lambda - a)$ determines a modulo $d_1^* d_2$ within at most 2 values. The same holds for b. Once a, b are fixed, c is similarly determined modulo d^* where d^* is the smallest number such that $d^*|d$ and $d|(d^*)^2$. We can refine the estimate for (6.14) as

$$\lambda^\epsilon \sum_{d \in \mathcal{D}} d \sum_{\substack{d = d_1 d_2 \\ \prod_{p | d_1} p | \lambda}} \sum_{\substack{\Delta | d \\ d | \Delta^2}} \sum_{\substack{d_1', d_1'' | d_1 \\ d_1 | (d_1')^2 . (d_1'')^2}} (\frac{\rho^2}{d_1' d_2} + 1)(\frac{\rho^2}{d_1'' d_2} + 1)(\frac{\rho^2}{\Delta} + 1)$$

$$\lesssim_\epsilon \rho^{2+\epsilon} \sum_{d \in \mathcal{D}} d^{1/2} \sum_{\substack{d = d_1 d_2 \\ \prod_{p | d_1} p | \lambda}} (\frac{\rho^4}{d_1 d_2^2} + \frac{\rho^2}{d_1^{1/2} d_2} + 1)$$

$$\lesssim_\epsilon \rho^{6+\epsilon} \lesssim_\epsilon \lambda^{1+\epsilon}.$$

This proves the conjectured bound (6.11) for $n = 4$.

References

1. J. Bourgain, C. Demeter, New bounds for the discrete Fourier restriction to the sphere in 4D and 5D. Int. Math. Res. Not. **2015**(11), 3150–3184 (2014)
2. J. Bourgain, C. Demeter, Mean value estimates for Weyl sums in two dimensions. J. Lond. Math. Soc. (2) **94**(3), 814–838 (2016)
3. J. Bourgain, Z. Rudnick, P. Sarnak, Spatial statistics for lattice points on the sphere I: individual results. Bull. Iranian Math. Soc. **43**(4), 361–386 (2017)
4. A. Greenleaf, A. Iosevich, B. Liu, E. Palsson, A group-theoretic viewpoint on Erdös-Falconer problems and the Mattila integral. Rev. Mat. Iberoamericana **31**(3), 799–810 (2015)
5. Y. Kitaoka, *Arithmetic of Quadratic Forms. Cambridge Tracts in Mathematics*, vol. 106 (Cambridge University Press, Cambridge, 1993)
6. L.C. Siegel, *Lectures on the Analytical Theory of Quadratic Forms: Second Term 1934–1935* The Institute for Advanced Study and Princeton University, Revised edition 1949, Reprinted January 1955

Chapter 7
Decouplings for Real Analytic Surfaces of Revolution

Jean Bourgain, Ciprian Demeter, and Dominique Kemp

Abstract We extend the decoupling results of the first two authors to the case of real analytic surfaces of revolution in $\mathbb{R}^3$. New examples of interest include the torus and the perturbed cone.

7.1 Background and the Main Result

Let

$$S = \{(\xi_1, \xi_2, g(\xi_1, \xi_2)) : (\xi_1, \xi_2) \in [-1, 1]^2\}$$

be a smooth, compact surface in $\mathbb{R}^3$, given by the graph of the function g. For each $0 < \delta < 1$ let $\mathcal{N}_\delta(S)$ be the δ-neighborhood of S.

Given a function $f : \mathbb{R}^3 \to \mathbb{C}$ and a set $\tau \subset \mathbb{R}^3$, we denote by f_τ the Fourier restriction of f to τ.

In [1, 2], the first two authors proved the following result.

Theorem 7.1 *Assume S has everywhere nonzero Gaussian curvature. Let $\mathcal{P}_\delta(S)$ be a partition of $\mathcal{N}_\delta(S)$ into near rectangular boxes τ of dimensions $\sim \delta^{1/2} \times \delta^{1/2} \times \delta$.*

The first two authors are partially supported by the Collaborative Research NSF grant DMS-1800305.

J. Bourgain
School of Mathematics, Institute for Advanced Study, Princeton, NJ, USA
e-mail: bourgain@math.ias.edu

C. Demeter (✉) · D. Kemp
Department of Mathematics, Indiana University, Bloomington, IN, USA
e-mail: demeterc@indiana.edu; dekemp@umail.iu.edu

© Springer Nature Switzerland AG 2020
B. Klartag, E. Milman (eds.), *Geometric Aspects of Functional Analysis*,
Lecture Notes in Mathematics 2256,
https://doi.org/10.1007/978-3-030-36020-7_7

Then for each f Fourier supported in $\mathcal{N}_\delta(S)$ and for $2 \le p \le 4$ we have

$$\|f\|_{L^p(\mathbb{R}^3)} \lesssim_\epsilon (\delta^{-1})^{\frac{1}{2}-\frac{1}{p}+\epsilon} (\sum_{\tau \in \mathcal{P}_\delta(S)} \|f_\tau\|_{L^p(\mathbb{R}^3)}^p)^{1/p}. \tag{7.1}$$

Moreover, if Gaussian curvature is positive then

$$\|f\|_{L^p(\mathbb{R}^3)} \lesssim_\epsilon \delta^{-\epsilon} (\sum_{\tau \in \mathcal{P}_\delta(S)} \|f_\tau\|_{L^p(\mathbb{R}^3)}^2)^{1/2}. \tag{7.2}$$

Inequality (7.2) is referred to as an l^2- decoupling. It is false for $p > 4$.

Inequality (7.1) is an l^p-decoupling. Since there are roughly δ^{-1} boxes in $\mathcal{P}_\delta(S)$, the l^p-decoupling follows from the l^2-decoupling and Hölder's inequality when S has positive curvature. However, if S has negative curvature, the stronger l^2-decoupling may fail. This is easiest to observe in the case of the hyperbolic paraboloid, corresponding to $g(\xi_1, \xi_2) = \xi_1^2 - \xi_2^2$. What rules out the l^2-decoupling here is the fact that this surface contains at least one line, and the following elementary principle (applied with $N \sim \delta^{-1/2}$).

Proposition 7.2 *Let L be a line segment in $\mathbb{R}^n$ of length $\sim$ 1. For each $0 \le \delta, N^{-1} < 1$, let $\mathcal{P}_{\delta,N}$ be a partition of the δ-neighborhood $\mathcal{N}_\delta(L)$ of L into $\sim N$ cylinders T with length N^{-1} and radius δ.*

For $p > 2$ let $D(\delta, N, p)$ be the smallest constant such that

$$\|f\|_{L^p(\mathbb{R}^n)} \le D(\delta, N, p)(\sum_{T \in \mathcal{P}_{\delta,N}} \|f_T\|_{L^p(\mathbb{R}^n)}^2)^{1/2} \tag{7.3}$$

holds for all f with Fourier transform supported on $\mathcal{N}_\delta(L)$. Then

$$D(\delta, N, p) \sim N^{\frac{1}{2}-\frac{1}{p}},$$

and (approximate) equality in (7.3) can be achieved by using a smooth approximation of $1_{\mathcal{N}_\delta(L)}$.

The implicit constants in (7.1) and (7.2) depend on ϵ, on the C^3 norm of g and on the lower bound for the Gaussian curvature. In [1] and [2], inequalities (7.2) and (7.1) are first proved for the model surfaces, the elliptic and hyperbolic paraboloid, respectively. The extension to the more general surfaces in Theorem 7.1 is then obtained via local approximation and induction on scales, using Taylor's formula with cubic error term. This is the reason why the third derivatives are also important, in addition to the first and second order ones.

The notable feature of the choice of the diameter $\delta^{1/2}$ of each $\tau \in \mathcal{P}_\delta(S)$ in Theorem 7.1 is that this is the largest scale for which τ can be thought of as being essentially flat. By that we mean that there is a rectangular box R_τ such that $R_\tau \subset \tau \subset 1000R_\tau$. This is of course a consequence of the nonzero curvature condition. The case when one of the principal curvatures is zero leads to new types of decoupling, that have been only partially explored (see also the last section). For future reference, we record the result from [1] for the cone

$$C^2 := \{(\xi_1, \xi_2, \sqrt{\xi_1^2 + \xi_2^2}) : \frac{1}{4} \leq \xi_1^2 + \xi_2^2 \leq 4\}$$

and the cylinder

$$Cyl^2 := \{(\xi_1, \xi_2, \xi_3) : \xi_1^2 + \xi_2^2 = 1, \ |\xi_3| \lesssim 1\}.$$

Theorem 7.3 *For S either C^2 or Cyl^2 we let $\mathcal{P}_\delta(S)$ be a partition of $\mathcal{N}_\delta(S)$ into roughly $\delta^{-1/2}$ essentially rectangular plates P with dimensions $\sim 1 \times \delta^{1/2} \times \delta$. Then for each $2 \leq p \leq 6$ and each f with Fourier transform supported in $\mathcal{N}_\delta(S)$ we have*

$$\|f\|_{L^p(\mathbb{R}^3)} \lesssim_\epsilon \delta^{-\epsilon} (\sum_{P \in \mathcal{P}_\delta(S)} \|f_P\|_{L^p(\mathbb{R}^3)}^2)^{1/2}.$$

The fact that we decouple using plates of length ~ 1 is enforced by Proposition 7.2. The range $[2, 6]$ here is larger than the range $[2, 4]$ from Theorem 7.1 because of subtle dimensionality considerations.

As an immediate corollary of Hölder's inequality, we get the following l^4 decoupling for $S = C^2, Cyl^2$, analogous to (7.1)

$$\|f\|_{L^4(\mathbb{R}^3)} \lesssim_\epsilon \delta^{-\epsilon - \frac{1}{8}} (\sum_{P \in \mathcal{P}_\delta(S)} \|f_P\|_{L^4(\mathbb{R}^3)}^4)^{1/4}. \tag{7.4}$$

We will refer to this inequality for the cylinder as *cylindrical decoupling*.

A natural step would be to try to extend Theorems 7.1 and 7.3 to the case of arbitrary real analytic surfaces S in $\mathbb{R}^3$, without any restriction on curvature. One of the issues is identifying the correct dimensions of the boxes in the partition of $\mathcal{P}_\delta(S)$. In analogy to the previous examples, we would like these boxes to be essentially flat. One possible way to formalize the question is recorded in the following conjecture.

Conjecture 7.4 If S is the graph of a nonconstant real analytic function $g : [-1, 1]^2 \to \mathbb{R}$ then for each $0 < \delta \leq 1$ there is a partition $\mathcal{P}_\delta(S)$ of $\mathcal{N}_\delta(S)$ into essentially flat boxes τ (of possibly different dimensions) such that for each f with Fourier transform supported in $\mathcal{N}_\delta(S)$ we have

$$\|f\|_{L^4(\mathbb{R}^3)} \lesssim_\epsilon \delta^{-\epsilon} |\mathcal{P}_\delta(S)|^{\frac{1}{4}} (\sum_{\tau \in \mathcal{P}_\delta(S)} \|f_\tau\|_{L^4(\mathbb{R}^3)}^4)^{1/4},$$

where $|\mathcal{P}_\delta(S)|$ refers to the cardinality of $\mathcal{P}_\delta(S)$.

In this generality, identifying such a partition seems to be a rather difficult task. We will limit our investigation to the class of surfaces of revolution, which as we shall soon see, is large enough to include some interesting new examples.

To get started, for each real analytic function $\gamma : [\frac{1}{2}, 2] \to \mathbb{R}$ we consider the associated surface of revolution

$$S_\gamma = \{(\xi_1, \xi_2, \gamma(\sqrt{\xi_1^2 + \xi_2^2})) : \frac{1}{4} \leq \xi_1^2 + \xi_2^2 \leq 4\}.$$

For example, the cone $\mathcal{C}^2$ corresponds to $\gamma(r) = r$. Our main result can be somewhat vaguely summarized as follows. We save the details about the precise definition of $\mathcal{P}_\delta(S)$ for the later sections. The interesting new feature of the partitions $\mathcal{P}_\delta(S)$ is that they will consist of boxes of different scales.

Theorem 7.5 (Main Result) *Conjecture 7.4 holds for all real analytic surfaces of revolution S_γ.*

As we shall soon see, the curvature of S_γ is zero exactly when either γ' or γ'' is zero. Let $r_1, \ldots, r_M$ be the zeros of $\gamma'\gamma''$ inside $[\frac{1}{2}, 2]$. The fact that there are only finitely many such zeros is a consequence of the real analyticity of γ. We consider pairwise disjoint intervals $I_i = (r_i - \Delta_i, r_i + \Delta_i)$, with Δ_i small enough such that the power series expansion of γ centered at r_i has radius of convergence $> \Delta_i$. Various other restrictions on the smallness of Δ_i will become apparent throughout the forthcoming argument. Note that the complement

$$[\frac{1}{2}, 2] \setminus \bigcup_{i=1}^{M} I_i = \bigcup J_i$$

is the union of at most $M + 1$ intervals J_i. The triangle inequality will allow us to separately consider the part of the surface corresponding to one such interval. On the intervals J_i the surface will have nonzero curvature, so Theorem 7.1 is applicable.

It remains to investigate the contribution from the intervals I_i. Let us fix such an interval. To simplify notation, we will assume it to be $(1 - \Delta, 1 + \Delta)$.

The partition $\mathcal{P}_\delta(S)$ and the type of analysis we will employ will depend on the derivatives of γ at 1. These derivatives encode all the necessary information concerning the size of the two principal curvatures of S_γ. This will be explored in more detail the next section.

7.2   A Case Analysis Based on Principal Curvatures

Differential geometry ties the notion of curvature of surfaces S in $\mathbb{R}^3$ to the change in the direction of the normal vector along curves in S. To be exact, it describes curvature by way of the derivative of the map $N : S \to \mathbb{S}^2$, whose value at p is the unit (outward) normal vector of S at p.

When S is given as the graph of a function g, this differential in local coordinates (ξ_1, ξ_2) has the form

$$(1 + (g_1)^2 + (g_2)^2)^{-\frac{3}{2}} \begin{pmatrix} g_{11}(1 + (g_2)^2) - g_1 g_2 g_{12} & g_{12}(1 + (g_2)^2) - g_1 g_2 g_{22} \\ \\ g_{12}(1 + (g_1)^2) - g_1 g_2 g_{11} & g_{22}(1 + (g_1)^2) - g_1 g_2 g_{12} \end{pmatrix} \tag{7.5}$$

where $g_i = \frac{\partial g}{\partial \xi_i}$ and $g_{ij} = \frac{\partial^2 g}{\partial \xi_i \partial \xi_j}$.

With a little algebra, the determinant (also known as the Gaussian curvature of S) at a point $(\xi_1, \xi_2, g(\xi_1, \xi_2))$ is found to be

$$K_S(\xi_1, \xi_2) = \frac{g_{11} g_{22} - (g_{12})^2}{(1 + (g_1)^2 + (g_2)^2)^2}. \tag{7.6}$$

The two eigenvalues λ_1, λ_2 are called principal curvatures. Their product equals the Gaussian curvature.

For a later convenience, we record the simplified version of (7.6) in the case that S is the surface of revolution S_γ. The Gaussian curvature along $\sqrt{(\xi_1)^2 + (\xi_2)^2} = r$ is

$$K(r) = \frac{\gamma'(r)\gamma''(r)}{r(1 + \gamma'(r)^2)^2}. \tag{7.7}$$

To motivate our intuition in the following sections, we also record the following well known formulae for the principal curvatures in the radial and angular directions

$$|\lambda_{rad}(r)| = \frac{|\gamma''(r)|}{(1 + (\gamma'(r))^2)^{3/2}}$$

$$|\lambda_{ang}(r)| = \frac{|\gamma'(r)|}{r(1 + (\gamma'(r))^2)^{1/2}}.$$

We will split our analysis into three cases.

Case 1 If $\gamma'(1) \neq 0$ and $\gamma^{(n)}(1) = 0$ for all $n \geq 2$, then we have in fact $\gamma(r) = \gamma'(1)r$. This is a cone, so it is covered by Theorem 7.3. The next two cases are new.

Case 2 If $\gamma'(1) = \ldots = \gamma^{(n-1)}(1) = 0$ and $\gamma^{(n)}(1) \neq 0$ for some $n \geq 2$, then the angular principal curvature is zero along the curve $r = 1$. We will refer to these manifolds as *quasi-tori* and will discuss them in Sect. 7.3.

The typical example to have in mind is the torus, corresponding to

$$\gamma(r) = (\frac{1}{4} - (r-1)^2)^{1/2} \tag{7.8}$$

defined on $(1 - \Delta, 1 + \Delta)$, $\Delta < \frac{1}{2}$.

Case 3 If $\gamma'(1) \neq 0$, $\gamma''(1) = \ldots = \gamma^{(n-1)}(1) = 0$ and $\gamma^{(n)}(1) \neq 0$ for some $n \geq 3$, then the radial principal curvature is zero along the curve $r = 1$. These manifolds can be thought of as perturbations of the cone and will be discussed in Sect. 7.4.

7.3   The Case of the Quasi-Torus

To simplify notation we will assume $\gamma(1) = 1$ and $\gamma^{(n)}(1) = n!$, so that

$$\gamma(r) = 1 + (r-1)^n + O((r-1)^{n+1}). \tag{7.9}$$

Fix δ. Our task is to describe the partition $\mathcal{P}_\delta(S)$. Recall that we want each element of $\mathcal{P}_\delta(S)$ to be an essentially rectangular box.

We start with a dyadic decomposition near 1

$$[1 - \Delta, 1 + \Delta] = [1 - \delta^{1/n}, 1 + \delta^{1/n}] \cup \cup_{k \geq 1} \{r : |r - 1| \in (2^{k-1}\delta^{1/n}, 2^k \delta^{1/n}]\}.$$

Note that k is restricted to $O(\log \frac{1}{\delta})$ values. Thus, since we can afford ϵ losses in Theorem 7.5 we may invoke again the triangle inequality and restrict our attention to a fixed k. Due to symmetry, we may further restrict attention to the right halves of these sets in the above decomposition, which we call U_k.

For $k \geq 0$ let us call S_k the part of the surface S_γ above the thin annulus

$$\mathbb{A}_k = \{(\xi_1, \xi_2) : (\xi_1^2 + \xi_2^2)^{1/2} \in U_k\}.$$

Figure 7.1 depicts S_0, S_1, S_2, with S_0 being the nearly horizontal circular strip at the top. The rationale for bringing in such a decomposition is that the two principal curvatures are essentially constant on each S_k

$$|\lambda_{rad}(r)| \sim (2^k \delta^{1/n})^{n-2}$$

$$|\lambda_{ang}(r)| \sim (2^k \delta^{1/n})^{n-1}.$$

We will first see how to deal with the surface S_0 corresponding to the interval $U_0 = [1, 1 + \delta^{1/n}]$. Note that $\mathcal{N}_\delta(S_0)$ sits inside the $C\delta^{1/n}$-neighborhood of the cylinder Cyl^2, with $C = O(1)$. We may thus apply cylindrical decoupling

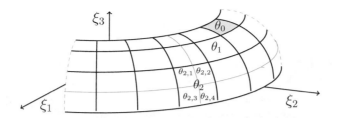

Fig. 7.1 The partition $\mathcal{P}_\delta(S)$

(Theorem 7.3) with δ replaced with $\delta^{1/n}$. Each vertical plate of dimensions $\sim 1 \times \delta^{\frac{1}{2n}} \times \delta^{\frac{1}{n}}$ will intersect S_0 in a cap θ_0 with dimensions $\sim \delta^{1/2n} \times \delta^{1/n}$. Note that for each such θ_0, the box $\mathcal{N}_\delta(\theta_0)$ is essentially flat.

Let $\mathcal{P}_\delta(S_0)$ be the partition consisting of all boxes $\tau_0 = \mathcal{N}_\delta(\theta_0)$. Invoking cylindrical decoupling, we find that whenever f has Fourier transform supported inside $\mathcal{N}_\delta(S_0)$ we have

$$\|f\|_{L^4(\mathbb{R}^3)} \lesssim_\epsilon \delta^{-\epsilon} |\mathcal{P}_\delta(S_0)|^{\frac{1}{4}} \Big(\sum_{\tau_0 \in \mathcal{P}_\delta(S_0)} \|f_{\tau_0}\|_{L^4(\mathbb{R}^3)}^4 \Big)^{1/4}.$$

The collection $\mathcal{P}_\delta(S_0)$ will provide the first elements of the final partition $\mathcal{P}_\delta(S)$.

Let us now investigate S_k, for $k \geq 1$. Fix f with Fourier transform supported inside $\mathcal{N}_\delta(S_k)$. There will be two steps needed in order to produce the desired partition $\mathcal{P}_\delta(S_k)$. The first step is very similar to the one we did for $k = 0$. Namely, we invoke cylindrical decoupling to write

$$\|f\|_{L^4(\mathbb{R}^3)} \lesssim_\epsilon \delta^{-\epsilon} |\tilde{\mathcal{P}}_\delta(S_k)|^{\frac{1}{4}} \Big(\sum_{\tau_k \in \tilde{\mathcal{P}}_\delta(S_k)} \|f_{\tau_k}\|_{L^4(\mathbb{R}^3)}^4 \Big)^{1/4}. \tag{7.10}$$

Each $\tau_k \in \tilde{\mathcal{P}}_\delta(S_k)$ is equal to $\mathcal{N}_\delta(\theta_k)$ for some cap θ_k of dimensions $\sim (2^k \delta^{1/n})^{1/2} \times (2^k \delta^{1/n})$.

It is not hard to see that τ_k is curved. This is because S_k has big radial curvature. More concretely, note that (7.9) forces the δ-neighborhood of the graph of γ on U_k to be a curved tube. This observation suggests that each f_{τ_k} can be further decoupled into smaller pieces. The principal curvatures of θ_k while nonzero, are very small. Consequently, Theorem 7.1 is not directly applicable. What compensates for the small curvatures is the fact that θ_k has tiny area. This will allow us to stretch it into a surface of scale ~ 1, whose principal curvatures are also ~ 1. To execute this strategy we use a linear transformation in the style of parabolic rescaling.

To simplify notation, let us denote by s_k the scale $2^k \delta^{1/n}$. It is also convenient to deal with θ_k sitting directly above the ξ_2 axis, so that a point $(\xi_1, \xi_2, \xi_3) \in \theta_k$ satisfies

$$|\xi_1| \lesssim s_k^{1/2}, \quad \xi_2 - 1 \sim s_k, \quad \xi_3 - 1 \sim s_k^n.$$

We will use the transformation

$$L_k(\xi_1, \xi_2, \xi_3) = (\frac{\xi_1}{s_k^{1/2}}, \frac{\xi_2 - 1}{s_k}, \frac{\xi_3 - 1}{s_k^n}).$$

Let us call $\theta_{k,new} = L_k(\theta_k)$. We make a few observations related to this new surface.

First, note that $L_k(\mathcal{N}_\delta(\theta_k)) \subset \mathcal{N}_{\frac{\delta}{s_k^n}}(\theta_{k,new})$. Thus the function f_{new} defined by

$$\widehat{f_{new}} = \widehat{f_{\tau_k}} \circ L_k^{-1}$$

has Fourier transform supported in $\mathcal{N}_{\frac{\delta}{s_k^n}}(\theta_{k,new})$.

Second, note that the equation of $\theta_{k,new}$ in the new coordinates η_1, η_2, η_3 is

$$\eta_3 = \frac{\gamma(\sqrt{1 + s_k(\eta_1^2 + 2\eta_2) + s_k^2 \eta_2^2}) - 1}{s_k^n}, \quad |\eta_1| \lesssim 1, \quad \eta_2 \sim 1.$$

Using (7.9) and the fact that $\sqrt{1 + r} = 1 + \frac{r}{2} + O(r^2)$ we may write

$$\eta_3 = \frac{1}{2^n s_k^n}(s_k(\eta_1^2 + 2\eta_2) + s_k^2 \eta_2^2)^n + O(s_k^{-n}(s_k(\eta_1^2 + 2\eta_2) + s_k^2 \eta_2^2)^{n+1})$$

$$= \frac{1}{2^n}(\eta_1^2 + 2\eta_2)^n + O(s_k)\Psi(\eta_1, \eta_2).$$

Here Ψ is a C^∞ function. Let S_{ref} be the surface

$$\{(\eta_1, \eta_2, \frac{1}{2^n}(\eta_1^2 + 2\eta_2)^n), \quad |\eta_1| \lesssim 1, \quad \eta_2 \sim 1\}.$$

The fact that $n \geq 2$ and the discussion from the previous section implies that S_{ref} has both principal curvatures ~ 1. The same remains true for $\theta_{k,new}$, as $s_k \ll 1$.

It is easy to see that $s_k = O(\frac{\delta}{s_k^n})$. This means that for some $C = O(1)$ we have $\mathcal{N}_{\frac{\delta}{s_k^n}}(\theta_{k,new}) \subset \mathcal{N}_{\frac{C\delta}{s_k^n}}(S_{ref})$. We can thus apply Theorem 7.1 to decouple f_{new} using N essentially flat boxes B of dimensions $\sim (\frac{\delta}{s_k^n})^{1/2} \times (\frac{\delta}{s_k^n})^{1/2} \times (\frac{\delta}{s_k^n})$

$$\|f_{new}\|_{L^4(\mathbb{R}^3)} \lesssim_\epsilon N^{1/4}\delta^{-\epsilon}(\sum_B \|f_{new,B}\|_{L^4(\mathbb{R}^3)}^4)^{1/4}. \tag{7.11}$$

Let us call $\tau_{k,l}$ the boxes $L_k^{-1}(B)$. These boxes partition $\mathcal{N}_\delta(\theta_k)$ and are essentially flat. Each $\tau_{k,l}$ is essentially the δ-neighborhood of some cap $\theta_{k,l} \subset \theta_k$. Figure 7.1 depicts the decomposition of some θ_2 into four smaller caps $\theta_{2,l}$.

Note that for each B

$$\widehat{f_{new,B}} = \widehat{f_{\tau_k, L^{-1}(B)}} \circ L_k^{-1}.$$

Thus, using a change of variables, (7.11) can be rewritten as

$$\|f_{\tau_k}\|_{L^4(\mathbb{R}^3)} \lesssim_\epsilon N^{1/4} \delta^{-\epsilon} (\sum_{\tau_{k,l}} \|f_{\tau_{k,l}}\|^4_{L^4(\mathbb{R}^3)})^{1/4}. \tag{7.12}$$

The number N is the same for each τ_k. We can now define the partition $\mathcal{P}_\delta(S_k)$ to consist of all $\tau_{k,l}$ with $\tau_k \in \tilde{\mathcal{P}}_\delta(S_k)$. Combining (7.10) with (7.12) we get the following decoupling for a function f with Fourier transform supported in $\mathcal{N}_\delta(S_k)$, $k \geq 1$

$$\|f\|_{L^4(\mathbb{R}^3)} \lesssim_\epsilon \delta^{-\epsilon} |\mathcal{P}_\delta(S_k)|^{\frac{1}{4}} (\sum_{\tau_{k,j} \in \mathcal{P}_\delta(S_k)} \|f_{\tau_{k,j}}\|^4_{L^4(\mathbb{R}^3)})^{1/4}.$$

The partition $\mathcal{P}_\delta(S)$ will be the union of all $\mathcal{P}_\delta(S_k)$, $k \geq 0$.

7.4   The Perturbed Cone

To simplify notation we will assume

$$\gamma(r) = r + (r-1)^n + O((r-1)^{n+1}). \tag{7.13}$$

We will use the decomposition into intervals U_k from the previous section

$$[1, 1 + \Delta) = [1, 1 + \delta^{1/n}] \cup \cup_{k \geq 1}[1 + 2^{k-1}\delta^{1/n}, 1 + 2^k \delta^{1/n}].$$

We continue to denote by S_k the part of S corresponding to U_k, and to write $s_k = 2^k \delta^{1/n}$.

Let us deal first with S_0. Note that $\mathcal{N}_\delta(S_0)$ sits inside $\mathcal{N}_{O(\delta)}(\mathcal{C}^2)$, so we can use the cone decoupling from Theorem 7.3 to produce the relevant partition $\mathcal{P}_\delta(S_0)$, consisting of essentially flat boxes of dimensions $\sim \delta^{1/n} \times \delta^{1/2} \times \delta$.

Next, we fix some $k \geq 1$ and assume f has Fourier transform supported inside $\mathcal{N}_\delta(S_k)$. We will decouple in two stages. The first one is similar to the case $k = 0$. More precisely, note that $\mathcal{N}_\delta(S_k) \subset \mathcal{N}_{O(s_k^n)}(\mathcal{C}^2)$. This allows us to run a cone decoupling

$$\|f\|_{L^4(\mathbb{R}^3)} \lesssim_\epsilon \delta^{-\epsilon} |\tilde{\mathcal{P}}_\delta(S_k)|^{\frac{1}{4}} \Big(\sum_{\tau_k \in \tilde{\mathcal{P}}_\delta(S_k)} \|f_{\tau_k}\|^4_{L^4(\mathbb{R}^3)} \Big)^{1/4}. \tag{7.14}$$

Each $\tau_k \in \tilde{\mathcal{P}}_\delta(S_k)$ is equal to $\mathcal{N}_\delta(\theta_k)$ for some cap θ_k of dimensions $\sim s_k^{n/2} \times s_k$. It is worth observing that cylindrical decoupling would have led to much wider caps of dimensions $\sim s_k^{1/2} \times s_k$. The caps θ_k however are small enough to behave well under rescaling.

We next perform a finer decoupling for each τ_k. It is convenient to deal with θ_k sitting directly above the ξ_2 axis, so that a point $(\xi_1, \xi_2, \xi_3) \in \theta_k$ satisfies

$$|\xi_1| \lesssim s_k^{n/2}, \qquad \xi_2 - 1 \sim s_k, \qquad \xi_3 - 1 \sim s_k.$$

To understand better how to rescale θ_k, we rotate it with $\frac{\pi}{4}$ about the ξ_1 axis and rescale the ξ_2 and ξ_3 variables by $\sqrt{2}$. Thus, the new coordinates satisfy

$$\begin{cases} \xi_1 & = \xi_1' \\ \xi_2 & = \xi_2' - \xi_3' \\ \xi_3 & = \xi_2' + \xi_3'. \end{cases}$$

Using (7.13), the equation of θ_k appears now in an implicit form

$$\xi_2' + \xi_3' = \sqrt{\xi_1'^2 + (\xi_2' - \xi_3')^2} + \Big(\sqrt{\xi_1'^2 + (\xi_2' - \xi_3')^2} - 1\Big)^n$$
$$+ O\Big(\big(\sqrt{\xi_1'^2 + (\xi_2' - \xi_3')^2} - 1\big)^{n+1}\Big),$$

with

$$|\xi_1'| \lesssim s_k^{n/2}, \qquad \xi_2' \sim 1 + s_k, \qquad \xi_3' \sim s_k^n. \tag{7.15}$$

This can be rearranged as follows

$$\xi_3' = \frac{\xi_1'^2}{4\xi_2'} + \frac{1}{4\xi_2'}\Big(\sqrt{\xi_1'^2 + (\xi_2' - \xi_3')^2} - 1\Big)^n + \frac{1}{4\xi_2'}O\Big(\big(\sqrt{\xi_1'^2 + (\xi_2' - \xi_3')^2} - 1\big)^{n+1}\Big).$$

Note that $\xi_3' = \frac{\xi_1'^2}{4\xi_2'}$ is the equation of the cone $\mathcal{C}^2$ in the new coordinates.

We will use the transformation

$$L_k(\xi_1', \xi_2', \xi_3') = (\eta_1, \eta_2, \eta_3) := \Big(\frac{\xi_1'}{s_k^{n/2}}, \frac{\xi_2' - 1}{s_k}, \frac{\xi_3' - 1}{s_k^n}\Big).$$

Let us call $\theta_{k,new} = L_k(\theta_k)$. If we define $\xi_3' = \psi(\xi_1', \xi_2')$ then the equation of $\theta_{k,new}$ in the coordinates η_1, η_2, η_3 becomes

$$\eta_3 = \psi_k(\eta_1, \eta_2) := \frac{1}{s_k^n}[\psi(s_k^{n/2}\eta_1, s_k\eta_2 + 1) - 1], \quad |\eta_1|, |\eta_2| \lesssim 1.$$

It remains to check that this surface satisfies the requirements in Theorem 7.1. More precisely, we have to show that the C^3 norm of ψ_k is $O(1)$, independent of k. Also, we need to show that the Gaussian curvature is away from zero, uniformly over k.

Lemma 7.6 *Write*

$$\phi(\xi_1', \xi_2') = \frac{1}{4\xi_2'}\left(\sqrt{\xi_1'^2 + (\xi_2' - \psi(\xi_1', \xi_2'))^2} - 1\right)^n$$

$$+ \frac{1}{4\xi_2'}O\left(\left(\sqrt{\xi_1'^2 + (\xi_2' - \psi(\xi_1', \xi_2'))^2} - 1\right)^{n+1}\right).$$

Then for each $p, q \geq 0$ with

$$|\xi_1'| \lesssim s_k^{n/2}, \quad \xi_2' - 1 \sim s_k \tag{7.16}$$

we have

$$|D_1^p D_2^q \phi(\xi_1', \xi_2')| \lesssim \min\{(s_k)^{n-p-q}, 1\}.$$

Consequently, for each $p, q \geq 0$

$$\sup_{|\eta_1|,|\eta_2|\lesssim 1} |D_1^p D_2^q \psi_k(\eta_1, \eta_2)| \lesssim 1,$$

with an implicit constant independent of k.

Proof It is clear that $\psi \in C^\infty$. Note that due to (7.15) we have

$$\xi_2' \sim 1$$

and

$$\left|\left(\sqrt{\xi_1'^2 + (\xi_2' - \psi(\xi_1', \xi_2'))^2} - 1\right)^m\right| \lesssim s_k^m$$

for each $0 \leq m \leq n$. The bound on the derivatives of ϕ is now quite immediate, using repeated differentiation.

Next, recall that

$$\psi(\xi_1', \xi_2') = \varphi(\xi_1', \xi_2') + \phi(\xi_1', \xi_2'),$$

where

$$\varphi(\xi_1', \xi_2') = \frac{\xi_1'^2}{4\xi_2'}.$$

Note that in the domain (7.16) we have

$$|D_1^p D_2^q \varphi(\xi_1', \xi_2')| \lesssim (s_k^{n/2})^{2-p}$$

for each $0 \le p \le 2, q \ge 0$ and the derivative becomes zero if $p \ge 3$.

Using all these observations, the desired bound on the derivatives of ψ_k is now immediate. $\qquad\square$

According to (7.6), the Gaussian curvature of $\theta_{k,new}$ is roughly

$$Hess(\psi_k) = det \begin{bmatrix} \psi_{11}(s_k^{n/2}\eta_1, s_k\eta_2 + 1) & \frac{(s_k)^{1+\frac{n}{2}}}{s_k^n} \psi_{12}(s_k^{n/2}\eta_1, s_k\eta_2 + 1) \\ \frac{(s_k)^{1+\frac{n}{2}}}{s_k^n} \psi_{12}(s_k^{n/2}\eta_1, s_k\eta_2 + 1) & (s_k)^{2-n}\psi_{22}(s_k^{n/2}\eta_1, s_k\eta_2 + 1) \end{bmatrix}.$$

It is immediate that

$$Hess(\psi_k) = (s_k)^{2-n} Hess(\psi).$$

Another application of (7.6) shows that $Hess(\psi)$ is roughly the Gaussian curvature of θ_k, in the coordinates (ξ_1', ξ_2', ξ_3'). This is in turn comparable to the Gaussian curvature of θ_k in the original coordinates (ξ_1, ξ_2, ξ_3). But (7.7) determines this curvature to be $\sim (s_k)^{n-2}$. We conclude that the curvature of $\theta_{k,new}$ is ~ 1, as desired.

Note that L_k maps $\mathcal{N}_\delta(\theta_k)$ inside $\mathcal{N}_{O(\frac{\delta}{s_k^n})}(\theta_{k,new})$. The rest of the argument is very similar to the one from the end of the previous section. We apply Theorem 7.1 to partition $\mathcal{N}_{O(\frac{\delta}{s_k^n})}(\theta_{k,new})$ into essentially flat boxes with dimensions $\sim (\frac{\delta}{s_k^n})^{1/2} \times (\frac{\delta}{s_k^n})^{1/2} \times (\frac{\delta}{s_k^n})$. Applying L_k^{-1}, this gives rise to a partition of $\mathcal{N}_\delta(\theta_k)$ into essentially flat boxes $\tau_{k,l} = \mathcal{N}_\delta(\theta_{k,l})$, where each $\theta_{k,l}$ has radial length $\sim s_k(\frac{\delta}{s_k^n})^{1/2}$ and angular length $\sim (s_k)^{n/2}(\frac{\delta}{s_k^n})^{1/2}$.

The desired partition $\mathcal{P}_\delta(S_k)$ will consist of all boxes $\tau_{k,l}$ corresponding to all $\theta_k \in \tilde{\mathcal{P}}_\delta(S_k)$. This concludes the analysis of the perturbed cone.

7.5 Final Remarks

There are various ways in which one could refine the analysis in this paper. We have only aimed to prove a universal l^4 decoupling on the space L^4. A more careful inspection of the argument will reveal that sometimes this can be naturally upgraded to an l^2 decoupling. For example, the torus (7.8) is positively curved on the outside $(r > 1)$ and negatively curved on the inside $(r < 1)$. Thus, the partition from Sect. 7.3 leads in fact to an l^2 decoupling, while the analogous partition for the inside part leads only to an l^4 decoupling.

Also, the boxes in our partitions are maximal, subject to the requirement of being essentially flat. Under this mild constraint some surfaces perform better than others. For example, we have seen earlier that the critical exponent for cone decoupling into plates is 6, rather than 4. Given a surface S, one may instead search for partitions consisting of boxes of smallest possible size, for which the l^4 decoupling holds. This issue seems to be much more delicate. For example, one of the most interesting open questions about surfaces in $\mathbb{R}^3$ is whether the cone can be decoupled into square-like caps. We conjecture the following.

Conjecture 7.7 Let $\mathcal{P}_\delta(\mathcal{C}^2)$ be a partition of $\mathcal{N}_\delta(\mathcal{C}^2)$ into roughly δ^{-1} near rectangular boxes τ of dimensions $\sim \delta^{1/2} \times \delta^{1/2} \times \delta$. Then for each f with Fourier transform supported in $\mathcal{N}_\delta(\mathcal{C}^2)$ and for $2 \leq p \leq 4$ we have

$$\|f\|_{L^p(\mathbb{R}^3)} \lesssim_\epsilon (\delta^{-1})^{\frac{1}{2} - \frac{1}{p} + \epsilon} \Big(\sum_{\tau \in \mathcal{P}_\delta(\mathcal{C}^2)} \|f_\tau\|^p_{L^p(\mathbb{R}^3)} \Big)^{1/p}.$$

The only range where this is known to hold is $2 \leq p \leq 3$, using trilinear restriction technology.

References

1. J. Bourgain, C. Demeter, The proof of the l^2 decoupling conjecture. Ann. Math. **182**(1), 351–389 (2015)
2. J. Bourgain, C. Demeter, Decouplings for curves and hypersurfaces with nonzero Gaussian curvature. J. Anal. Math. **133**, 279–311 (2017)

Chapter 8
On Discrete Hardy–Littlewood Maximal Functions over the Balls in $\mathbb{Z}^d$: Dimension-Free Estimates

Jean Bourgain, Mariusz Mirek, Elias M. Stein, and Błażej Wróbel

Abstract We show that the discrete Hardy–Littlewood maximal functions associated with the Euclidean balls in $\mathbb{Z}^d$ with dyadic radii have bounds independent of the dimension on $\ell^p(\mathbb{Z}^d)$ for $p \in [2, \infty]$.

Jean Bourgain was supported by NSF grant DMS-1800640. Mariusz Mirek was partially supported by the Schmidt Fellowship and the IAS School of Mathematics and by the National Science Center, Poland grant DEC-2015/19/B/ST1/01149. Elias M. Stein was partially supported by NSF grant DMS-1265524. Błażej Wróbel was partially supported by the National Science Centre, Poland grant Opus 2018/31/B/ST1/00204.

J. Bourgain
School of Mathematics, Institute for Advanced Study, Princeton, NJ, USA
e-mail: bourgain@math.ias.edu

M. Mirek (✉)
Department of Mathematics, Rutgers University, Piscataway, NJ, USA

Instytut Matematyczny, Uniwersytet Wrocławski, Wrocław, Poland
e-mail: mariusz.mirek@rutgers.edu

E. M. Stein
Department of Mathematics, Princeton University, Princeton, NJ, USA
e-mail: stein@math.princeton.edu

B. Wróbel
Instytut Matematyczny, Uniwersytet Wrocławski, Wrocław, Poland
e-mail: blazej.wrobel@math.uni.wroc.pl

© Springer Nature Switzerland AG 2020
B. Klartag, E. Milman (eds.), *Geometric Aspects of Functional Analysis*,
Lecture Notes in Mathematics 2256,
https://doi.org/10.1007/978-3-030-36020-7_8

8.1   Introduction

8.1.1   Motivations and Statement of the Results

Let G be a convex centrally symmetric body in $\mathbb{R}^d$, which is simply a bounded closed and centrally symmetric convex subset of $\mathbb{R}^d$ with non-empty interior. An important class of convex symmetric bodies in $\mathbb{R}^d$ are q-balls

$$
B^q = \left\{ x \in \mathbb{R}^d : |x|_q = \left(\sum_{1 \le k \le d} |x_k|^q \right)^{1/q} \le 1 \right\} \quad \text{for} \quad q \in [1, \infty),
$$

$$
B^\infty = \{ x \in \mathbb{R}^d : |x|_\infty = \max_{1 \le k \le d} |x_k| \le 1 \}.
$$

(8.1.1)

For every $t > 0$ and for every $x \in \mathbb{R}^d$ we define the integral Hardy–Littlewood averaging operator

$$
M_t^G f(x) = \frac{1}{|G_t|} \int_{G_t} f(x - y) \mathrm{d}y \quad \text{for} \quad f \in L_{\mathrm{loc}}^1(\mathbb{R}^d),
$$

(8.1.2)

where $G_t = \{ y \in \mathbb{R}^d : t^{-1}y \in G \}$. For $p \in (1, \infty]$, let $C_p(d, G) > 0$ be the best constant such that the following inequality

$$
\left\| \sup_{t>0} |M_t^G f| \right\|_{L^p(\mathbb{R}^d)} \le C_p(d, G) \| f \|_{L^p(\mathbb{R}^d)}
$$

(8.1.3)

holds for every $f \in L^p(\mathbb{R}^d)$. If $p = \infty$, then (8.1.3) holds with $C_p(d, G) = 1$, since M_t^G is the averaging operator. By appealing to the real interpolation and a covering argument for $p = 1$, it is not difficult to see that $C_p(d, G) < \infty$ for every $p \in (1, \infty)$ and for every convex symmetric body $G \subset \mathbb{R}^d$.

In the case of the Euclidean balls $G = B^2$ the theory of spherical maximal functions was used [16] to show that $C_p(d, B^2)$ is bounded independently of the dimension for every $p \in (1, \infty]$. Not long afterwards it was shown, in [1] for $p = 2$, and in [2, 7] for $p \in (3/2, \infty]$, that $C_p(d, G)$ is bounded by an absolute constant, which is independent of the underlying convex symmetric body $G \subset \mathbb{R}^d$. However, if the supremum in (8.1.3) is taken over a dyadic set, i.e. $t \in \mathbb{D} = \{2^n : n \in \mathbb{N} \cup \{0\}\}$, then (8.1.3) holds for all $p \in (1, \infty]$ and $C_p(d, G)$ is independent of the body $G \subset \mathbb{R}^d$ as well.

It is conjectured that the inequality in (8.1.3) holds for all $p \in (1, \infty]$ and for all convex symmetric bodies $G \subset \mathbb{R}^d$ with $C_p(d, G)$ independent of d. It is reasonable to believe that this is true, since it was verified for a large class of convex symmetric bodies. Namely, for the q-balls $G = B^q$ the full range $p \in (1, \infty]$ of dimension-free estimates for $C_p(d, B^q)$ was established in [13] (for $q \in [1, \infty)$) and in [3] (for cubes $q = \infty$) with constants depending only on q. The general case is beyond our reach at this point. We refer also to the survey article [8] for a very careful and exhaustive exposition of the subject, and see also [4] and [11, 12] for extensions of dimension-free estimates to r-variational and jump inequalities.

However, similar questions have been recently investigated by the authors [6] for the discrete analogues of the operators M_t^G in $\mathbb{Z}^d$. The aim of the present article is to continue the investigations in this direction.

For every $t > 0$ and for every $x \in \mathbb{Z}^d$ we define the discrete Hardy–Littlewood averaging operator

$$\mathcal{M}_t^G f(x) = \frac{1}{|G_t \cap \mathbb{Z}^d|} \sum_{y \in G_t \cap \mathbb{Z}^d} f(x - y) \quad \text{for} \quad f \in \ell^1(\mathbb{Z}^d). \tag{8.1.4}$$

We note that the operator $\mathcal{M}_t^G$ is a discrete analogue of (8.1.2).

For $p \in (1, \infty]$, let $\mathcal{C}_p(d, G) > 0$ be the best constant such that the following inequality

$$\left\| \sup_{t > 0} |\mathcal{M}_t^G f| \right\|_{\ell^p(\mathbb{Z}^d)} \le \mathcal{C}_p(d, G) \| f \|_{\ell^p(\mathbb{Z}^d)} \tag{8.1.5}$$

holds for every $f \in \ell^p(\mathbb{Z}^d)$. Arguing in a similar way as in (8.1.3) we conclude that $\mathcal{C}_p(d, G) < \infty$ for every $p \in (1, \infty]$ and for every convex symmetric body $G \subset \mathbb{R}^d$.

The question now is to decide whether $\mathcal{C}_p(d, G)$ can be bounded independently of the dimension d for every $p \in (1, \infty)$. In [6] the authors examined this question in the case of the discrete cubes $B^\infty \cap \mathbb{Z}^d$, and showed that for every $p \in (3/2, \infty]$ there is a constant $C_p > 0$ independent of the dimension such that $\mathcal{C}_p(d, B^\infty) \le C_p$. It was also shown in [6] that if the supremum in (8.1.5) is restricted to the dyadic set $\mathbb{D}$, then (8.1.5) holds for all $p \in (1, \infty]$ and $\mathcal{C}_p(d, G)$ is independent of the dimension.

On the other hand, we constructed in [6] a simple example of a convex symmetric body in $\mathbb{Z}^d$ for which maximal estimate (8.1.5) on $\ell^p(\mathbb{Z}^d)$ involves the smallest constant $\mathcal{C}_p(d, G) > 0$ unbounded in d for every $p \in (1, \infty)$. In order to carry out the construction it suffices to fix a sequence $1 \le \lambda_1 < \ldots < \lambda_d < \ldots < \sqrt{2}$ and consider the ellipsoid

$$E(d) = \left\{ x \in \mathbb{R}^d : \sum_{k=1}^d \lambda_k^2 x_k^2 \le 1 \right\}.$$

Then one can prove that for every $p \in (1, \infty)$ there is $C_p > 0$ such that for every $d \in \mathbb{N}$ one has

$$\mathcal{C}_p(d, E(d)) \ge C_p (\log d)^{1/p}. \tag{8.1.6}$$

Inequality (8.1.6) shows that the dimension-free phenomenon for the Hardy–Littlewood maximal functions in the discrete setting is much more delicate and not as broad as in the continuous case. All these results give us strong motivation to understand the situation more generally, in particular in the case of q-balls $G = B^q$ where $q \in [1, \infty)$, see (8.1.1), which is well understood in the continuous setup.

The main purpose of this work is to prove a dyadic variant of inequality (8.1.5) for (8.1.4) with $G = B^2$.

Theorem 1.1 *For every $p \in [2, \infty]$ there exists a constant $C_p > 0$ independent of $d \in \mathbb{N}$ such that for every $f \in \ell^p(\mathbb{Z}^d)$ we have*

$$\Big\| \sup_{N \in \mathbb{D}} |\mathcal{M}_N^{B^2} f| \Big\|_{\ell^p(\mathbb{Z}^d)} \leq C_p \|f\|_{\ell^p(\mathbb{Z}^d)}. \tag{8.1.7}$$

We shall briefly outline the strategy for proving Theorem 1.1. By a simple interpolation the proof of inequality (8.1.7) is only interesting for $p = 2$, and it will consist of three steps. In the consecutive steps, we shall consider maximal functions corresponding to the operators $\mathcal{M}_N^{B^2}$ in which the supremum is restricted respectively to the sets:

1. $\mathbb{D}_{C_3, \infty} = \{N \in \mathbb{D} : N \geq C_3 d\}$, the large-scale case;
2. $\mathbb{D}_{C_1, C_2} = \{N \in \mathbb{D} : C_1 d^{1/2} \leq N \leq C_2 d\}$, the intermediate-scale case;
3. $\mathbb{D}_{C_0} = \{N \in \mathbb{D} : N \leq C_0 d^{1/2}\}$, the small-scale case;

for some universal constants $C_0, C_1, C_2, C_3 > 0$. Since we are working with the dyadic numbers $\mathbb{D}$ the exact values of C_0, C_1, C_2, C_3 will never play a role as long as they are absolute constants. Moreover, the implied constants will be always allowed to depend on C_0, C_1, C_2, C_3.

8.1.2   The Large-Scale Case

In this step, we will appeal to the comparison principle from [5], where it was shown that there are absolute constants $C, c > 0$ such that for every $p \in (1, \infty)$ and for every $f \in \ell^p(\mathbb{Z}^d)$ we have

$$\Big\| \sup_{t \geq cd} |\mathcal{M}_t^{B^2} f| \Big\|_{\ell^p(\mathbb{Z}^d)} \leq C C_p(d, B^2) \|f\|_{\ell^p(\mathbb{Z}^d)}. \tag{8.1.8}$$

Inequality (8.1.8) combined with the dimension-free estimates for $C_p(d, B^2)$ from (8.1.3) yield the estimates for the full maximal function in the large-scale case. This is the easiest case, the remaining two cases, where it will be important that we are working with the dyadic numbers $\mathbb{D}$, are much more challenging.

8.1.3 The Intermediate-Scale Case

This case will be discussed in Sect. 8.2, where we shall bound the maximal function corresponding to $\mathcal{M}_N^{B^2}$ with the supremum taken over the set $\mathbb{D}_{C_1,C_2}$, see Theorem 2.2. In this case, by a comparison $\mathcal{M}_N^{B^2}$ with a suitable semigroup P_t (see (8.2.11), and also [6]), the proof will be reduced, using a standard square function argument, to estimates of the multipliers $\mathfrak{m}_N^{B^2}$ associated with the operators $\mathcal{M}_N^{B^2}$.

The main objective of Sect. 8.2 is to show that there is a constant $C > 0$ independent of $d \in \mathbb{N}$ such that for every $\xi \in \mathbb{T}^d$ we have

$$|\mathfrak{m}_N^{B^2}(\xi) - 1| \le C(\kappa(d, N)\|\xi\|)^2, \tag{8.1.9}$$

$$|\mathfrak{m}_N^{B^2}(\xi)| \le C\big((\kappa(d, N)\|\xi\|)^{-1} + \kappa(d, N)^{-\frac{1}{7}}\big), \tag{8.1.10}$$

where $\|\xi\|^2 = \sum_{i=1}^d \|\xi_i\|^2$, and $\|\xi_i\| = \text{dist}(\xi_i, \mathbb{Z})$ for all $i \in \mathbb{N}_d = \{1, 2, \ldots, d\}$, and $\kappa(d, N) = Nd^{-1/2}$ is the proportionality factor, which can be identified with the isotropic constant corresponding to the Euclidean ball B_N^2 with radius $N > 0$, see (8.2.30) and Lemma 2.11. We also refer to [1] for more details.

The proof of inequality (8.1.9) is given in Proposition 2.1 and relies on the invariance of $B_N^2 \cap \mathbb{Z}^d$ under the permutation group of $\mathbb{N}_d$. These invariance properties of $B_N^2 \cap \mathbb{Z}^d$ play important roles in the whole article and allow us to exploit probabilistic arguments on the permutation groups.

The proof of inequality (8.1.10) is given in Proposition 2.2 and it requires a more sophisticated analysis, and in particular three tools that we now highlight:

1. Lemma 2.4, which tells us, to some extent, that a significant amount of mass of $B_N^2 \cap \mathbb{Z}^d$, like in the continuous setup, is concentrated near the boundary of $B_N^2 \cap \mathbb{Z}^d$. This lemma combined with Lemma 2.5, which is a variant of a concentration inequality for the hypergeometric distribution, leads us to a decrease dimension trick described in Lemma 2.7.
2. Lemma 2.8, which is an outgrowth of the idea implicit in Lemma 2.7, permits us to control the multiplier $\mathfrak{m}_N^{B^2}$ by multipliers corresponding to the averages associated with balls in lower dimensional spaces, and consequently exploit the estimates for the multipliers corresponding to the operators $M_t^{B^2}$ in the continuous setting, see Theorem 2.3 and Lemma 2.12.
3. A convexity lemma described in Lemma 2.6, which is essential in the proof of inequality (8.1.10).

Let us remark that if we could prove the inequality

$$|\mathfrak{m}_N^{B^2}(\xi)| \le C(\kappa(d, N)\|\xi\|)^{-1}, \tag{8.1.11}$$

instead of (8.1.10), then we would be able to extend inequality (8.1.7) with N restricted to the set $\mathbb{D}_{C_1,C_2}$ for all $p \in (1,\infty]$. However, this will surely require new methods.

8.1.4   The Small-Scale Case

This case will be discussed in Sect. 8.3, where we shall be bounding the maximal function corresponding to $\mathcal{M}_N^{B^2}$ with the supremum taken over the set $\mathbb{D}_{C_0}$, see Theorem 3.1. Our strategy will be much the same as for the proof in the previous case. We shall find suitable approximating multipliers and reduce the matters to the square function estimates using Proposition 3.1. However, this case will require a more sophisticated analysis, due to its different nature that becomes apparent in Lemma 3.2, which says, to a certain degree, that a large percentage of mass of $B_N^2 \cap \mathbb{Z}^d$ is concentrated on the set $\{-1, 0, 1\}^d$. This observation allows us to employ the properties of the Krawtchouk polynomials (8.3.17), as in [9], to prove Proposition 3.3, which is the core of the proof of Proposition 3.1. Using a uniform bound for the Krawtchouk polynomials (see Property 5 in Theorem 3.2) we are able to deduce a decay of the multipliers $\mathfrak{m}_N^{B^2}$ at infinity. Namely, we show (see Proposition 3.3) that there are absolute constants $C, c > 0$ such that for every $\xi \in \mathbb{T}^d$ we have

$$|\mathfrak{m}_N^{B^2}(\xi)| \le Ce^{-\frac{c\kappa(d,N)^2}{100}\sum_{i=1}^d \sin^2(\pi\xi_i)} + Ce^{-\frac{c\kappa(d,N)^2}{100}\sum_{i=1}^d \cos^2(\pi\xi_i)}. \tag{8.1.12}$$

As it was proven in Proposition 3.1, inequality (8.1.12), while different from (8.1.10) or (8.1.11), is good enough to provide $\ell^2(\mathbb{Z}^d)$ theory for the maximal function associated with $\mathcal{M}_N^{B^2}$ in which the supremum is restricted the set $\mathbb{D}_{C_0}$. However, it is not clear at this moment whether (8.1.12) can be used to give an extension of (8.1.7) for some $p \in (1, 2)$ in the small-scale case.

Finally some comments are in order. Currently our methods are limited to $\ell^2(\mathbb{Z}^d)$ theory for the dyadic maximal function $\sup_{N \in \mathbb{D}} |\mathcal{M}_N^{B^2} f|$. It is clear that more information must be provided, if one thinks about an extension of (8.1.7) to the full maximal inequality as in (8.1.5) with $G = B^2$, even for $p = 2$. If we knew that (8.1.11) holds and additionally we could control the difference of $\mathfrak{m}_N^{B^2}$, let us say, in the following sense: that is for every $N \in \mathbb{N}$ and $\xi \in \mathbb{T}^d$ we would have

$$|\mathfrak{m}_{N+1}^{B^2}(\xi) - \mathfrak{m}_N^{B^2}(\xi)| \le CN^{-1} \tag{8.1.13}$$

for some constant independent of the dimension $d \in \mathbb{N}$. Then using the methods from [6] or [12], and taking into account (8.1.9), (8.1.11) and (8.1.13) we would obtain that for every $p \in (3/2, \infty]$ there is a constant $C_p > 0$ independent of the dimension such that $C_p(d, B^2) \leq C_p$. We hope to return to these questions in the near future.

8.1.5 Notation

1. From now on we shall use abbreviated notation and we will write $B_t = B_t^2$, and $Q_t = B_t^\infty$, and $Q = Q_{1/2} = [-1/2, 1/2]^d$, and also $|x| = |x|_2$ for any $x \in \mathbb{R}^d$. Moreover, $M_t = M_t^{B^2}$, and $\mathcal{M}_t = \mathcal{M}_t^{B^2}$ for any $t > 0$.
2. Throughout the whole paper $d \in \mathbb{N}$ will denote the dimension and $C, c, C_0, C_1, \ldots > 0$ will be absolute constants which do not depend on the dimension, however their values may vary from line to line. We will use the convention that $A \lesssim_\delta B$ ($A \gtrsim_\delta B$) to say that there is an absolute constant $C_\delta > 0$ (which possibly depends on $\delta > 0$) such that $A \leq C_\delta B$ ($A \geq C_\delta B$). We will write $A \simeq_\delta B$ when $A \lesssim_\delta B$ and $A \gtrsim_\delta B$ hold simultaneously.
3. Let $\mathbb{N} = \{1, 2, \ldots\}$ be the set of positive integers and $\mathbb{N}_0 = \mathbb{N} \cup \{0\}$, and $\mathbb{D} = \{2^n : n \in \mathbb{N}_0\}$ will denote the set of all dyadic numbers. We set $\mathbb{N}_N = \{1, 2, \ldots, N\}$ for any $N \in \mathbb{N}$.
4. The Euclidean space $\mathbb{R}^d$ is endowed with the standard inner product

$$x \cdot \xi = \langle x, \xi \rangle = \sum_{k=1}^{d} x_k \xi_k$$

for every $x = (x_1, \ldots, x_d)$ and $\xi = (\xi_1, \ldots, \xi_d) \in \mathbb{R}^d$.
5. For a countable set $\mathcal{Z}$ endowed with the counting measure we will write for any $p \in [1, \infty]$ that

$$\ell^p(\mathcal{Z}) = \{f : \mathcal{Z} \to \mathbb{C} : \|f\|_{\ell^p(\mathcal{Z})} < \infty\},$$

where for any $p \in [1, \infty)$ we have

$$\|f\|_{\ell^p(\mathcal{Z})} = \left(\sum_{m \in \mathcal{Z}} |f(m)|^p\right)^{1/p} \quad \text{and} \quad \|f\|_{\ell^\infty(\mathcal{Z})} = \sup_{m \in \mathcal{Z}} |f(m)|.$$

In our case usually $\mathcal{Z} = \mathbb{Z}^d$. We will also abbreviate $\|\cdot\|_{\ell^p(\mathbb{Z}^d)}$ to $\|\cdot\|_{\ell^p}$.
6. Let $\mathcal{F}$ denote the Fourier transform on $\mathbb{R}^d$ defined for any function $f \in L^1(\mathbb{R}^d)$ as

$$\mathcal{F}f(\xi) = \int_{\mathbb{R}^d} f(x) e^{2\pi i \xi \cdot x} \, dx \quad \text{for any} \quad \xi \in \mathbb{R}^d.$$

If $f \in \ell^1(\mathbb{Z}^d)$ we define the discrete Fourier transform by setting

$$\hat{f}(\xi) = \sum_{x \in \mathbb{Z}^d} f(x) e^{2\pi i \xi \cdot x} \quad \text{for any} \quad \xi \in \mathbb{T}^d,$$

where $\mathbb{T}^d$ denote d-dimensional torus, which will be identified with $Q = [-1/2, 1/2]^d$. To simplify notation we will denote by $\mathcal{F}^{-1}$ the inverse Fourier transform on $\mathbb{R}^d$ or the inverse Fourier transform (Fourier coefficient) on the torus $\mathbb{T}^d$. It will cause no confusions and it will be always clear from the context.

8.2 Estimates for the Dyadic Maximal Function: Intermediate Scales

This section is intended to provide bounds independent of the dimension for the dyadic maximal function with supremum taken over all dyadic numbers N such that $d^{1/2} \lesssim N \lesssim d$. Since, as we discussed in the introduction the estimate on $\ell^2(\mathbb{Z}^d)$ for the maximal function $\sup_{N \in \mathbb{D}_{C_3,\infty}} |\mathcal{M}_N f|$ is covered by inequality (8.2.1), which was proved in [5].

Theorem 2.1 *For every* $p \in (1, \infty)$ *there is a constant* $C_p > 0$ *independent of* $d \in \mathbb{N}$ *such that for every* $f \in \ell^p(\mathbb{Z}^d)$ *we have*

$$\left\| \sup_{N \geq cd} |\mathcal{M}_N f| \right\|_{\ell^p} \leq C_p \|f\|_{\ell^p}, \tag{8.2.1}$$

for an absolute large constant $c > 0$.

Now, in view of Theorem 2.1 our aim is to prove Theorem 2.2.

Theorem 2.2 *Let* $C_1, C_2 > 0$ *and define* $\mathbb{D}_{C_1,C_2} = \{N \in \mathbb{D} : C_1 d^{1/2} \leq N \leq C_2 d\}$. *Then there exists a constant* $C > 0$ *independent of dimension such that for every* $f \in \ell^2(\mathbb{Z}^d)$ *we have*

$$\left\| \sup_{N \in \mathbb{D}_{C_1,C_2}} \mathcal{M}_N f \right\|_{\ell^2} \leq C \|f\|_{\ell^2}. \tag{8.2.2}$$

The operator $\mathcal{M}_N$ is a convolution operator with the kernel

$$\mathcal{K}_N(x) = \frac{1}{|B_N \cap \mathbb{Z}^d|} \sum_{y \in B_N \cap \mathbb{Z}^d} \delta_y(x),$$

where δ_y is the Dirac's delta at $y \in \mathbb{Z}^d$. In what follows for any $\xi \in \mathbb{T}^d \equiv [-1/2, 1/2)^d$ we will consider the multipliers corresponding to the operators $\mathcal{M}_N$, which are exponential sums given by

$$\mathfrak{m}_N(\xi) = \hat{\mathcal{K}}_N(\xi) = \frac{1}{|B_N \cap \mathbb{Z}^d|} \sum_{x \in B_N \cap \mathbb{Z}^d} e^{2\pi i \xi \cdot x}. \tag{8.2.3}$$

For $\xi \in \mathbb{T}^d$ we will write $\|\xi\|^2 = \|\xi_1\|^2 + \ldots + \|\xi_d\|^2$, where $\|\xi_j\| = \mathrm{dist}(\xi_j, \mathbb{Z})$ for any $j \in \mathbb{N}_d$. Since we identify $\mathbb{T}^d$ with $[-1/2, 1/2)^d$ hence the norm $\| \cdot \|$ coincides with the Euclidean norm $| \cdot |$ restricted to $[-1/2, 1/2)^d$. Moreover, for every $\eta \in \mathbb{T}$ we know that $\|\eta\| \simeq |\sin(\pi \eta)|$, since $|\sin(\pi \eta)| = \sin(\pi \|\eta\|)$ and for $0 \le |\eta| \le 1/2$ we have

$$2|\eta| \le |\sin(\pi \eta)| \le \pi |\eta|. \tag{8.2.4}$$

The proof of Theorem 2.2 will be based on Proposition 2.1, which provides estimates of the multiplier $\mathfrak{m}_N(\xi)$ at the origin, and on Proposition 2.2, which provides estimates of the multiplier $\mathfrak{m}_N(\xi)$ at infinity. Both of the estimates will be described in terms of a proportionality constant

$$\kappa(d, N) = N d^{-1/2}. \tag{8.2.5}$$

Proposition 2.1 *For every $d, N \in \mathbb{N}$ and for every $\xi \in \mathbb{T}^d$ we have*

$$|\mathfrak{m}_N(\xi) - 1| \le 2\pi^2 \kappa(d, N)^2 \|\xi\|^2. \tag{8.2.6}$$

Proof Exploiting the symmetries of $B_N \cap \mathbb{Z}^d$ we have

$$\mathfrak{m}_N(\xi) = \frac{1}{|B_N \cap \mathbb{Z}^d|} \sum_{x \in B_N \cap \mathbb{Z}^d} \prod_{j=1}^d e^{2\pi i x_j \xi_j}$$

$$= \frac{1}{|B_N \cap \mathbb{Z}^d|} \sum_{x \in B_N \cap \mathbb{Z}^d} \prod_{j=1}^d \cos(2\pi x_j \xi_j). \tag{8.2.7}$$

Recall that for any sequence $(a_j : j \in \mathbb{N}_d) \subseteq \mathbb{C}$ and $(b_j : j \in \mathbb{N}_d) \subseteq \mathbb{C}$, if $\sup_{j \in \mathbb{N}_d} |a_j| \le 1$ and $\sup_{j \in \mathbb{N}_d} |b_j| \le 1$ then we have

$$\left| \prod_{j=1}^d a_j - \prod_{j=1}^d b_j \right| \le \sum_{j=1}^d |a_j - b_j|. \tag{8.2.8}$$

Therefore, using (8.2.8) and the formula $\cos(2x) = \cos^2 x - \sin^2 x = 1 - 2\sin^2 x$, we obtain

$$|m_N(\xi) - 1| \leq \frac{1}{|B_N \cap \mathbb{Z}^d|} \sum_{x \in B_N \cap \mathbb{Z}^d} \left| \prod_{j=1}^d \cos(2\pi x_j \xi_j) - 1 \right|$$

$$\leq \frac{1}{|B_N \cap \mathbb{Z}^d|} \sum_{x \in B_N \cap \mathbb{Z}^d} \sum_{j=1}^d |\cos(2\pi x_j \xi_j) - 1|$$

$$\leq \frac{2}{|B_N \cap \mathbb{Z}^d|} \sum_{x \in B_N \cap \mathbb{Z}^d} \sum_{j=1}^d \sin^2(\pi x_j \xi_j).$$

Observe that $|\sin(\pi xy)| \leq |x||\sin(\pi y)|$ for every $x \in \mathbb{Z}$ and $y \in \mathbb{R}$, and observe also that for every $i \neq j$ one has

$$\sum_{x \in B_N \cap \mathbb{Z}^d} x_i^2 = \sum_{x \in B_N \cap \mathbb{Z}^d} x_j^2 = \frac{1}{d} \sum_{x \in B_N \cap \mathbb{Z}^d} |x|^2.$$

Thus, taking into account these observations and changing the order of summations we obtain

$$|m_N(\xi) - 1| \leq \frac{2}{|B_N \cap \mathbb{Z}^d|} \sum_{x \in B_N \cap \mathbb{Z}^d} \sum_{j=1}^d \sin^2(\pi x_j \xi_j)$$

$$\leq \frac{2}{|B_N \cap \mathbb{Z}^d|} \sum_{j=1}^d \sin^2(\pi \xi_j) \sum_{x \in B_N \cap \mathbb{Z}^d} x_j^2 \tag{8.2.9}$$

$$\leq \frac{2}{|B_N \cap \mathbb{Z}^d|} \sum_{j=1}^d \sin^2(\pi \xi_j) \frac{1}{d} \sum_{x \in B_N \cap \mathbb{Z}^d} |x|^2$$

$$\leq 2\pi^2 \kappa(d, N)^2 \|\xi\|^2$$

and (8.2.6) is justified.   □

The rest of this section is devoted to prove of Proposition 2.2.

Proposition 2.2 *There is a constant $C > 0$ such that for any $d, N \in \mathbb{N}$ if $10 \leq \kappa(d, N) \leq 50d^{1/2}$ then for all $\xi \in \mathbb{T}^d$ we have*

$$|m_N(\xi)| \leq C\big((\kappa(d, N)\|\xi\|)^{-1} + \kappa(d, N)^{-\frac{1}{7}}\big). \tag{8.2.10}$$

Proposition 2.2 is essential in the proof of Theorem 2.2. Assume momentarily that Proposition 2.2 has been proven. We show how the inequalities (8.2.6) and (8.2.10) can be used to deduce (8.2.2).

Proof of Theorem 2.2 Since $\mathbb{D}_{C_1,C_2}$ is a subset of the dyadic set $\mathbb{D}$ we can assume, without loss of generality, that $C_1 = C_2 = 1$ and (8.2.10) is valid when $N \in \mathbb{D}_{1,1}$. To complete the proof we shall compare the averages $\mathcal{M}_N$ with a symmetric diffusion semigroup on $\mathbb{Z}^d$. Namely, for every $t > 0$ let P_t be the semigroup with the multiplier

$$\mathfrak{p}_t(\xi) = e^{-t \sum_{i=1}^d \sin^2(\pi \xi_i)} \quad \text{for} \quad \xi \in \mathbb{T}^d. \tag{8.2.11}$$

It follows from a general theory for symmetric diffusion semigroups [15], (see also [6] for more details) that for every $p \in (1, \infty)$ there is $C_p > 0$ independent of $d \in \mathbb{N}$ such that for every $f \in \ell^p(\mathbb{Z}^d)$ we have

$$\left\| \sup_{t>0} |P_t f| \right\|_{\ell^p} \le C_p \|f\|_{\ell^p}. \tag{8.2.12}$$

Hence (8.2.12) reduces the proof of (8.2.2) to the dimension-free estimate on $\ell^2(\mathbb{Z}^d)$ for the following square function

$$Sf(x) = \left(\sum_{N \in \mathbb{D}_{C_1,C_2}} |\mathcal{M}_N f(x) - P_{N^2/d} f(x)|^2 \right)^{1/2} \quad \text{for} \quad x \in \mathbb{Z}^d.$$

By Plancherel's formula, (8.2.6) and (8.2.10), we have

$$\|S(f)\|_{\ell^2}^2 \le \int_{\mathbb{T}^d} \left(\sum_{\substack{m \in \mathbb{Z}: \\ d^{1/2} \le 2^m \le d}} \min \left\{ 2^{2m} \|\xi\|^2/d, (2^{2m} \|\xi\|^2/d)^{-1} \right\} \right.$$

$$\left. + d^{1/7} \sum_{\substack{m \in \mathbb{Z}: \\ d^{1/2} \le 2^m \le d}} 2^{-2m/7} \right) |\hat{f}(\xi)|^2 \mathrm{d}\xi \le C \|f\|_{\ell^2}^2.$$

This completes the proof of Theorem 2.2. □

8.2.1   Some Preparatory Estimates

The proof of Proposition 2.2 will require some bunch of lemmas, which will be based on the following precise version of Stirling's formula [14]. For every $m \in \mathbb{N}$ one has

$$\sqrt{2\pi}\, m^{m+1/2} e^{-m} e^{\frac{1}{12m+1}} \le m! \le \sqrt{2\pi}\, m^{m+1/2} e^{-m} e^{\frac{1}{12m}}. \tag{8.2.13}$$

We shall need the following crude size estimates for the number of lattice points in B_N.

Lemma 2.3 *For all* $d, N \in \mathbb{N}$ *we have*

$$(2\lfloor \kappa(d, N)\rfloor + 1)^d \leq |B_N \cap \mathbb{Z}^d| \leq (2\pi e)^{d/2}(\kappa(d, N)^2 + 1/4)^{d/2}.$$

Proof The lower bound follows from the inclusion $[-\kappa(d, N), \kappa(d, N)]^d \cap \mathbb{Z}^d \subseteq B_N \cap \mathbb{Z}^d$. To prove the upper bound we use [5, Lemma 5.1] to obtain that

$$|B_N \cap \mathbb{Z}^d| \leq 2|B_{(N^2+d/4)^{1/2}}| = \frac{2\pi^{d/2}}{\Gamma(d/2+1)}(N^2 + d/4)^{d/2}$$

$$= \frac{2\pi^{d/2}d^{d/2}}{\Gamma(d/2+1)}(\kappa(d, N)^2 + 1/4)^{d/2}.$$

Assume that $d = 2m$ is even and note that by (8.2.13) we have

$$\frac{\pi^{d/2}}{\Gamma(d/2+1)} = \frac{\pi^m}{m!} \leq \frac{\pi^m e^m}{(2\pi)^{1/2}m^{m+1/2}} \leq \frac{(2\pi e)^{d/2}}{2d^{d/2}}.$$

If $d = 1$ then $\frac{\pi^{1/2}}{\Gamma(3/2)} = 2 \leq (2\pi e)^{1/2}/2$. Assume now that $d = 2m + 1 \geq 3$ is odd and note

$$\frac{\pi^{d/2}}{\Gamma(d/2+1)} = \frac{2^{2m+1}\pi^m m!}{(2m+1)!} \leq \frac{2^d \pi^{(d-1)/2}m^{m+1/2}e^{d-m}e^{1/12}}{d^{d+1/2}}$$

$$\leq \left(\frac{e^{1/6}}{3\pi}\right)^{1/2}\frac{(2\pi e)^{d/2}}{d^{d/2}} \leq \frac{(2\pi e)^{d/2}}{2d^{d/2}}.$$

The proof of the lemma is completed. $\qquad\qquad\qquad\qquad\qquad\qquad\qquad\square$

We shall also need balls in lower dimensions. For every $r \in \mathbb{N}_d$ let $B_R^{(r)}$ denote the Euclidean ball in $\mathbb{R}^r$ centered at the origin with radius $R > 0$. We now use Lemma 2.3 to control size of certain error subsets of $B_N \cap \mathbb{Z}^d$, which tells us, to some extent, that a significant amount of mass of $B_N \cap \mathbb{Z}^d$ is concentrated near its boundary, like in the continuous setup.

Lemma 2.4 *Given* $\varepsilon_1, \varepsilon_2 \in (0, 1]$ *we define for every* $d, N \in \mathbb{N}$ *the set*

$$E = \{x \in B_N \cap \mathbb{Z}^d : |\{i \in \mathbb{N}_d : |x_i| \geq \varepsilon_2\kappa(d, N)\}| \leq \varepsilon_1 d\}.$$

If $\varepsilon_1, \varepsilon_2 \in (0, 1/10]$ *and* $\kappa(d, N) \geq 10$, *then we have*

$$|E| \leq 2e^{-\frac{d}{10}}|B_N \cap \mathbb{Z}^d|. \qquad\qquad\qquad (8.2.14)$$

Proof Note that if $x \in E$, then there is $I \subseteq \mathbb{N}_d$ such that $|I| \leq \varepsilon_1 d$ and $|x_i| \geq \varepsilon_2 \kappa(d, N)$ precisely when $i \in I$. Therefore, we have

$$E \subseteq \bigcup_{I \subseteq \mathbb{N}_d : |I| \leq \varepsilon_1 d} \{x \in B_N \cap \mathbb{Z}^d : |x_i| \geq \varepsilon_2 \kappa(d, N) \text{ precisely when } i \in I\}.$$

For I as above we have $|x_i| \leq \varepsilon_2 \kappa(d, N)$ for every $i \in \mathbb{N}_d \setminus I$ and consequently

$$|\{x \in B_N \cap \mathbb{Z}^d : |x_i| \geq \varepsilon_2 \kappa(d, N) \text{ precisely when } i \in I\}|$$

$$\leq (2\varepsilon_2 \kappa(d, N) + 1)^{d-|I|} |B_N^{(|I|)} \cap \mathbb{Z}^{|I|}|.$$

Hence

$$|E| \leq \sum_{I \subseteq \mathbb{N}_d : |I| \leq \varepsilon_1 d} (2\varepsilon_2 \kappa(d, N) + 1)^{d-|I|} |B_N^{(|I|)} \cap \mathbb{Z}^{|I|}|. \tag{8.2.15}$$

Clearly, there are $\binom{d}{m}$ subsets I of $\mathbb{N}_d$ of size m. We use the upper bound from Lemma 2.3 (with $d = m$) to estimate (8.2.15) and obtain

$$|E| \leq (2\varepsilon_2 \kappa(d, N) + 1)^d + \sum_{1 \leq m \leq \varepsilon_1 d} \binom{d}{m} (2\varepsilon_2 \kappa(d, N) + 1)^{d-m} |B_N^{(m)} \cap \mathbb{Z}^m|$$

$$\leq (2\varepsilon_2 \kappa(d, N) + 1)^d + \sum_{1 \leq m \leq \varepsilon_1 d} \frac{d^m}{m!} (2\varepsilon_2 \kappa(d, N) + 1)^{d-m} (2\pi e)^{m/2}$$

$$\times \left(\kappa(d, N)^2 d/m + 1/4\right)^{m/2}$$

$$\leq (2\varepsilon_2 \kappa(d, N) + 1)^d + \sum_{1 \leq m \leq \varepsilon_1 d} \left(\frac{d}{m}\right)^{3m/2} (2\varepsilon_2 \kappa(d, N) + 1)^{d-m} (2\pi e^3)^{m/2}$$

$$\times \left(\kappa(d, N)^2 + m/(4d)\right)^{m/2},$$

where in the last line we have used that $\frac{1}{m!} \leq \frac{e^m}{m^m}$. Therefore, using the lower bound from Lemma 2.3, we obtain

$$|E| \leq (2\varepsilon_2 \kappa(d, N) + 1)^d + \sum_{1 \leq m \leq \varepsilon_1 d} \left(\frac{2^{1/3} ed}{m}\right)^{3m/2} (2\varepsilon_2 \kappa(d, N) + 1)^{d-m}$$

$$\times \left(\pi^{1/2} \kappa(d, N) + 1\right)^m$$

$$\leq \left(\left(\frac{2\varepsilon_2 \kappa(d, N) + 1}{2\lfloor \kappa(d, N) \rfloor + 1} \right)^d + \sum_{1 \leq m \leq \varepsilon_1 d} \left(\frac{2^{1/3} ed}{m} \right)^{3m/2} \right.$$

$$\left. \times \left(\frac{2\varepsilon_2 \kappa(d, N) + 1}{2\lfloor \kappa(d, N) \rfloor + 1} \right)^{d-m} \right) |B_N \cap \mathbb{Z}^d|$$

$$\leq \left(e^{-\frac{16d}{19}} + e^{-\frac{72d}{95}} \sum_{m=1}^{d_0} e^{\varphi(m)} \right) |B_N \cap \mathbb{Z}^d|, \tag{8.2.16}$$

where $d_0 = \lfloor \varepsilon_1 d \rfloor$ and $\varphi(x) = \frac{3x}{2} \log \left(\frac{2^{1/3} ed}{x} \right)$. In the last inequality we have used that $\kappa(d, N) \geq 10$ and $\varepsilon_1, \varepsilon_2 \leq 1/10$ and the following bound $\frac{2\varepsilon_2 \kappa(d,N)+1}{2\lfloor \kappa(d,N) \rfloor + 1} \leq \frac{2\varepsilon_2 \kappa(d,N)+1}{2\kappa(d,N)-1} = 1 - \frac{2(1-\varepsilon_2)\kappa(d,N)-2}{2\kappa(d,N)-1} \leq e^{-\frac{16}{19}}$. Note that $(0, d/10] \ni x \mapsto \varphi(x) = \frac{3x}{2} \log \left(\frac{2^{1/3} ed}{x} \right)$ is increasing, since

$$\varphi'(x) = \frac{3}{2} \log \left(\frac{2^{1/3} ed}{x} \right) - \frac{3}{2} \geq \log 3.$$

Thus

$$\sum_{m=1}^{d_0} e^{\varphi(m)} \leq e^{\varphi(d_0)} \sum_{m=0}^{d_0} e^{-(d_0-m) \log 3} \leq \frac{3}{2} e^{\varphi(d_0)} \leq \frac{3}{2} e^{\frac{3 \log(2^{1/3} \cdot 10 \cdot e)d}{20}} \leq \frac{3}{2} e^{\frac{3d}{5}},$$

and consequently

$$e^{-\frac{16d}{19}} + e^{-\frac{72d}{95}} \sum_{m=1}^{d_0} e^{\varphi(m)} \leq e^{-\frac{16d}{19}} + \frac{3}{2} e^{-\frac{3d}{19}} \leq e^{-\frac{d}{10}} \left(e^{-\frac{141}{190}} + \frac{3}{2} \right) \leq 2 e^{-\frac{d}{10}}. \tag{8.2.17}$$

Combining (8.2.16) with (8.2.17) we obtain (8.2.14).   □

8.2.2   Analysis on Permutation Groups

We have to fix more notation and terminology. Let $\mathrm{Sym}(d)$ be the permutation group on $\mathbb{N}_d$. We will write $\sigma \cdot x = (x_{\sigma(1)}, \ldots, x_{\sigma(d)})$ for every $x \in \mathbb{R}^d$ and $\sigma \in \mathrm{Sym}(d)$. Later on $\mathbb{P}$ will denote the uniform distribution on the symmetry group $\mathrm{Sym}(d)$, i.e. $\mathbb{P}(A) = |A|/d!$ for any $A \subseteq \mathrm{Sym}(d)$, since we know that $|\mathrm{Sym}(d)| = d!$. The expectation $\mathbb{E}$ will be always taken with respect to the uniform distribution $\mathbb{P}$ on the symmetry group $\mathrm{Sym}(d)$. The next few lemmas will rely on the properties of the permutation group.

Lemma 2.5 *Assume that* $I, J \subseteq \mathbb{N}_d$ *and* $|J| = r$ *for some* $0 \leq r \leq d$. *Then*

$$\mathbb{P}[\{\sigma \in \mathrm{Sym}(d) : |\sigma(I) \cap J| \leq r|I|/(5d)\}] \leq e^{-\frac{r|I|}{10d}}. \tag{8.2.18}$$

In particular, if $\delta_1, \delta_2 \in (0, 1]$ *satisfy* $5\delta_2 \leq \delta_1$ *and* $\delta_1 d \leq |I| \leq d$, *then we have*

$$\mathbb{P}[\{\sigma \in \mathrm{Sym}(d) : |\sigma(I) \cap J| \leq \delta_2 r\}] \leq e^{-\frac{\delta_1 r}{10}}. \tag{8.2.19}$$

Proof Inequality (8.2.19) is a consequence of (8.2.18). To prove (8.2.18) we fix $0 \leq r \leq d$ and $I, J \subseteq \mathbb{N}_d$ such that $|J| = r$. It is not difficult to see that

$$\mathbb{P}[\{\sigma \in \mathrm{Sym}(d) : |\sigma(I) \cap J| = k\}] = \binom{r}{k} \binom{d-r}{|I|-k} \binom{d}{|I|}^{-1},$$

which means that the random variable $X = |\sigma(I) \cap J|$ has the hypergeometric distribution. Appealing to the Hoeffding type inequality [10, Theorem 2.10 and inequality (2.6)], we obtain for every $\tau \in (0, 1)$ that

$$\mathbb{P}[\{X \leq (1 - \tau)\mathbb{E}[X]\}] \leq e^{-\tau^2 \mathbb{E}[X]/2}.$$

Taking $\tau = 4/5$ in this inequality and noting that $\mathbb{E}[X] = r|I|/d$ we conclude that

$$\mathbb{P}[\{\sigma \in \mathrm{Sym}(d) : |\sigma(I) \cap J| \leq r|I|/(5d)\}] \leq e^{-\frac{8r|I|}{25d}}.$$

This completes the proof of the lemma.   □

Lemma 2.6 *Assume that we have a finite decreasing sequence* $0 \leq u_d \leq \dots \leq u_2 \leq u_1 \leq (1 - \delta_0)/2$ *for some* $\delta_0 \in (0, 1)$. *Suppose that* $I \subseteq \mathbb{N}_d$ *satisfies* $\delta_1 d \leq |I| \leq d$ *for some* $\delta_1 \in (0, 1]$. *Then for every* $J = (d_0, d] \cap \mathbb{Z}$ *with* $0 \leq d_0 \leq d$ *we have*

$$\mathbb{E}\left[\exp\left(-\sum_{j \in \sigma(I) \cap J} u_j \right) \right] \leq 3 \exp\left(-\frac{\delta_0 \delta_1}{20} \sum_{j \in J} u_j \right). \tag{8.2.20}$$

Proof Let us define $\bar{m} = \sum_{j \in J} u_j$, and without loss of generality, we may assume that $\bar{m} \geq 1$, since otherwise (8.2.20) is obvious. We now take $m \in \mathbb{N}$ such that $m \leq \bar{m} < m + 1$, with this choice of m, we define

$$U_k = (d - (k+1)m, d - km]$$

for every $0 \leq k \leq k_0$, where k_0 is the maximal integer such that $U_{k_0} \cap J \neq \emptyset$.
Observe that

$$\sum_{j \in \sigma(I) \cap J} u_j \geq \sum_{k=0}^{k_0-1} \sum_{j \in \sigma(I) \cap J \cap U_k} u_j \geq \sum_{k=0}^{k_0-1} u_{d-km} |\sigma(I) \cap J \cap U_k|. \qquad (8.2.21)$$

We shall prove that

$$\delta_0 \leq \sum_{k=0}^{k_0-1} u_{d-km} \leq 2. \qquad (8.2.22)$$

Indeed, on the one hand, we have

$$\sum_{k=0}^{k_0-1} u_{d-km} \geq \frac{1}{m} \sum_{k=1}^{k_0-1} \sum_{j \in U_{k-1}} u_j \geq \frac{1}{m} \left(\bar{m} - \sum_{j \in U_{k_0}} u_j - \sum_{j \in U_{k_0-1}} u_j \right)$$

$$\geq \frac{1}{m} \left(m - (1 - \delta_0)m \right) \geq \delta_0.$$

On the other hand, we have

$$\sum_{k=0}^{k_0-1} u_{d-km} \leq \frac{1}{m} \sum_{k=0}^{k_0-1} \sum_{j \in U_k} u_j \leq \frac{\bar{m}}{m} \leq 2.$$

which proves (8.2.22). Let $s = \sum_{k=0}^{k_0-1} u_{d-km}$ and we note that (8.2.21) yields

$$\mathbb{E}\left[\exp\left(- \sum_{j \in \sigma(I) \cap J} u_j \right) \right] \leq \mathbb{E}\left[\exp\left(-s \sum_{k=0}^{k_0-1} \frac{u_{d-km}}{s} |\sigma(I) \cap J \cap U_k| \right) \right]$$

$$\leq \sum_{k=0}^{k_0-1} \frac{u_{d-km}}{s} \mathbb{E}\left[e^{-s|\sigma(I) \cap J \cap U_k|} \right]$$

$$\leq \sup_{0 \leq k \leq k_0-1} \mathbb{E}\left[e^{-s|\sigma(I) \cap J \cap U_k|} \right],$$

$$(8.2.23)$$

where in the second inequality of (8.2.23) we have used convexity. Take $\delta_2 \in (0, 1)$ such that $\delta_2 = \delta_1/5$ and define $A_m = \{ \sigma \in \mathrm{Sym}(d) \colon |\{\sigma(I) \cap J \cap U_k\}| \leq \delta_2 m \}$. Invoking (8.2.19), with $U_k \cap J$ in place of J and $r = m$, since $U_k \cap J = U_k$ for any

$0 \le k \le k_0 - 1$, we see that

$$\mathbb{E}\big[e^{-s|\sigma(I) \cap J \cap U_k|}\big] \le \mathbb{E}\big[e^{-s|\sigma(I) \cap J \cap U_k|}\mathbb{1}_{A_m^c}\big] + \mathbb{E}\big[e^{-s|\sigma(I) \cap J \cap U_k|}\mathbb{1}_{A_m}\big]$$

$$\le e^{-s\delta_2 m} + \mathbb{P}[\{\sigma \in \mathrm{Sym}(d) \colon |\{\sigma(I) \cap J \cap U_k\}| \le \delta_2 m\}]$$

$$\le 3\exp\left(-\frac{\delta_0\delta_1}{20}\sum_{j \in J} u_j\right)$$

due to (8.2.22) and $m \ge \frac{1}{2}\sum_{j \in J} u_j$. This completes the proof of Lemma 2.6.   □

8.2.3   A Decrease Dimension Trick

Choosing r large, but sufficiently small compared with $\kappa(d, N)$, one may then perform finer estimates by exploiting the distribution of lattice points in the ball $B_{\sqrt{l}}^{(r)}$ with $l \gtrsim \kappa(d, N)^2 r$. The next two lemmas will allow us to reduce the matters to the lower dimensional case, where things are simpler. For $r \in \mathbb{N}$ let $S_{\sqrt{R}}^r = \{x \in \mathbb{R}^r \colon |x|^2 = R\}$ be the sphere in $\mathbb{R}^r$ centered at the origin and radius $\sqrt{R} > 0$.

Lemma 2.7 *For $d, N \in \mathbb{N}$, $\varepsilon \in (0, 1/50]$ and an integer $1 \le r \le d$ we define*

$$E = \{x \in B_N \cap \mathbb{Z}^d \colon \sum_{i=1}^{r} x_i^2 < \varepsilon^3 \kappa(d, N)^2 r\}.$$

If $\kappa(d, N) \ge 10$, then we have

$$|E| \le 4e^{-\frac{\varepsilon r}{10}}|B_N \cap \mathbb{Z}^d|. \tag{8.2.24}$$

As a consequence, there exists $E' \subseteq E$, such that

$$B_N \cap \mathbb{Z}^d = \bigcup_{l \ge \varepsilon^3 \kappa(d, N)^2 r} \left(B_{\sqrt{l}}^{(r)} \cap \mathbb{Z}^r\right) \times \left(S_{\sqrt{n-l}}^{d-r} \cap \mathbb{Z}^{d-r}\right) \cup E' \quad \text{with} \quad n = N^2.$$

$$\tag{8.2.25}$$

Proof Let $\delta_1 \in (0, 1/10]$ be such that $\delta_1 \ge 5\varepsilon$, and define $I_x = \{i \in \mathbb{N}_d \colon |x_i| \ge \varepsilon\kappa(d, N)\}$. We have $E \subseteq E_1 \cup E_2$, where

$$E_1 = \{x \in B_N \cap \mathbb{Z}^d \colon \sum_{i \in I_x \cap \mathbb{N}_r} x_i^2 < \varepsilon^3 \kappa(d, N)^2 r \text{ and } |I_x| \ge \delta_1 d\},$$

$$E_2 = \{x \in B_N \cap \mathbb{Z}^d \colon |I_x| < \delta_1 d\}.$$

By Lemma 2.4 (with $\varepsilon_1 = \delta_1$ and $\varepsilon_2 = \varepsilon$) we have $|E_2| \leq 2e^{-\frac{d}{10}}|B_N \cap \mathbb{Z}^d|$, provided that $\kappa(d, N) \geq 10$. Observe that

$$
\begin{aligned}
|E_1| &= \sum_{x \in B_N \cap \mathbb{Z}^d} \frac{1}{d!} \sum_{\sigma \in \mathrm{Sym}(d)} \mathbb{1}_{E_1}(\sigma^{-1} \cdot x) \\
&= \sum_{x \in B_N \cap \mathbb{Z}^d} \mathbb{P}[\{\sigma \in \mathrm{Sym}(d) : \sum_{i \in \sigma(I_x) \cap \mathbb{N}_r} x_{\sigma^{-1}(i)}^2 \\
&\qquad < \varepsilon^3 \kappa(d, N)^2 r \ \text{and} \ |\sigma(I_x)| \geq \delta_1 d\}],
\end{aligned}
$$

since $I_{\sigma^{-1} \cdot x} = \sigma(I_x)$. Now by Lemma 2.5 (with $J = \mathbb{N}_r$, $\delta_2 = \frac{\delta_1}{5}$ and δ_1 as above) we obtain, for every $x \in B_N \cap \mathbb{Z}^d$, that

$$
\begin{aligned}
&\mathbb{P}[\{\sigma \in \mathrm{Sym}(d) : \sum_{i \in \sigma(I_x) \cap \mathbb{N}_r} x_{\sigma^{-1}(i)}^2 < \varepsilon^3 \kappa(d, N)^2 r \ \text{and} \ |\sigma(I_x)| \geq \delta_1 d\}] \\
&\leq \mathbb{P}[\{\sigma \in \mathrm{Sym}(d) : |\sigma(I_x) \cap \mathbb{N}_r| \leq \delta_2 r\}] \leq 2e^{-\frac{\delta_1 r}{10}},
\end{aligned}
$$

since

$$
\begin{aligned}
\{\sigma \in \mathrm{Sym}(d) : &\sum_{i \in \sigma(I_x) \cap \mathbb{N}_r} x_{\sigma^{-1}(i)}^2 < \varepsilon^3 \kappa(d, N)^2 r \ \text{and} \ |\sigma(I_x)| \\
&\geq \delta_1 d \ \text{and} \ |\sigma(I_x) \cap \mathbb{N}_r| > \delta_2 r\} = \emptyset.
\end{aligned}
$$

Thus $|E_1| \leq 2e^{-\frac{\varepsilon r}{2}}$, which proves (8.2.24). To prove (8.2.25) we write

$$
B_N \cap \mathbb{Z}^d = \bigcup_{l=0}^{n} \left(B_{\sqrt{l}}^{(r)} \cap \mathbb{Z}^r \right) \times \left(S_{\sqrt{n-l}}^{d-r} \cap \mathbb{Z}^{d-r} \right).
$$

Then we see that

$$
\begin{aligned}
\left(\bigcup_{l=0}^{n} \left(B_{\sqrt{l}}^{(r)} \cap \mathbb{Z}^r \right) \times \left(S_{\sqrt{n-l}}^{d-r} \cap \mathbb{Z}^{d-r} \right) \right) \cap E^{\mathbf{c}} &= \Bigg(\bigcup_{l \geq \varepsilon^3 \kappa(d,N)^2 r} \left(B_{\sqrt{l}}^{(r)} \cap \mathbb{Z}^r \right) \\
&\quad \times \left(S_{\sqrt{n-l}}^{d-r} \cap \mathbb{Z}^{d-r} \right) \Bigg) \cap E^{\mathbf{c}},
\end{aligned}
$$

where $n = N^2$, and consequently we obtain (8.2.25) with some $E' \subseteq E$. The proof is completed. $\square$

We shall need the lower dimensional multipliers

$$\mathfrak{m}_R^{(r)}(\eta) = \frac{1}{|B_R^{(r)} \cap \mathbb{Z}^d|} \sum_{x \in B_R^{(r)} \cap \mathbb{Z}^d} e^{2\pi i \eta \cdot x}, \qquad \eta \in \mathbb{T}^r, \tag{8.2.26}$$

where $r \in \mathbb{N}$ and $R > 0$.

Lemma 2.8 *For* $d, N \in \mathbb{N}$ *and* $\varepsilon \in (0, 1/50]$ *if* $\kappa(d, N) \geq 10$, *then for every* $1 \leq r \leq d$ *and* $\xi \in \mathbb{T}^d$ *we have*

$$|\mathfrak{m}_N(\xi)| \leq \sup_{l \geq \varepsilon^3 \kappa(d,N)^2 r} |\mathfrak{m}_{\sqrt{l}}^{(r)}(\xi_1, \dots, \xi_r)| + 4e^{-\frac{\varepsilon r}{10}}. \tag{8.2.27}$$

Proof We identify $\mathbb{R}^d \equiv \mathbb{R}^r \times \mathbb{R}^{d-r}$ and $\mathbb{T}^d \equiv \mathbb{T}^r \times \mathbb{T}^{d-r}$ and we will write $\mathbb{R}^d \ni x = (x^1, x^2) \in \mathbb{R}^r \times \mathbb{R}^{d-r}$ and $\mathbb{T}^d \ni \xi = (\xi^1, \xi^2) \in \mathbb{T}^r \times \mathbb{T}^{d-r}$ respectively. Invoking (8.2.25) one obtains

$$|\mathfrak{m}_N(\xi)| \leq \frac{1}{|B_N \cap \mathbb{Z}^d|} \sum_{l \geq \varepsilon^3 \kappa(d,N)^2 r} \sum_{x^2 \in S_{\sqrt{n-l}}^{d-r} \cap \mathbb{Z}^{d-r}} |B_{\sqrt{l}}^{(r)} \cap \mathbb{Z}^r| \frac{1}{|B_{\sqrt{l}}^{(r)} \cap \mathbb{Z}^r|}$$

$$\times \left| \sum_{x^1 \in B_{\sqrt{l}}^{(r)} \cap \mathbb{Z}^r} e^{2\pi i \xi^1 \cdot x^1} \right| + 4e^{-\frac{\varepsilon r}{10}}$$

$$\leq \sup_{l \geq \varepsilon^3 \kappa(d,N)^2 r} |\mathfrak{m}_{\sqrt{l}}^{(r)}(\xi_1, \dots, \xi_r)| + 4e^{-\frac{\varepsilon r}{10}}.$$

In the last inequality the disjointness in the decomposition from (8.2.25) has been used.   □

Lemma 2.8 will play an essential role in the proof of Proposition 2.2. The decrease of the dimension will allow us to approximate the resulting multiplier (8.2.27) by its continuous counterpart with a dimension-free error term. In order to control the error term efficiently we will need the following two simple lemmas.

Lemma 2.9 *Let* $R \geq 1$ *and let* $r \in \mathbb{N}$ *be such that* $r \leq R^\delta$ *for some* $\delta \in (0, 2/3)$. *Then for every* $z \in \mathbb{R}^r$ *we have*

$$\left| |(z + B_R^{(r)}) \cap \mathbb{Z}^r| - |B_R^{(r)}| \right| \leq |B_R^{(r)}| r^{3/2} R^{-1} e^{r^{3/2}/R} \leq e |B_R^{(r)}| R^{-1+3\delta/2}. \tag{8.2.28}$$

Proof Throughout the proof we abbreviate $B_R = B_R^{(r)}$. Observe that

$$|(z + B_R) \cap \mathbb{Z}^r| = \sum_{x \in \mathbb{Z}^r} \int_Q \mathbb{1}_{z+B_R}(x) dy \leq \sum_{x \in \mathbb{Z}^r} \int_Q \mathbb{1}_{z+B_{R+r^{1/2}}}(x+y) dy$$

$$= |z + B_{R+r^{1/2}}|,$$

and

$$|B_{R+r^{1/2}}| = |B_R| \left(1 + \frac{r^{1/2}}{R} \right)^r \leq e^{r^{3/2}/R} |B_R| \leq |B_R| \left(1 + r^{3/2} R^{-1} e^{r^{3/2}/R} \right),$$

since $e^x \leq (1 + x e^x)$. Arguing in a similar way we obtain

$$|(z + B_R) \cap \mathbb{Z}^r| = \sum_{x \in \mathbb{Z}^r} \int_Q \mathbb{1}_{z+B_R}(x) dy \geq \sum_{x \in \mathbb{Z}^r} \int_Q \mathbb{1}_{z+B_{R-r^{1/2}}}(x+y) dy$$

$$= |z + B_{R-r^{1/2}}|,$$

and

$$|B_{R-r^{1/2}}| = |B_R| \left(1 - \frac{r^{1/2}}{R} \right)^r \geq |B_R| \left(1 - r^{3/2} R^{-1} \right).$$

These inequalities imply (8.2.28), since $r \leq R^\delta$. □

Lemma 2.10 *Let $R \geq 1$ and let $r \in \mathbb{N}$ be such that $r \leq R^\delta$ for some $\delta \in (0, 2/3)$. Then for every $z \in \mathbb{R}^r$ we have*

$$\left| \left(B_R^{(r)} \cap \mathbb{Z}^r \right) \triangle \left((z + B_R^{(r)}) \cap \mathbb{Z}^r \right) \right|$$

$$\leq 4e \left(r|z|R^{-1} e^{r|z|R^{-1}} + e^{r|z|R^{-1}} R^{-1+3\delta/2} \right) |B_R^{(r)}|$$

$$\leq 4e \left(|z|R^{-1+\delta} e^{|z|R^{-1+\delta}} + e^{|z|R^{-1+\delta}} R^{-1+3\delta/2} \right) |B_R^{(r)}|. \tag{8.2.29}$$

Proof We again abbreviate $B_R = B_R^{(r)}$. Observe that

$$\left(B_R \cap \mathbb{Z}^r \right) \setminus \left((z + B_R) \cap \mathbb{Z}^r \right) \subseteq \left((z + B_{R+|z|}) \cap \mathbb{Z}^r \right) \setminus \left((z + B_R) \cap \mathbb{Z}^r \right).$$

Thus by Lemma 2.9 one has

$$\left| \left((z + B_{R+|z|}) \cap \mathbb{Z}^r \right) \setminus \left((z + B_R) \cap \mathbb{Z}^r \right) \right|$$

$$= \left| \left((z + B_{R+|z|}) \cap \mathbb{Z}^r \right) \right| - \left| \left((z + B_R) \cap \mathbb{Z}^r \right) \right|$$

$$\leq |B_{R+|z|}| - |B_R| + e|B_{R+|z|}|(R + |z|)^{-1+3\delta/2} + e|B_R|R^{-1+3\delta/2}$$

$$\leq |B_{R+|z|}| - |B_R| + 2e|B_{R+|z|}|R^{-1+3\delta/2}$$

$$\leq |B_{R+|z|}| - |B_R| + 2e|B_R|e^{r|z|R^{-1}}R^{-1+3\delta/2}.$$

We also see

$$|B_{R+|z|}| - |B_R| = \big((1 + |z|R^{-1})^r - 1\big)|B_R|$$

$$\leq r|z|R^{-1}(1 + |z|R^{-1})^r|B_R|$$

$$\leq r|z|R^{-1}e^{r|z|R^{-1}}|B_R|.$$

Hence

$$\big|(B_R \cap \mathbb{Z}^r) \setminus \big((z + B_R) \cap \mathbb{Z}^r\big)\big| \leq \big(r|z|R^{-1}e^{r|z|R^{-1}} + 2e \cdot e^{r|z|R^{-1}}R^{-1+3\delta/2}\big)|B_R|.$$

We obtain the same bound for $\big|\big((z + B_R) \cap \mathbb{Z}^r\big) \setminus \big(B_R \cap \mathbb{Z}^r\big)\big|$ and this gives (8.2.29).

□

We now recall the dimension-free estimates of the Fourier transform for the multiplies associated with averaging operators (8.1.2) in $\mathbb{R}^d$. For a symmetric convex body $G \subset \mathbb{R}^d$ we let the multipliers

$$m^G(\xi) = \frac{1}{|G|}\mathcal{F}(\mathbb{1}_G)(\xi) \quad \text{for} \quad \xi \in \mathbb{R}^d.$$

It is easy to see that $m^G(t\xi)$ is the multiplier corresponding to the operator M_t^G from (8.1.2).

Assume that $|G| = 1$ and that G is in the isotropic position, which means that there is an isotropic constant $L = L(G) > 0$ such that for every unit vector $\xi \in \mathbb{R}^d$ we have

$$\int_G (\xi \cdot x)^2 dx = L(G)^2. \qquad (8.2.30)$$

Then the kernel of the averaging operator (8.1.2) satisfies

$$|G_R|^{-1}\mathbb{1}_{G_R}(x) = R^{-d}\mathbb{1}_G(R^{-1}x)$$

for all $R > 0$, and the multiplier satisfies

$$\mathcal{F}(|G_R|^{-1}\mathbb{1}_{G_R})(\xi) = m^G(R\xi) = \mathcal{F}(\mathbb{1}_G)(R\xi).$$

The isotropic position of G allows us to provide dimension-free estimates for the multiplier m^G.

Theorem 2.3 ([1, eq. (10), (11), (12)]) *Given a symmetric convex body $G \subset \mathbb{R}^d$ with volume one, which is in the isotropic position, there exists a constant $C > 0$ such that for every $\xi \in \mathbb{R}^d$ we have*

$$|m^G(\xi)| \leq C(L|\xi|)^{-1}, \qquad |m^G(\xi) - 1| \leq CL|\xi|, \qquad |\langle \xi, \nabla m^G(\xi) \rangle| \leq C. \tag{8.2.31}$$

The constant $L = L(G)$ is defined in (8.2.30), while C is a universal constant which does not depend on d.

The following lemma is a simple consequence of Theorem 2.3.

Lemma 2.11 *For every $q \in [1, \infty]$ there is a constant $c_q > 0$ independent of d and such that for every $R > 0$ and $\xi \in \mathbb{R}^d$ we have*

$$|m^{B_R^q}(\xi)| \leq C(c_q R d^{-1/q}|\xi|)^{-1}, \quad \text{and} \quad |m^{B_R^q}(\xi) - 1| \leq C(c_q R d^{-1/q}|\xi|), \tag{8.2.32}$$

with the implied constant $C > 0$ independent of d.

Proof Let $s = s(d, q)$ be such that $|B_s^q| = s^d|B^q| = 1$. It is justified in [13, Section 3] that $s \simeq a_q d^{1/q}$ and that B_s^q is in isotropic position with $L(B_s^q) \simeq A_q$. Here the constants a_q and A_q depend only on the parameter q. Therefore, from (8.2.31) it follows that

$$|m^{B_s^q}(\xi)| \leq C(A_q|\xi|)^{-1}, \quad \text{and} \quad |m^{B_s^q}(\xi) - 1| \leq C(A_q|\xi|). \tag{8.2.33}$$

Changing variables we see that

$$m^{B_R^q}(\xi) = \frac{1}{|B_R^q|} \int_{B_R^q} e^{2\pi i x \cdot \xi} \mathrm{d}x = \frac{1}{|B_s^q|} \int_{B_s^q} e^{2\pi i y \cdot (Rs^{-1}\xi)} \mathrm{d}y = m^{B_s^q}(Rs^{-1}\xi).$$

In view of the above equality, recalling that $s^{-1} \simeq a_q^{-1}d^{-1/q}$ and using (8.2.33) we are lead to (8.2.32) with $c_q = A_q a_q^{-1}$ and the proof is completed. $\square$

Lemma 2.12 *There exists a constant $C > 0$ such that for every $\delta \in (0, 1/2)$ and for all $r \in \mathbb{N}$ and $R > 0$ satisfying $1 \leq r \leq R^\delta$ we have*

$$|\mathfrak{m}_R^{(r)}(\eta)| \leq C\left(\kappa(r, R)^{-\frac{1}{3} + \frac{2\delta}{3}} + r\kappa(r, R)^{-\frac{1+\delta}{3}} + \left(\kappa(r, R)\|\eta\|\right)^{-1}\right)$$

for every $\eta \in \mathbb{T}^r$, where $\mathfrak{m}_R^{(r)}(\eta)$ is the multiplier from (8.2.26).

Proof The inequality is obvious when $R \leq 16$, hence it suffices to consider $R > 16$.

Firstly, we assume that $\max\{\|\eta_1\|, \ldots, \|\eta_r\|\} > \kappa(r, R)^{-\frac{1+\delta}{3}}$. Let $M = \lfloor \kappa(r, R)^{\frac{2-\delta}{3}} \rfloor$ and assume without loss of generality that $\|\eta_1\| > \kappa(r, R)^{-\frac{1+\delta}{3}}$. Then

$$
|\mathfrak{m}_R^{(r)}(\eta)| \leq \frac{1}{|B_R^{(r)} \cap \mathbb{Z}^r|} \sum_{x \in B_R^{(r)} \cap \mathbb{Z}^r} \frac{1}{M} \Big| \sum_{s=1}^{M} e^{2\pi i (x+se_1)\cdot\eta} \Big|
$$
$$
+ \frac{1}{M} \sum_{s=1}^{M} \frac{1}{|B_R^{(r)} \cap \mathbb{Z}^r|} \Big| \sum_{x \in B_R^{(r)} \cap \mathbb{Z}^r} e^{2\pi i x\cdot\eta} - e^{2\pi i (x+se_1)\cdot\eta} \Big|.
$$
(8.2.34)

Since $\kappa(r, R) \geq 1$ we now see that

$$
\frac{1}{M} \Big| \sum_{s=1}^{M} e^{2\pi i (x+se_1)\cdot\eta} \Big| \leq M^{-1} \|\eta_1\|^{-1} \leq 2\kappa(r, R)^{-\frac{1}{3}+\frac{2\delta}{3}}.
$$
(8.2.35)

We have assumed that $r \leq R^\delta$, thus by Lemma 2.10, with $z = se_1$ and $s \leq M \leq \kappa(r, R)^{\frac{2-\delta}{3}}$, we get

$$
\frac{1}{|B_R^{(r)} \cap \mathbb{Z}^r|} \Big| \sum_{x \in B_R^{(r)} \cap \mathbb{Z}^r} e^{2\pi i x\cdot\eta} - e^{2\pi i (x+se_1)\cdot\eta} \Big|
$$
$$
\leq \frac{1}{|B_R^{(r)} \cap \mathbb{Z}^r|} \Big| (B_R^{(r)} \cap \mathbb{Z}^r) \triangle ((se_1 + B_R^{(r)}) \cap \mathbb{Z}^r) \Big|
$$
$$
\leq 8e \big(sr R^{-1} e^{sr R^{-1}} + e^{sr R^{-1}} R^{-1+3\delta/2} \big)
$$
$$
\leq 16e^2 \kappa(r, R)^{-\frac{1}{3}+\frac{2\delta}{3}},
$$
(8.2.36)

since $sr R^{-1} \leq \kappa(r, R)^{\frac{2-\delta}{3}} R^{-1+\delta} \leq \kappa(r, R)^{-\frac{1}{3}+\frac{2\delta}{3}} \leq 1$, and $R^{-1+3\delta/2} \leq R^{-\frac{1}{3}+\frac{2\delta}{3}}$, we have, for $R > 16$,

$$
|B_R^{(r)} \cap \mathbb{Z}^r| \geq |B_{R-r^{1/2}}^{(r)}| = |B_R^{(r)}| \Big(1 - \frac{r^{1/2}}{R}\Big)^r \geq |B_R^{(r)}| \big(1 - r^{3/2} R^{-1}\big) \geq |B_R^{(r)}|/2.
$$

Combining (8.2.34) with (8.2.35) and (8.2.36) we obtain

$$
|\mathfrak{m}_R^{(r)}(\eta)| \leq (16e^2 + 2)\kappa(r, R)^{-\frac{1}{3}+\frac{2\delta}{3}}.
$$

Secondly, we assume that $\max\{\|\eta_1\|, \ldots, \|\eta_r\|\} \le \kappa(r, R)^{-\frac{1+\delta}{3}}$. Observe that by (8.2.28) we have

$$\left| \frac{1}{|B_R^{(r)} \cap \mathbb{Z}^r|} - \frac{1}{|B_R^{(r)}|} \right| \le \frac{e R^{-1+3\delta/2}}{|B_R^{(r)} \cap \mathbb{Z}^r|} \le \frac{2e\kappa(r, R)^{-\frac{1}{3}+\frac{2\delta}{3}}}{|B_R^{(r)}|}.$$

Then

$$|\mathrm{m}_R^{(r)}(\eta)| \le \left| \mathrm{m}_R^{(r)}(\eta) - \frac{1}{|B_R^{(r)}|} \mathcal{F}(\mathbb{1}_{B_R^{(r)}})(\eta) \right| + \frac{1}{|B_R^{(r)}|} \left| \mathcal{F}(\mathbb{1}_{B_R^{(r)}})(\eta) \right|$$

$$\le 2e\kappa(r, R)^{-\frac{1}{3}+\frac{2\delta}{3}} + \frac{1}{|B_R^{(r)} \cap \mathbb{Z}^r|} \left| \sum_{x \in B_R^{(r)} \cap \mathbb{Z}^r} e^{2\pi i x \cdot \eta} - \int_{B_R^{(r)}} e^{2\pi i y \cdot \eta} dy \right|$$

$$+ \frac{1}{|B_R^{(r)}|} \left| \mathcal{F}(\mathbb{1}_{B_R^{(r)}})(\eta) \right|. \tag{8.2.37}$$

Let $Q^{(r)} = [-1/2, 1/2]^r$ and note that by Lemma 2.10, with $z = t \in Q^{(r)}$ we get

$$\frac{1}{|B_R^{(r)} \cap \mathbb{Z}^r|} \left| \sum_{x \in B_R^{(r)} \cap \mathbb{Z}^r} e^{2\pi i x \cdot \eta} - \int_{B_R^{(r)}} e^{2\pi i y \cdot \eta} dy \right|]$$

$$= \frac{1}{|B_R^{(r)} \cap \mathbb{Z}^r|} \left| \sum_{x \in \mathbb{Z}^r} \int_{Q^{(r)}} e^{2\pi i x \cdot \eta} \mathbb{1}_{B_R^{(r)}}(x) - e^{2\pi i (x+t) \cdot \eta} \mathbb{1}_{B_R^{(r)}}(x + t) dt \right|$$

$$\le \frac{1}{|B_R^{(r)} \cap \mathbb{Z}^r|} \int_{Q^{(r)}} \left| (B_R^{(r)} \cap \mathbb{Z}^r) \triangle \left((t + B_R^{(r)}) \cap \mathbb{Z}^r \right) \right| dt$$

$$+ \frac{1}{|B_R^{(r)} \cap \mathbb{Z}^r|} \sum_{x \in \mathbb{Z}^r} \mathbb{1}_{B_R^{(r)}}(x) \int_{Q^{(r)}} |e^{2\pi i x \cdot \eta} - e^{2\pi i (x+t) \cdot \eta}| dt$$

$$\le 16 e^2 \kappa(r, R)^{-\frac{1}{3}+\frac{2\delta}{3}} + 2\pi \left(\|\eta_1\| + \ldots + \|\eta_r\| \right)$$

$$\le 16 e^2 \kappa(r, R)^{-\frac{1}{3}+\frac{2\delta}{3}} + 2\pi r \kappa(r, R)^{-\frac{1+\delta}{3}}. \tag{8.2.38}$$

Finally, by Lemma 2.11 we obtain

$$\frac{1}{|B_R^{(r)}|} |\mathcal{F}(\mathbb{1}_{B_R^{(r)}})(\eta)| \le C \left(\kappa(r, R) \|\eta\| \right)^{-1}. \tag{8.2.39}$$

Combining (8.2.37) with (8.2.38) and (8.2.39) we obtain that

$$|\mathrm{m}_R^{(r)}(\eta)| \le (16 e^2 + 2e) \kappa(r, R)^{-\frac{1}{3}+\frac{2\delta}{3}} + 2\pi r \kappa(r, R)^{-\frac{1+\delta}{3}} + C \left(\kappa(r, R) \|\eta\| \right)^{-1},$$

which completes the proof. $\qquad\square$

Lemma 2.13 *For every $\delta \in (0, 1/2)$ and $\varepsilon \in (0, 1/50]$ there is a constant $C_{\delta,\varepsilon} > 0$ such that for every $d, N \in \mathbb{N}$, if r is an integer such that $1 \le r \le d$ and $\max\{1, \varepsilon^{\frac{3\delta}{2}} \kappa(d, N)^\delta / 2\} \le r \le \max\{1, \varepsilon^{\frac{3\delta}{2}} \kappa(d, N)^\delta\}$, then for every $\xi = (\xi_1, \ldots, \xi_d) \in \mathbb{T}^d$ we have*

$$|\mathfrak{m}_N(\xi)| \le C_{\delta,\varepsilon} \big(\kappa(d, N)^{-\frac{1}{3} + \frac{2\delta}{3}} + (\kappa(d, N) \|\eta\|)^{-1} \big),$$

where $\eta = (\xi_1, \ldots, \xi_r)$.

Proof If $\kappa(d, N) \le \varepsilon^{-\frac{3}{2}}$, then there is nothing to do, since the implied constant in question is allowed to depend on δ and ε. We will assume that $\kappa(d, N) \ge \varepsilon^{-\frac{3}{2}}$, which ensures that $\kappa(d, N) \ge 10$. In view of Lemma 2.8 we have

$$|\mathfrak{m}_N(\xi)| \le \sup_{R \ge \varepsilon^{3/2} \kappa(d,N) r^{1/2}} |\mathfrak{m}_R^{(r)}(\eta)| + 4 e^{-\frac{\varepsilon r}{10}}, \tag{8.2.40}$$

where $\eta = (\xi_1, \ldots, \xi_r)$. By Lemma 2.12, since $r \le \varepsilon^{\frac{3\delta}{2}} \kappa(d, N)^\delta \le \kappa(r, R)^\delta \le R^\delta$, we obtain

$$|\mathfrak{m}_R^{(r)}(\eta)| \lesssim \kappa(r, R)^{-\frac{1}{3} + \frac{2\delta}{3}} + r \kappa(r, R)^{-\frac{1+\delta}{3}} + \big(\kappa(r, R) \|\eta\| \big)^{-1}. \tag{8.2.41}$$

Combining (8.2.40) and (8.2.41) with our assumptions we obtain the desired claim. $\qquad\square$

8.2.4 All Together

We have prepared all necessary tools to prove inequality (8.2.10). We shall be working under the assumptions of Lemma 2.13 with $\delta = 2/7$.

Proof of Proposition 2.2 Assume that $\varepsilon = 1/50$. If $\kappa(d, N) \le 2 \cdot 50^{\frac{3}{7}}$ then clearly (8.2.10) holds. Therefore, we can assume that $d, N \in \mathbb{N}$ satisfy $\kappa(d, N) \ge 2 \cdot 50^{\frac{3}{7}}$. We choose an integer $1 \le r \le d$ satisfying $50^{-\frac{3}{7}} \kappa(d, N)^{\frac{2}{7}} / 2 \le r \le 50^{-\frac{3}{7}} \kappa(d, N)^{\frac{2}{7}}$, it is possible since $50^{-\frac{3}{7}} \kappa(d, N)^{\frac{2}{7}} \le 50^{-\frac{1}{7}} d^{\frac{1}{7}} \le 2d/3$. We will also assume that $\|\xi_1\| \ge \ldots \ge \|\xi_d\|$ and we shall distinguish two cases. Suppose first that

$$\|\xi_1\|^2 + \ldots + \|\xi_r\|^2 \ge \frac{1}{4} \|\xi\|^2. \tag{8.2.42}$$

Then in view of Lemma 2.13 (with $\delta = 2/7$ and $r \simeq \kappa(d, N)^{\frac{2}{7}}$) and (8.2.42) we obtain

$$|\mathfrak{m}_N(\xi)| \le C \big(\kappa(d, N)^{-\frac{1}{7}} + (\kappa(d, N) \|\xi\|)^{-1} \big),$$

and we are done. So we can assume that

$$\|\xi_1\|^2 + \ldots + \|\xi_r\|^2 \leq \frac{1}{4}\|\xi\|^2. \tag{8.2.43}$$

Let $\varepsilon_1 = 1/10$ and assume first that

$$\|\xi_j\| \leq \frac{\varepsilon_1^{1/2}}{10\kappa(d, N)} \quad \text{for all} \quad r \leq j \leq d. \tag{8.2.44}$$

Using (8.2.7) and the Cauchy–Schwarz inequality we have

$$|\mathbf{m}_N(\xi)|^2 \leq \frac{1}{|B_N \cap \mathbb{Z}^d|} \sum_{x \in B_N \cap \mathbb{Z}^d} \prod_{j=1}^{d} \cos^2(2\pi x_j \xi_j)$$

$$\leq \frac{1}{|B_N \cap \mathbb{Z}^d|} \sum_{x \in B_N \cap \mathbb{Z}^d} \prod_{j=1}^{d}(1 - \sin^2(2\pi x_j \xi_j)) \tag{8.2.45}$$

$$\leq \frac{1}{|B_N \cap \mathbb{Z}^d|} \sum_{x \in B_N \cap \mathbb{Z}^d} \exp\left(-\sum_{j=r+1}^{d} \sin^2(2\pi x_j \xi_j)\right).$$

For $x \in B_N \cap \mathbb{Z}^d$ we define

$$I_x = \{i \in \mathbb{N}_d : \varepsilon\kappa(d, N) \leq |x_i| \leq 2\varepsilon_1^{-1/2}\kappa(d, N)\},$$

$$I_x' = \{i \in \mathbb{N}_d : 2\varepsilon_1^{-1/2}\kappa(d, N) < |x_i|\},$$

$$I_x'' = \{i \in \mathbb{N}_d : \varepsilon\kappa(d, N) \leq |x_i|\} = I_x \cup I_x'.$$

and

$$E = \{x \in B_N \cap \mathbb{Z}^d : |I_x| \geq \varepsilon_1 d/2\}.$$

Observe that

$$E^c = \{x \in B_N \cap \mathbb{Z}^d : |I_x| < \varepsilon_1 d/2\} = \{x \in B_N \cap \mathbb{Z}^d : |I_x''| < \varepsilon_1 d/2 + |I_x'|\}$$

$$\subseteq \{x \in B_N \cap \mathbb{Z}^d : |I_x''| < \varepsilon_1 d/2 + |I_x'| \text{ and } |I_x'| \leq \varepsilon_1 d/2\}$$

$$\cup \{x \in B_N \cap \mathbb{Z}^d : |I_x'| > \varepsilon_1 d/2\}.$$

Then it is not difficult to see that

$$E^c \subseteq \{x \in B_N \cap \mathbb{Z}^d : |I_x''| < \varepsilon_1 d\},$$

since $\{x \in B_N \cap \mathbb{Z}^d : |I_x'| > \varepsilon_1 d/2\} = \emptyset$. Then by Lemma 2.4 with $\varepsilon_2 = \varepsilon$, we obtain

$$|E^{\mathsf{c}}| \leq |\{x \in B_N \cap \mathbb{Z}^d : |I_x''| < \varepsilon_1 d\}| \leq 2e^{-\frac{d}{10}}|B_N \cap \mathbb{Z}^d|.$$

Therefore, by (8.2.45) we have

$$|\mathfrak{m}_N(\xi)|^2 \leq \frac{1}{|B_N \cap \mathbb{Z}^d|} \sum_{x \in B_N \cap \mathbb{Z}^d} \exp\Big(- \sum_{j \in I_x \cap J_r} \sin^2(2\pi x_j \xi_j)\Big)\mathbb{1}_E(x) + 2e^{-\frac{d}{10}},$$

(8.2.46)

where $J_r = \{r+1, \ldots, d\}$. Using (8.2.4) and definition I_x we have

$$\sin^2(2\pi x_j \xi_j) \geq 16|x_j|^2\|\xi_j\|^2 \geq 16\varepsilon^2\kappa(d, N)^2\|\xi_j\|^2,$$

since $2|x_j|\|\xi_j\| \leq 1/2$ by (8.2.44), and consequently we estimate (8.2.46) and obtain for some $C, c > 0$ that

$$\frac{1}{|B_N \cap \mathbb{Z}^d|} \sum_{x \in B_N \cap \mathbb{Z}^d} \exp\Big(- \sum_{j \in I_x \cap J_r} \sin^2(2\pi x_j \xi_j)\Big)\mathbb{1}_E(x)$$

$$\leq \frac{1}{|B_N \cap \mathbb{Z}^d|} \sum_{x \in B_N \cap \mathbb{Z}^d \cap E} \exp\Big(- 16\varepsilon^2\kappa(d, N)^2 \sum_{j \in I_x \cap J_r} \|\xi_j\|^2\Big)$$

$$\leq Ce^{-c\kappa(d,N)^2\|\xi\|^2}.$$

(8.2.47)

In order to obtain the last inequality in (8.2.47) observe that

$$\frac{1}{|B_N \cap \mathbb{Z}^d|} \sum_{x \in B_N \cap \mathbb{Z}^d \cap E} \exp\Big(- 16\varepsilon^2\kappa(d, N)^2 \sum_{j \in I_x \cap J_r} \|\xi_j\|^2\Big)$$

$$= \frac{1}{|B_N \cap \mathbb{Z}^d|} \sum_{x \in B_N \cap \mathbb{Z}^d \cap E} \frac{1}{d!} \sum_{\sigma \in \mathrm{Sym}(d)} \exp\Big(- 16\varepsilon^2\kappa(d, N)^2 \sum_{j \in \sigma(I_x) \cap J_r} \|\xi_j\|^2\Big)$$

$$= \frac{1}{|B_N \cap \mathbb{Z}^d|} \sum_{x \in B_N \cap \mathbb{Z}^d \cap E} \mathbb{E}\Big[\exp\Big(- 16\varepsilon^2\kappa(d, N)^2 \sum_{j \in \sigma(I_x) \cap J_r} \|\xi_j\|^2\Big)\Big],$$

since $\sigma \cdot (B_N \cap \mathbb{Z}^d \cap E) = B_N \cap \mathbb{Z}^d \cap E$ for every $\sigma \in \mathrm{Sym}(d)$. Appealing now to Lemma 2.6 with $\delta_1 = \varepsilon_1/2$, $d_0 = r$, $I = I_x$ and $\delta_0 = 3/5$, we conclude that

$$\mathbb{E}\Big[\exp\Big(- 16\varepsilon^2\kappa(d, N)^2 \sum_{j \in \sigma(I_x) \cap J_r} \|\xi_j\|^2\Big)\Big] \leq C\exp\Big(- c'\kappa(d, N)^2 \sum_{j=r+1}^{d} \|\xi_j\|^2\Big),$$

for some $c' > 0$ and for all $x \in B_N \cap \mathbb{Z}^d \cap E$. This proves (8.2.47) since by (8.2.43) we obtain

$$\exp\left(-c'\kappa(d,N)^2 \sum_{j=r+1}^{d} \|\xi_j\|^2\right) \leq \exp\left(-\frac{c'\kappa(d,N)^2}{4} \sum_{j=1}^{d} \|\xi_j\|^2\right).$$

Assume now that (8.2.44) does not hold. Then

$$\|\xi_j\| \geq \frac{\varepsilon_1^{1/2}}{10\kappa(d,N)} \quad \text{for all} \quad 1 \leq j \leq r. \tag{8.2.48}$$

Hence (8.2.48) gives that

$$\|\xi_1\|^2 + \ldots + \|\xi_r\|^2 \geq \frac{\varepsilon_1 r}{100\kappa(d,N)^2}.$$

Therefore, we invoke Lemma 2.13 with $\eta = (\xi_1, \ldots, \xi_r)$ again and obtain

$$|\mathfrak{m}_N(\xi)| \lesssim \kappa(d,N)^{-\frac{1}{7}} + (\kappa(d,N)\|\eta\|)^{-1}$$

$$\lesssim \kappa(d,N)^{-\frac{1}{7}},$$

since $r \simeq \kappa(d,N)^{\frac{2}{7}}$. This completes the proof of Proposition 2.2. $\qquad \square$

8.3   Estimates for the Dyadic Maximal Function: Small Scales

This section is intended to provide bounds independent of the dimension for the dyadic maximal function with supremum taken over all dyadic numbers N such that $1 \lesssim N \lesssim d^{1/2}$. Theorem 3.1 combined with Theorem 2.2 from the previous section implies our main result Theorem 1.1.

Theorem 3.1 *Let $C_0 > 0$ and define $\mathbb{D}_{C_0} = \{N \in \mathbb{D} : 1 \leq N \leq C_0 d^{1/2}\}$. Then there exists a constant $C > 0$ independent of dimension such that for every $f \in \ell^2(\mathbb{Z}^d)$ we have*

$$\left\| \sup_{N \in \mathbb{D}_{C_0}} \mathcal{M}_N f \right\|_{\ell^2} \leq C\|f\|_{\ell^2}. \tag{8.3.1}$$

The strategy of the proof of Theorem 3.1 is much the same as for the proof of Theorem 2.2. However, the approximating multiplies are different and they depend on the size of the set

$$V_\xi = \{i \in \mathbb{N}_d : \cos(2\pi\xi_i) < 0\} = \{i \in \mathbb{N}_d : 1/4 < |\xi_i| \leq 1/2\} \quad \text{for} \quad \xi \in \mathbb{T}^d. \tag{8.3.2}$$

We will approximate the maximal function from (8.3.1) by the maximal functions associated with the following multipliers

$$\lambda_N^1(\xi) = e^{-\kappa(d,N)^2 \sum_{i=1}^d \sin^2(\pi\xi_i)} \qquad \text{if } |V_\xi| \leq d/2,$$

$$(8.3.3)$$

$$\lambda_N^2(\xi) = \frac{1}{|B_N \cap \mathbb{Z}^d|}\left(\sum_{x \in B_N \cap \mathbb{Z}^d}(-1)^{\sum_{i=1}^d x_i}\right)e^{-\kappa(d,N)^2 \sum_{i=1}^d \cos^2(\pi\xi_i)} \qquad \text{if } |V_\xi| \geq d/2.$$

$$(8.3.4)$$

In Proposition 3.1, which is the main results of this section, we show that multiplies (8.3.3) and (8.3.4) are close to the multiplier $\mathfrak{m}_N$ defined in (8.2.3) in the sense of inequalities (8.3.5) and (8.3.6) respectively.

Proposition 3.1 *For every $d, N \in \mathbb{N}$, if $N \geq 2^{9/2}$ and $\kappa(d, N) \leq 1/5$, then for every $\xi \in \mathbb{T}^d$ we have the following bounds with the constant $c \in (0, 1)$ as in (8.3.18). Namely,*

1. if $|V_\xi| \leq d/2$, then

$$|\mathfrak{m}_N(\xi) - \lambda_N^1(\xi)| \leq 17 \min\left\{e^{-\frac{c\kappa(d,N)^2}{400}\sum_{i=1}^d \sin^2(\pi\xi_i)}, \kappa(d, N)^2 \sum_{i=1}^d \sin^2(\pi\xi_i)\right\},$$

$$(8.3.5)$$

2. if $|V_\xi| \geq d/2$, then

$$|\mathfrak{m}_N(\xi) - \lambda_N^2(\xi)| \leq 17 \min\left\{e^{-\frac{c\kappa(d,N)^2}{400}\sum_{i=1}^d \cos^2(\pi\xi_i)}, \kappa(d, N)^2 \sum_{i=1}^d \cos^2(\pi\xi_i)\right\}.$$

$$(8.3.6)$$

Throughout this section all of the estimates will be also described in terms of the proportionality constant $\kappa(d, N)$ from (8.2.5). Assume momentarily that Proposition 3.1 has been proven and let us deduce Theorem 3.1.

Proof of Theorem 3.1 Let $f \in \ell^2(\mathbb{Z}^d)$ and we write $f = f_1 + f_2$, where $\hat{f}_1(\xi) = \hat{f}(\xi)\mathbb{1}_{\{\eta \in \mathbb{T}^d \,:\, |V_\eta| \leq d/2\}}(\xi)$. Then

$$\left\|\sup_{N \in \mathbb{D}_{C_0}}|\mathcal{F}^{-1}(\mathfrak{m}_N \hat{f})|\right\|_{\ell^2} \leq \sum_{i=1}^2 \left\|\sup_{N \in \mathbb{D}_{C_0}}|\mathcal{F}^{-1}(\mathfrak{m}_N \hat{f}_i)|\right\|_{\ell^2}$$

$$\leq \sum_{i=1}^2 \left\|\sup_{N \in \mathbb{D}_{C_0}}|\mathcal{F}^{-1}(\lambda_N^i \hat{f}_i)|\right\|_{\ell^2}$$

$$+ \sum_{i=1}^2 \left\|\left(\sum_{N \in \mathbb{D}_{C_0}}|\mathcal{F}^{-1}((\mathfrak{m}_N - \lambda_N^i)\hat{f}_i)|^2\right)\right\|_{\ell^2}.$$

The usual square function argument permits therefore to reduce the problem to controlling the maximal functions associated with the multipliers λ_N^1 and λ_N^2. Since by Proposition 3.1 we have

$$\sum_{i=1}^{2} \left\| \left(\sum_{N\in\mathbb{D}_{C_0}} |\mathcal{F}^{-1}((\mathfrak{m}_N - \lambda_N^i)\hat{f}_i)|^2 \right) \right\|_{\ell^2} \lesssim \|f\|_{\ell^2}.$$

We only have to bound the maximal functions. Observe that

$$\left\| \sup_{N\in\mathbb{D}_{C_0}} |\mathcal{F}^{-1}(\lambda_N^1 \hat{f}_1)| \right\|_{\ell^2} \lesssim \|f\|_{\ell^2}, \tag{8.3.7}$$

since the multiplier λ_N^1 corresponds to the semi-group of contractions P_t, which was defined in the proof of Theorem 2.2. The multiplier $\mathfrak{p}_t$ corresponding to P_t satisfies $\lambda_N^1(\xi) = \mathfrak{p}_{\kappa(d,N)^2}(\xi)$. Therefore, by (8.2.12) we obtain (8.3.7) as desired. It is also not difficult to see that

$$\left\| \sup_{N\in\mathbb{D}_{C_0}} |\mathcal{F}^{-1}(\lambda_N^2 \hat{f}_2)| \right\|_{\ell^2} \lesssim \|f\|_{\ell^2}. \tag{8.3.8}$$

In fact (8.3.8) can be deduced from (8.3.7). For this purpose we denote $F_2(x) = (-1)^{\sum_{j=1}^{d} x_j} f_2(x)$, hence

$$\hat{F}_2(\xi) = \sum_{x\in\mathbb{Z}^d} e^{2\pi i x\cdot\xi}(-1)^{\sum_{j=1}^{d} x_j} f_2(x) = \sum_{x\in\mathbb{Z}^d} e^{2\pi i x\cdot(\xi+1/2)} f_2(x) = \hat{f}_2(\xi + 1/2).$$

Thus we write

$$\left\| \sup_{N\in\mathbb{D}_{C_0}} |\mathcal{F}^{-1}(\lambda_N^2 \hat{f}_2)| \right\|_{\ell^2}$$

$$\leq \left\| \sup_{N\in\mathbb{D}_{C_0}} \left| \int_{\mathbb{T}^d} e^{-\kappa(d,N)^2 \sum_{j=1}^{d} \cos^2(\pi\xi_j)} \hat{f}_2(\xi) e^{-2\pi i x\cdot\xi} \, d\xi \right| \right\|_{\ell^2}$$

$$\leq \left\| \sup_{N\in\mathbb{D}_{C_0}} \left| \int_{\mathbb{T}^d} e^{-\kappa(d,N)^2 \sum_{j=1}^{d} \sin^2(\pi\xi_j)} \hat{F}_2(\xi) e^{-2\pi i x\cdot\xi} \, d\xi \right| \right\|_{\ell^2}$$

$$\leq \|f\|_{\ell^2},$$

since $\|F_2\|_{\ell^2} = \|f_2\|_{\ell^2} \leq \|f\|_{\ell^2}$. $\qquad\square$

8.3.1 Some Preparatory Estimates

We begin with a very useful lemma, which will allow us to control efficiently sizes of certain error sets, which say, to some extent, that a large amount of mass of $B_N \cap \mathbb{Z}^d$ is concentrated on the set $\{-1, 0, 1\}^d$.

Lemma 3.2 *For every $d, N \in \mathbb{N}$, if $n = N^2$ and $\kappa(d, N) \le 1/5$ and $n \ge k \ge 2^9 \max\{1, \kappa(d, N)^6 n\}$, then*

$$|\{x \in B_N \cap \mathbb{Z}^d : |\{i \in \mathbb{N}_d : x_i = \pm 1\}| \le n - k\}| \le 2^{-k+1}|B_N \cap \mathbb{Z}^d|. \qquad (8.3.9)$$

In particular, we have

$$\left|\left\{x \in B_N \cap \mathbb{Z}^d : \sum_{\substack{i=1 \\ |x_i| \ge 2}}^d x_i^2 > k\right\}\right| \le 2^{-k+1}|B_N \cap \mathbb{Z}^d|. \qquad (8.3.10)$$

Proof If $n = N^2$ then $nd^{-1} = \kappa(d, N)^2$. It is easy to see that (8.3.10) follows from (8.3.9), since

$$\left|\left\{x \in B_N \cap \mathbb{Z}^d : \sum_{\substack{i=1 \\ |x_i| \ge 2}}^d x_i^2 > k\right\}\right| \le \left|\left\{x \in B_N \cap \mathbb{Z}^d : \sum_{\substack{i=1 \\ |x_i|=1}}^d x_i^2 \le n - k\right\}\right|$$

$$= |\{x \in B_N \cap \mathbb{Z}^d : |\{i \in \mathbb{N}_d : x_i = \pm 1\}| \le n - k\}|.$$

To prove (8.3.9) let us define for every $m \in \mathbb{N}_n$ the set

$$E_m = \left\{x \in B_N \cap \mathbb{Z}^d : \sum_{\substack{i=1 \\ |x_i|=1}}^d x_i^2 = n - m\right\},$$

and observe that

$$\left|\left\{x \in B_N \cap \mathbb{Z}^d : \sum_{\substack{i=1 \\ |x_i|=1}}^d x_i^2 \le n - k\right\}\right| \le \sum_{m=k}^n |E_m|.$$

In view of this bound it suffices to show, for all $m \in \mathbb{N}_n \setminus \mathbb{N}_{k-1}$, that

$$|E_m| \le 2^{-m}|B_N \cap \mathbb{Z}^d|. \qquad (8.3.11)$$

Our task now is to prove (8.3.11). If $x \in E_m$ then

$$\sum_{\substack{i=1 \\ |x_i| \geq 2}}^{d} x_i^2 \leq m$$

and consequently we obtain that

$$|\{i \in \mathbb{N}_d : |x_i| \geq 2\}| \leq \lfloor m/4 \rfloor. \tag{8.3.12}$$

We now establish the following upper bound for the sets E_m

$$|E_m| \leq 2^{n-m} \binom{d}{n-m} \binom{d-n+m}{\lfloor m/4 \rfloor} |B_{\sqrt{m}}^{(\lfloor m/4 \rfloor)} \cap \mathbb{Z}^{\lfloor m/4 \rfloor}|. \tag{8.3.13}$$

If $x = (x_1, \ldots, x_d) \in E_m$ then precisely $n - m$ of its coordinates are ± 1. Due to (8.3.12) at most $\lfloor m/4 \rfloor$ coordinates of x have an absolute value at least 2. Finally, the remaining $d - n + m - \lfloor m/4 \rfloor$ coordinates vanish. This allows us to justify (8.3.13). There are 2^{n-m} sequences of length $n - m$ whose elements are ± 1, the elements of these sequences can be placed into $\binom{d}{n-m}$ spots. We have also at most $|B_{\sqrt{m}}^{(\lfloor m/4 \rfloor)} \cap \mathbb{Z}^{\lfloor m/4 \rfloor}|$ sequences with elements whose absolute value is at least 2, these elements can be placed into $\binom{d-n+m}{\lfloor m/4 \rfloor}$ spots. Finally there is only one way to put zeros into the remaining $d - n + m - \lfloor m/4 \rfloor$ spots.

By Lemma 2.3 we get for $m \geq 8$ that

$$\begin{aligned}
|B_{\sqrt{m}}^{(\lfloor m/4 \rfloor)} \cap \mathbb{Z}^{\lfloor m/4 \rfloor}| &\leq (2\pi e)^{m/8} \left(\frac{m}{m/4 - 1} + \frac{1}{4} \right)^{m/8} \\
&\leq (2\pi e)^{m/8} \left(\frac{4m}{m - 4} + \frac{1}{4} \right)^{m/8} \\
&\leq (17\pi e)^{m/8}.
\end{aligned} \tag{8.3.14}$$

Moreover, we obtain

$$\begin{aligned}
\binom{d}{n-m} \binom{d-n+m}{\lfloor m/4 \rfloor} \binom{d}{n}^{-1} &= \frac{n!(d-n)!}{(n-m)!(d-n+m-\lfloor m/4 \rfloor)!\lfloor m/4 \rfloor!} \\
&\leq \frac{n^{\lfloor m/4 \rfloor}}{\lfloor m/4 \rfloor!} \left(\frac{n}{d-n} \right)^{m-\lfloor m/4 \rfloor} \\
&\leq \left(\frac{en}{\lfloor m/4 \rfloor} \right)^{\lfloor m/4 \rfloor} \left(\frac{n}{d-n} \right)^{m-\lfloor m/4 \rfloor} \\
&\leq \left(\frac{4en}{m} \right)^{m/4} \left(\frac{\kappa(d,N)^2}{1 - \kappa(d,N)^2} \right)^{3m/4},
\end{aligned} \tag{8.3.15}$$

since $\lfloor m/4 \rfloor! \geq (\lfloor m/4 \rfloor/e)^{\lfloor m/4 \rfloor}$ and for every $a, b > 0$ the function $(0, a/e] \ni t \mapsto \left(\frac{a}{t}\right)^{bt}$ is increasing.

We also have

$$2^n \binom{d}{n} \leq |B_N \cap \mathbb{Z}^d|. \tag{8.3.16}$$

Thus taking into account (8.3.13), (8.3.14), (8.3.15) and (8.3.16), we obtain

$$
\begin{aligned}
|E_m| &\leq 2^n \binom{d}{n} 2^{-m} (17\pi e)^{m/8} \left(\frac{4en}{m}\right)^{m/4} \left(\frac{\kappa(d, N)^2}{1 - \kappa(d, N)^2}\right)^{3m/4} \\
&\leq \left(\frac{272\pi e^3}{256}\right)^{m/8} \left(\frac{n}{m}\right)^{m/4} \left(\frac{\kappa(d, N)^2}{1 - \kappa(d, N)^2}\right)^{3m/4} |B_N \cap \mathbb{Z}^d| \\
&\leq 2^m \left(\frac{n}{m}\right)^{m/4} \left(\frac{\kappa(d, N)^2}{1 - \kappa(d, N)^2}\right)^{3m/4} |B_N \cap \mathbb{Z}^d| \\
&\leq 2^m \left(\frac{25}{24}\right)^{3m/4} \left(\frac{n}{m}\right)^{m/4} \kappa(d, N)^{3m/2} |B_N \cap \mathbb{Z}^d| \\
&\leq 2^{-m} |B_N \cap \mathbb{Z}^d|,
\end{aligned}
$$

since $m \geq 2^9 \kappa(d, N)^6 n$. The proof now is completed. □

8.3.2 Analysis Exploiting the Krawtchouk Polynomials

The proof of Proposition 3.3 will rely on the properties of the Krawtchouk polynomials. We need to introduce some definitions and formulate basic facts. For every $n \in \mathbb{N}_0$ and integers $x, k \in [0, n]$ we define the k-th Krawtchouk polynomial

$$\Bbbk_k^{(n)}(x) = \frac{1}{\binom{n}{k}} \sum_{j=0}^{k} (-1)^j \binom{x}{j} \binom{n-x}{k-j}. \tag{8.3.17}$$

We gather some facts about the Krawtchouk polynomials in the theorem stated below.

Theorem 3.2 For every $n \in \mathbb{N}_0$ and integers $x, k \in [0, n]$ we have

1. Symmetry: $\Bbbk_k^{(n)}(x) = \Bbbk_x^{(n)}(k)$.
2. Reflection symmetry: $\Bbbk_k^{(n)}(n - x) = (-1)^k \Bbbk_k^{(n)}(x)$.

3. *Orthogonality: for every* $k, m \in [0, n]$

$$\sum_{x=0}^{n} \binom{n}{x} \Bbbk_k^{(n)}(x) \Bbbk_m^{(n)}(x) = 2^n \binom{n}{k}^{-1} \delta_k(m).$$

4. *Roots: the roots of* $\Bbbk_k^{(n)}(x)$ *are real, distinct, and lie in* $\left[\frac{n}{2} - \sqrt{k(n-k)}, \frac{n}{2} + \sqrt{k(n-k)} \right]$.
5. *A uniform bound: there exists a constant* $c \in (0, 1)$ *such that for all* $n \in \mathbb{N}_0$ *the following inequality*

$$\left| \Bbbk_k^{(n)}(x) \right| \leq e^{-\frac{ckx}{n}} \tag{8.3.18}$$

holds for all integers $0 \leq x, k \leq n/2$.

The proof of Theorem 3.2 can be found in [9], see also the references therein. In Proposition 3.3, using Theorem 3.2, we will be able to describe the decay at infinity of the multipliers corresponding to our averages. In fact, we will only use properties 1, 2 and 5. Properties 3 and 4 are only provided to give an idea about the objects we are going to work with. The support of $x \in \mathbb{R}^d$ will be denoted by $\operatorname{supp} x = \{ i \in \mathbb{N}_d : x_i \neq 0 \}$.

Proposition 3.3 *For every* $d, n \in \mathbb{N}$, *if* $N \geq 2^{9/2}$ *and* $\kappa(d, N) \leq 1/5$ *then for all* $\xi \in \mathbb{T}^d$ *we have the following estimate*

$$|\mathfrak{m}_N(\xi)| \leq 8 e^{-\frac{c\kappa(d,N)^2}{100} \sum_{i=1}^{d} \sin^2(\pi \xi_i)} + 8 e^{-\frac{c\kappa(d,N)^2}{100} \sum_{i=1}^{d} \cos^2(\pi \xi_i)}, \tag{8.3.19}$$

where $c \in (0, 1)$ *is the constant as in* (8.3.18).

Proof We let $n = N^2 \geq 2^9$ and for any $x \in \mathbb{Z}^d$ we define the set $I_x = \{ i \in \mathbb{N}_d : x_i = \pm 1 \}$ and the set $E = \{ x \in B_N \cap \mathbb{Z}^d : |I_x| > n/2 \}$. By Lemma 3.2, with $k = \lfloor n/2 \rfloor$, we see that

$$|E^{\mathbf{c}}| \leq 2^{-n/2+2} |B_N \cap \mathbb{Z}^d| \leq 2^{-\frac{\kappa(d,N)^2 d}{2}+2} |B_N \cap \mathbb{Z}^d| \leq 4 e^{-\frac{\kappa(d,N)^2 d}{4}} |B_N \cap \mathbb{Z}^d|.$$

In view of this estimate it now suffices to show that

$$\frac{1}{|B_N \cap \mathbb{Z}^d|} \left| \sum_{x \in B_N \cap \mathbb{Z}^d \cap E} \prod_{j=1}^{d} \cos(2\pi x_j \xi_j) \right| \leq 6 e^{-\frac{c\kappa(d,N)^2}{100} \sum_{i=1}^{d} \sin^2(\pi \xi_i)}$$

$$+ 6 e^{-\frac{c\kappa(d,N)^2}{100} \sum_{i=1}^{d} \cos^2(\pi \xi_i)}. \tag{8.3.20}$$

For this purpose we decompose every $x \in B(N) = B_N \cap \mathbb{Z}^d \cap E = E$ uniquely as $x = Y(x) + Z(x)$, where $Z(x) = (Z_1(x), \ldots, Z_d(x))$ is given by

$$Z_j(x) = \begin{cases} x_j, & \text{if } |x_j| \geq 2 \\ 0, & \text{if } |x_j| \leq 1, \end{cases}$$

and we shall exploit the following disjoint decomposition

$$B(N) = \bigcup_{z \in Z(B(N))} z + Y(N, z), \tag{8.3.21}$$

where

$$Y(N, z) = \left\{ y \in \{-1, 0, 1\}^d : z + y \in B(N), \text{ and } \operatorname{supp} y \subseteq (\operatorname{supp} z)^{\mathbf{c}} \right\}.$$

Note that then $|Y(N, z)|$ depends only on N and $|\operatorname{supp} z|$. We abbreviate $\operatorname{supp} z$ to S_z and using (8.3.21) we write

$$\frac{1}{|B_N \cap \mathbb{Z}^d|} \left| \sum_{x \in B_N \cap \mathbb{Z}^d \cap E} \prod_{j=1}^d \cos(2\pi x_j \xi_j) \right|$$

$$= \frac{1}{|B_N \cap \mathbb{Z}^d|} \left| \sum_{z \in Z(B(N))} \prod_{j \in S_z} \cos(2\pi z_j \xi_j) \sum_{y \in Y(N,z)} \prod_{j \in \mathbb{N}_d \setminus S_z} \cos(2\pi y_j \xi_j) \right|.$$

We claim that for every $z \in Z(B(N))$, whenever $Y(N, z)$ is non-empty, then

$$\frac{1}{|Y(N, z)|} \left| \sum_{y \in Y(N,z)} \prod_{j \in \mathbb{N}_d \setminus S_z} \cos(2\pi y_j \xi_j) \right|$$

$$\leq 2 e^{-\frac{c\kappa(d,N)^2}{4} \sum_{i \in \mathbb{N}_d \setminus S_z} \sin^2(\pi \xi_i)} + 2 e^{-\frac{c\kappa(d,N)^2}{4} \sum_{i \in \mathbb{N}_d \setminus S_z} \cos^2(\pi \xi_i)}. \tag{8.3.22}$$

Assuming momentarily (8.3.22) and exploiting symmetries we obtain (8.3.20) as follows

$$\frac{1}{|B_N \cap \mathbb{Z}^d|} \left| \sum_{x \in B_N \cap \mathbb{Z}^d \cap E} \prod_{j=1}^d \cos(2\pi x_j \xi_j) \right|$$

$$\leq \frac{2}{|B_N \cap \mathbb{Z}^d|} \sum_{z \in Z(B(N))} |Y(N, z)| e^{-\frac{c\kappa(d,N)^2}{4} \sum_{i \in \mathbb{N}_d \setminus S_z} \sin^2(\pi \xi_i)}$$

$$+ \frac{2}{|B_N \cap \mathbb{Z}^d|} \sum_{z \in Z(B(N))} |Y(N, z)| e^{-\frac{c\kappa(d,N)^2}{4} \sum_{i \in \mathbb{N}_d \setminus S_z} \cos^2(\pi \xi_i)}$$

$$= \frac{2}{|B_N \cap \mathbb{Z}^d|} \sum_{z \in Z(B(N))} |Y(N, z)| \mathbb{E}\left[e^{-\frac{c\kappa(d,N)^2}{4} \sum_{i \in \sigma(\mathbb{N}_d \setminus S_z)} \sin^2(\pi \xi_i)} \right]$$

$$+ \frac{2}{|B_N \cap \mathbb{Z}^d|} \sum_{z \in Z(B(N))} |Y(N, z)| \mathbb{E}\left[e^{-\frac{c\kappa(d,N)^2}{4} \sum_{i \in \sigma(\mathbb{N}_d \setminus S_z)} \cos^2(\pi \xi_i)} \right],$$

where in the last two lines we used the fact that $Z(B(N))$ is $\text{Sym}(d)$ invariant and $|Y(N, \sigma \cdot z)| = |Y(N, z)|$ for every $\sigma \in \text{Sym}(d)$. Using the fact that $|Y(N, z)| = |z + Y(N, z)|$ and decomposition (8.3.21) it suffices to show, for every $z \in Z(B(N))$, that

$$\mathbb{E}\left[e^{-\frac{c\kappa(d,N)^2}{4} \sum_{i \in \sigma(\mathbb{N}_d \setminus S_z)} \sin^2(\pi \xi_i)} \right] \le 3 e^{-\frac{c\kappa(d,N)^2}{100} \sum_{i=1}^d \sin^2(\pi \xi_i)} \qquad (8.3.23)$$

$$\mathbb{E}\left[e^{-\frac{c\kappa(d,N)^2}{4} \sum_{i \in \sigma(\mathbb{N}_d \setminus S_z)} \cos^2(\pi \xi_i)} \right] \le 3 e^{-\frac{c\kappa(d,N)^2}{100} \sum_{i=1}^d \cos^2(\pi \xi_i)} \qquad (8.3.24)$$

with the constant $c \in (0, 1)$ as in (8.3.18). We shall only focus on estimating (8.3.23), as the bound for (8.3.24) is completely analogous. For this purpose we will apply Lemma 2.6. Observe that for every $z \in Z(B(N))$ we have $|S_z| \le n/4$, and consequently $|\mathbb{N}_d \setminus S_z| = d - |S_z| \ge d - n/4 \ge d(1 - \kappa(d, N)^2/4) \ge 99d/100$. Invoking Lemma 2.6, with $d_0 = 0$, $I = \mathbb{N}_d \setminus S_z$, $\delta_0 = 49/50$ and $\delta_1 = 99/100$, we conclude

$$\mathbb{E}\left[e^{-\frac{c\kappa(d,N)^2}{4} \sum_{i \in \sigma(\mathbb{N}_d \setminus S_z)} \sin^2(\pi \xi_i)} \right] \le 3 e^{-\frac{c\kappa(d,N)^2}{100} \sum_{i=1}^d \sin^2(\pi \xi_i)}.$$

We now return to the proof of (8.3.22). Throughout the proof of (8.3.22) we fix $z \in Z(B(N))$ and assume that $Y(N, z)$ is non-empty. We let $I = \mathbb{N}_d \setminus S_z$ and $m = |I| \ge 99d/100$. We shall exploit the properties of the symmetry group restricted to the set I, i.e. $\text{Sym}(I) = \{\sigma \in \text{Sym}(d): \sigma(y) = y \text{ for every } y \in S_z\}$. Since $\sigma \cdot Y(N, z) = \sigma \cdot Y(N, \sigma^{-1} \cdot z) = Y(N, z)$ for every $\sigma \in \text{Sym}(I)$ then we have

$$\frac{1}{|Y(N, z)|} \sum_{y \in Y(N,z)} \prod_{j \in \mathbb{N}_d \setminus S_z} \cos(2\pi y_j \xi_j) = \frac{1}{|Y(N, z)|} \sum_{y \in Y(N,z)} \frac{1}{m!}$$

$$\times \sum_{\sigma \in \text{Sym}(I)} \prod_{j \in I} \cos(2\pi y_{\sigma^{-1}(j)} \xi_j).$$

Our aim is to show that for every $y \in Y(N, z)$ we have

$$\frac{1}{m!} \left| \sum_{\sigma \in \text{Sym}(I)} \prod_{j \in I} \cos(2\pi y_{\sigma^{-1}(j)} \xi_j) \right| \le 2 e^{-\frac{c\kappa(d,N)^2}{4} \sum_{i \in \mathbb{N}_d \setminus S_z} \sin^2(\pi \xi_i)}$$

$$+ 2 e^{-\frac{c\kappa(d,N)^2}{4} \sum_{i \in \mathbb{N}_d \setminus S_z} \cos^2(\pi \xi_i)} \qquad (8.3.25)$$

with the constant $c \in (0, 1)$ as in (8.3.18). We fix $y \in Y(N, z)$, we set $l = |\operatorname{supp} y|$, and write

$$\operatorname{Sym}(I) = \bigcup_{\substack{J \subseteq I \\ |J| = l}} \{\sigma \in \operatorname{Sym}(I) : |y_{\sigma^{-1}(j)}| = 1 \text{ exactly for } j \in J\}.$$

Using this decomposition we obtain

$$\frac{1}{m!} \sum_{\sigma \in \operatorname{Sym}(I)} \prod_{j \in I} \cos(2\pi y_{\sigma^{-1}(j)} \xi_j) = \frac{1}{\binom{m}{l}} \sum_{\substack{J \subseteq I \\ |J| = l}} \prod_{j \in J} \cos(2\pi \xi_j).$$

We remark that $n/2 < l \le n$, since $y + z \in E$, and $m \ge 99d/100 \ge n$, since $\kappa(d, N) \le 1/5$. Thus $\binom{m}{l}$ is well defined. For $S \subseteq I$ we denote

$$a_S(\xi) = \prod_{j \in I \setminus S} \cos^2(\pi \xi_j) \cdot \prod_{i \in S} \sin^2(\pi \xi_i), \tag{8.3.26}$$

so that

$$\sum_{S \subseteq I} a_S = \prod_{j \in I} \left(\cos^2(\pi \xi_j) + \sin^2(\pi \xi_j) \right) = 1.$$

Taking $\varepsilon(J) \in \{-1, 1\}^d$ such that $\varepsilon(J)_j = -1$ precisely for $j \in J$ we may rewrite

$$\prod_{j \in J} \cos(2\pi \xi_j) = \prod_{j \in I} \left(\frac{1 + \cos(2\pi \xi_j)}{2} + \varepsilon(J)_j \frac{1 - \cos(2\pi \xi_j)}{2} \right)$$

$$= \prod_{j \in I} \left(\cos^2(\pi \xi_j) + \varepsilon(J)_j \sin^2(\pi \xi_j) \right)$$

$$= \sum_{S \subseteq I} a_S(\xi) w_S(\varepsilon(J)),$$

where we have defined $w_S : \{-1, 1\}^d \to \{-1, 1\}$ by setting $w_S(\varepsilon) = \prod_{j \in S} \varepsilon_j$. Changing the order of summation we thus have

$$\frac{1}{m!} \sum_{\sigma \in \operatorname{Sym}(I)} \prod_{j \in I} \cos(2\pi y_{\sigma^{-1}(j)} \xi_j) = \frac{1}{\binom{m}{l}} \sum_{\substack{J \subseteq I \\ |J| = l}} \prod_{j \in J} \cos(2\pi \xi_j)$$

$$= \sum_{S \subseteq I} a_S(\xi) \frac{1}{\binom{m}{l}} \sum_{\substack{J \subseteq I \\ |J| = l}} w_S(\varepsilon(J)).$$

Now, a direct computation shows that

$$\frac{1}{\binom{m}{l}} \sum_{\substack{J \subseteq I \\ |J|=l}} w_S(\varepsilon(J)) = \frac{1}{\binom{m}{l}} \sum_{j=0}^{l} (-1)^j \binom{|S|}{j} \binom{m-|S|}{l-j}$$

$$= \Bbbk_l^{(m)}(|S|),$$

(8.3.27)

where $\Bbbk_l^{(m)}$ is the Krawtchouk polynomial from (8.3.17). The first equality in (8.3.27) can be deduced from the disjoint splitting

$$\{J \subseteq I : |J| = l\} = \bigcup_{j=0}^{l} \{J \subseteq I : |J| = l, \text{ and } |J \cap S| = j\}.$$

Indeed, if $|J \cap S| = j$ then $w_S(\varepsilon(J)) = (-1)^j$ and thus using the above and recalling that $|I| = m$ we write

$$\sum_{\substack{J \subseteq I \\ |J|=l}} w_S(\varepsilon(J)) = \sum_{j=0}^{l} \sum_{\substack{J \subseteq I \\ |J|=l, |J \cap S|=j}} (-1)^j = \sum_{j=0}^{l} (-1)^j \binom{|S|}{j} \binom{m-|S|}{l-j}.$$

Finally, we obtain

$$\frac{1}{m!} \sum_{\sigma \in \mathrm{Sym}(I)} \prod_{j \in I} \cos(2\pi y_{\sigma^{-1}(j)} \xi_j) = \sum_{S \subseteq I} a_S(\xi) \Bbbk_l^{(m)}(|S|).$$

(8.3.28)

By the uniform estimates for Krawtchouk polynomials (8.3.18), if $0 \le l, |S| \le m/2$, we obtain

$$|\Bbbk_l^{(m)}(|S|)| \le e^{-\frac{cl|S|}{m}}.$$

(8.3.29)

Otherwise, if $|S| > m/2$ and $l \le m/2$, then we use symmetries and (8.3.18) and obtain

$$|\Bbbk_l^{(m)}(|S|)| = |\Bbbk_l^{(m)}(m-|S|)| \le e^{-\frac{cl(m-|S|)}{m}}.$$

(8.3.30)

Using (8.3.28), (8.3.29) and (8.3.30) we obtain

$$\frac{1}{m!} \left| \sum_{\sigma \in \mathrm{Sym}(I)} \prod_{j \in I} \cos(2\pi y_{\sigma^{-1}(j)} \xi_j) \right| \le \sum_{\substack{S \subseteq I \\ \emptyset \ne S \ne I}} a_S(\xi) e^{-\frac{cl|S|}{m}} + \sum_{\substack{S \subseteq I \\ \emptyset \ne S \ne I}} a_S(\xi) e^{-\frac{cl(m-|S|)}{m}}$$

$$+ e^{-\sum_{i \in I} \sin^2(\pi \xi_i)} + e^{-\sum_{i \in I} \cos^2(\pi \xi_i)},$$

where $c \in (0, 1)$ is the constant as in (8.3.18). Recalling the definition of a_S from (8.3.26) we have

$$\sum_{\substack{S \subseteq I \\ \emptyset \neq S \neq I}} a_S(\xi) e^{-\frac{cl|S|}{m}} + \sum_{\substack{S \subseteq I \\ \emptyset \neq S \neq I}} a_S(\xi) e^{-\frac{cl(m-|S|)}{m}}$$

$$\leq \sum_{S \subseteq I} \prod_{j \in I \setminus S} \cos^2(\pi \xi_j) \cdot \prod_{j \in S} e^{-\frac{cl}{m}} \sin^2(\pi \xi_j)$$

$$+ \sum_{S \subseteq I} \prod_{j \in I \setminus S} e^{-\frac{cl}{m}} \cos^2(\pi \xi_j) \cdot \prod_{j \in S} \sin^2(\pi \xi_j)$$

$$= \prod_{j \in I} \left(\cos^2(\pi \xi_j) + e^{-\frac{cl}{m}} \sin^2(\pi \xi_j) \right) + \prod_{j \in I} \left(e^{-\frac{cl}{m}} \cos^2(\pi \xi_j) + \sin^2(\pi \xi_j) \right)$$

$$= \prod_{j \in I} \left(1 - (1 - e^{-\frac{cl}{m}}) \sin^2(\pi \xi_j) \right) + \prod_{j \in I} \left(1 - (1 - e^{-\frac{cl}{m}}) \cos^2(\pi \xi_j) \right)$$

$$\leq e^{-\frac{c\kappa(d,N)^2}{4} \sum_{i \in I} \sin^2(\pi \xi_i)} + e^{-\frac{c\kappa(d,N)^2}{4} \sum_{i \in I} \cos^2(\pi \xi_i)},$$

where in the last inequality we have applied two simple inequalities $1 - x \leq e^{-x}$ and $xe^{-x/2} \leq 1 - e^{-x}$ for all $x \geq 0$ and we used the fact that $\frac{1}{2}\kappa(d, N)^2 \leq \frac{l}{m} \leq 2\kappa(d, N)^2$. This completes the proof of (8.3.25), since $I = \mathbb{N}_d \setminus S_z$, and consequently we complete the proof of Proposition 3.3. □

8.3.3 All Together

To prove Proposition 3.1 we will need the following lemma.

Lemma 3.4 *For every $d, N \in \mathbb{N}$ and for every $\xi \in \mathbb{T}^d$ we have*

$$\left| m_N(\xi) - \frac{1}{|B_N \cap \mathbb{Z}^d|} \sum_{x \in B_N \cap \mathbb{Z}^d} (-1)^{\sum_{i \in V_\xi} x_i} \right| \leq 2\kappa(d, N)^2 \sum_{i=1}^{d} \cos^2(\pi \xi_i).$$

$$(8.3.31)$$

Proof For every $\xi \in \mathbb{T}^d$, let $\xi' \in \mathbb{T}^d$ be defined as follows

$$\xi'_i = \begin{cases} \xi_i & \text{if } i \notin V_\xi, \\ \xi_i - \frac{1}{2} & \text{if } i \in V_\xi \text{ and } \frac{1}{4} < \xi_i \leq \frac{1}{2}, \\ \xi_i + \frac{1}{2} & \text{if } i \in V_\xi \text{ and } -\frac{1}{2} \leq \xi_i < -\frac{1}{4}, \end{cases} \qquad (8.3.32)$$

where V_ξ is the set defined in (8.3.2). Hence by (8.3.32) we see that $|\xi'_i| \leq 1/4$ and

$$\sin^2(\pi \xi'_i) = \cos^2(\pi \xi_i) \quad \text{for} \quad i \in V_\xi. \tag{8.3.33}$$

Moreover,

$$\begin{aligned}
\mathfrak{m}_N(\xi) &= \frac{1}{|B_N \cap \mathbb{Z}^d|} \sum_{x \in B_N \cap \mathbb{Z}^d} \prod_{j=1}^{d} \cos(2\pi x_j \xi_j) \\
&= \frac{1}{|B_N \cap \mathbb{Z}^d|} \sum_{x \in B_N \cap \mathbb{Z}^d} \left((-1)^{\sum_{i \in V_\xi} x_i} \prod_{j=1}^{d} \cos(2\pi x_j \xi'_j) \right),
\end{aligned} \tag{8.3.34}$$

since $\cos(2\pi x_j \xi_j) = \cos(2\pi x_j \xi'_j) \cos(\pi x_j) = (-1)^{x_j} \cos(2\pi x_j \xi'_j)$.

Arguing in a similar way as in the proof of Proposition 2.1 we see, by (8.3.34), that

$$\left| \mathfrak{m}_N(\xi) - \frac{1}{|B_N \cap \mathbb{Z}^d|} \sum_{x \in B_N \cap \mathbb{Z}^d} (-1)^{\sum_{i \in V_\xi} x_i} \right|$$

$$\leq \frac{1}{|B_N \cap \mathbb{Z}^d|} \sum_{x \in B_N \cap \mathbb{Z}^d} \left(1 - \prod_{j=1}^{d} \cos(2\pi x_j \xi'_j) \right)$$

$$\leq \frac{2}{|B_N \cap \mathbb{Z}^d|} \sum_{x \in B_N \cap \mathbb{Z}^d} \sum_{j=1}^{d} \sin^2(\pi x_j \xi'_j)$$

$$\leq \frac{2}{|B_N \cap \mathbb{Z}^d|} \sum_{x \in B_N \cap \mathbb{Z}^d} \sum_{j=1}^{d} x_j^2 \sin^2(\pi \xi'_j)$$

$$\leq \frac{2N^2}{d} \sum_{j=1}^{d} \sin^2(\pi \xi'_j)$$

$$\leq 2\kappa(d, N)^2 \sum_{j=1}^{d} \cos^2(\pi \xi_j),$$

since by (8.3.33) we obtain

$$\sum_{j \in V_\xi} \sin^2(\pi \xi'_j) = \sum_{j \in V_\xi} \cos^2(\pi \xi_j), \quad \text{and} \quad \sum_{j \in \mathbb{N}_d \setminus V_\xi} \sin^2(\pi \xi'_j) \leq \sum_{j \in \mathbb{N}_d \setminus V_\xi} \cos^2(\pi \xi'_j).$$

The proof of (8.3.31) is completed.   □

Proof of Proposition 3.1 Firstly, we assume that $|V_\xi| \leq d/2$, then $|\mathbb{N}_d \setminus V_\xi| = d - |V_\xi| \geq d/2$ and (8.3.19) implies that

$$|\mathsf{m}_N(\xi)| \leq 16e^{-\frac{c\kappa(d,N)^2}{400}\sum_{i=1}^d \sin^2(\pi\xi_i)},$$

since $\sum_{i=1}^d \cos^2(\pi\xi_i) \geq d/4 \geq 1/4\sum_{i=1}^d \sin^2(\pi\xi_i)$. Thus

$$\left| \mathsf{m}_N(\xi) - e^{-\kappa(d,N)^2\sum_{i=1}^d \sin^2(\pi\xi_i)} \right| \leq 17e^{-\frac{c\kappa(d,N)^2}{400}\sum_{i=1}^d \sin^2(\pi\xi_i)}. \tag{8.3.35}$$

On the other hand, using (8.2.9) from the proof of Proposition 2.1 we obtain

$$\left| \mathsf{m}_N(\xi) - e^{-\kappa(d,N)^2\sum_{i=1}^d \sin^2(\pi\xi_i)} \right| \leq 3\kappa(d,N)^2 \sum_{i=1}^d \sin^2(\pi\xi_i). \tag{8.3.36}$$

We now see that (8.3.35) and (8.3.36) imply (8.3.5).

Secondly, we assume that $|V_\xi| \geq d/2$, then (8.3.19) implies that

$$|\mathsf{m}_N(\xi)| \leq 16e^{-\frac{c\kappa(d,N)^2}{400}\sum_{i=1}^d \cos^2(\pi\xi_i)},$$

since $\sum_{i=1}^d \sin^2(\pi\xi_i) \geq d/4 \geq 1/4\sum_{i=1}^d \cos^2(\pi\xi_i)$. Thus

$$\left| \mathsf{m}_N(\xi) - \frac{1}{|B_N \cap \mathbb{Z}^d|}\left(\sum_{x \in B_N \cap \mathbb{Z}^d} (-1)^{\sum_{i=1}^d x_i} \right)e^{-\kappa(d,N)^2\sum_{i=1}^d \cos^2(\pi\xi_i)} \right|$$

$$\leq 17e^{-\frac{c\kappa(d,N)^2}{400}\sum_{i=1}^d \cos^2(\pi\xi_i)}. \tag{8.3.37}$$

On the other hand, by Lemma 3.4 we obtain

$$\left| \mathsf{m}_N(\xi) - \frac{1}{|B_N \cap \mathbb{Z}^d|}\left(\sum_{x \in B_N \cap \mathbb{Z}^d} (-1)^{\sum_{i \in V_\xi} x_i} \right)e^{-\kappa(d,N)^2\sum_{i=1}^d \cos^2(\pi\xi_i)} \right|$$

$$\leq 3\kappa(d,N)^2 \sum_{i=1}^d \cos^2(\pi\xi_i). \tag{8.3.38}$$

Moreover, arguing in a similar way as in the proof of Proposition 2.1 we see that

$$\left| \frac{1}{|B_N \cap \mathbb{Z}^d|} \sum_{x \in B_N \cap \mathbb{Z}^d} (-1)^{\sum_{i \in V_\xi} x_i} - \frac{1}{|B_N \cap \mathbb{Z}^d|} \sum_{x \in B_N \cap \mathbb{Z}^d} (-1)^{\sum_{i=1}^d x_i} \right|$$

$$\leq \frac{1}{|B_N \cap \mathbb{Z}^d|} \sum_{\substack{x \in B_N \cap \mathbb{Z}^d \\ (\mathbb{N}_d \setminus V_\xi) \cap \mathrm{supp}\, x \neq \emptyset}} \left| (-1)^{\sum_{i \in \mathbb{N}_d \setminus V_\xi} x_i} - 1 \right|$$

$$\leq \frac{2}{|B_N \cap \mathbb{Z}^d|} \sum_{x \in B_N \cap \mathbb{Z}^d} \sum_{i \in \mathbb{N}_d \setminus V_\xi} x_i^2$$

$$\leq 2\kappa(d, N)^2 (d - |V_\xi|)$$

$$\leq 4\kappa(d, N)^2 \sum_{i=1}^{d} \cos^2(\pi \xi_i), \tag{8.3.39}$$

since $1/2 \leq \cos^2(\pi \xi_i)$ for any $i \in \mathbb{N}_d \setminus V_\xi$ and consequently $d - |V_\xi| \leq 2 \sum_{i \in \mathbb{N}_d \setminus V_\xi} \cos^2(\pi \xi_i)$. Combining (8.3.38) and (8.3.39) we obtain

$$\left| \mathsf{m}_N(\xi) - \frac{1}{|B_N \cap \mathbb{Z}^d|} \left(\sum_{x \in B_N \cap \mathbb{Z}^d} (-1)^{\sum_{i=1}^{d} x_i} \right) e^{-\kappa(d,N)^2 \sum_{i=1}^{d} \cos^2(\pi \xi_i)} \right|$$

$$\leq 7\kappa(d, N)^2 \sum_{i=1}^{d} \cos^2(\pi \xi_i). \tag{8.3.40}$$

We now see that (8.3.37) and (8.3.40) imply (8.3.6). This completes the proof. □

Acknowledgements The authors are grateful to the referees for careful reading of the manuscript and useful remarks that led to the improvement of the presentation. We also thank for pointing out a simple proof of Lemma 2.5.

References

1. J. Bourgain, On high dimensional maximal functions associated to convex bodies. Am. J. Math. **108**, 1467–1476 (1986)
2. J. Bourgain, On L^p bounds for maximal functions associated to convex bodies in $\mathbb{R}^n$. Israel J. Math. **54**, 257–265 (1986)
3. J. Bourgain, On the Hardy-Littlewood maximal function for the cube. Israel J. Math. **203**, 275–293 (2014)
4. J. Bourgain, M. Mirek, E.M. Stein, B. Wróbel, Dimension-free variational estimates on $L^p(\mathbb{R}^d)$ for symmetric convex bodies. Geom. Funct. Anal. **28**(1), 58–99 (2018)
5. J. Bourgain, M. Mirek, E. M. Stein, B. Wróbel, On the Hardy–Littlewood maximal functions in high dimensions: continuous and discrete perspective, in *Geometric Aspects of Harmonic Analysis. A conference proceedings on the Occasion of Fulvio Ricci's 70th Birthday.* Springe INdAM Series (Cortona, 2018)
6. J. Bourgain, M. Mirek, E.M. Stein, B. Wróbel, Dimension-free estimates for discrete Hardy–Littlewood averaging operators over the cubes in $\mathbb{Z}^d$. Am. J. Math. **141**(4), 857–905 (2019)
7. A. Carbery, An almost-orthogonality principle with applications to maximal functions associated to convex bodies. Bull. Am. Math. Soc. **14**(2), 269–274 (1986)
8. L. Delaval, O. Guédon, B. Maurey, Dimension-free bounds for the Hardy-Littlewood maximal operator associated to convex sets. Ann. Fac. Sci. Toulouse Math. **27**(1), 1–198 (2018)
9. A.W. Harrow, A. Kolla, L.J. Schulman, Dimension-free L_2 maximal inequality for spherical means in the hypercube. Theory Comput. **10**(3), 55–75 (2014)

10. S. Janson, T. Łuczak, A. Ruciński, *Random Graphs* (Wiley, New York, 2000), pp. 1–348
11. M. Mirek, E.M. Stein, P. Zorin–Kranich, Jump inequalities via real interpolation. Math. Ann. (2019). arXiv:1808.04592
12. M. Mirek, E.M. Stein, P. Zorin–Kranich, A bootstrapping approach to jump inequalities and their applications (2019). arXiv:1808.09048
13. D. Müller, A geometric bound for maximal functions associated to convex bodies. Pac. J. Math. **142**(2), 297–312 (1990)
14. H. Robbins, A Remark on Stirling's formula. Am. Math. Month. **62**(1), 26–29 (1955)
15. E.M. Stein, *Topics in Harmonic Analysis Related to the Littlewood–Paley Theory*. Annals of Mathematics Studies (Princeton University Press, Princeton, 1970), pp. 1–157
16. E.M. Stein, The development of square functions in the work of A. Zygmund. Bull. Am. Math. Soc., **7**, 359–376 (1982)

Chapter 9
On the Poincaré Constant
of Log-Concave Measures

Patrick Cattiaux and Arnaud Guillin

Abstract The goal of this paper is to push forward the study of those properties of log-concave measures that help to estimate their Poincaré constant. First we revisit E. Milman's result (Invent Math 177:1–43, 2009) on the link between weak (Poincaré or concentration) inequalities and Cheeger's inequality in the log-concave cases, in particular extending localization ideas and a result of Latala, as well as providing a simpler proof of the nice Poincaré (dimensional) bound in the unconditional case. Then we prove alternative transference principle by concentration or using various distances (total variation, Wasserstein). A mollification procedure is also introduced enabling, in the log-concave case, to reduce to the case of the Poincaré inequality for the mollified measure. We finally complete the transference section by the comparison of various probability metrics (Fortet-Mourier, bounded-Lipschitz, . . .) under a log-concavity assumption.

9.1 Introduction and Overview

Let ν be a Probability measure defined on $\mathbb{R}^n$. For a real valued function f, $\nu(f)$ and $m_\nu(f)$ will denote respectively the ν mean and a ν median of f, when these quantities exist. We also denote by

$$\mathrm{Var}_\nu(f) = \nu(f^2) - \nu^2(f)$$

the ν variance of f.

P. Cattiaux (✉)
Institut de Mathématiques de Toulouse, Université de Toulouse, CNRS UMR 5219, Toulouse, France
e-mail: cattiaux@math.univ-toulouse.fr

A. Guillin
Laboratoire de Mathématiques, CNRS UMR 6620, Université Clermont-Auvergne, Aubière, France
e-mail: guillin@math.univ-bpclermont.fr

© Springer Nature Switzerland AG 2020
B. Klartag, E. Milman (eds.), *Geometric Aspects of Functional Analysis*,
Lecture Notes in Mathematics 2256,
https://doi.org/10.1007/978-3-030-36020-7_9

The Poincaré constant $C_P(\nu)$ of ν is defined as the best constant such that

$$\mathrm{Var}_\nu(f) \le C_P(\nu)\, \nu(|\nabla f|^2)\,.$$

In all the paper, we shall denote equally $C_P(Z)$ or $C_P(\nu)$ the Poincaré constant for a random variable Z with distribution ν. We say that ν satisfies a Poincaré inequality when $C_P(\nu)$ is finite. Note also that in the whole paper, for $x \in \mathbb{R}^n$, $|x| = \left(\sum_{i=1}^{n} x_i^2\right)^{\frac{1}{2}}$ denotes the Euclidean norm of x, and a function f is said to be K-Lipschitz if $\sup_{|x-y|>0} \frac{|f(x)-f(y)|}{|x-y|} \le K$.

It is well known that, as soon as a Poincaré inequality is satisfied, the tails of ν are exponentially small, i.e. $\nu(|x| > R) \le C\, e^{-cR/\sqrt{C_P(\nu)}}$ for some universal c and C (see [14]), giving a very useful necessary condition for this inequality to hold. Conversely, during the last 80 years a lot of sufficient conditions have been given for a Poincaré inequality to hold. In 1976, Brascamp and Lieb [15] connected Poincaré inequality to convexity by proving the following: if $\nu(dx) = e^{-V(x)}\, dx$, then

$$\mathrm{Var}_\nu(f) \le \int {}^t\nabla f\, (\mathrm{Hess}^{-1}(V))\, \nabla f\, d\nu$$

where $\mathrm{Hess}(V)$ denotes the Hessian matrix of V. Consequently, if V is uniformly convex, i.e. $\inf_x {}^t\xi\, \mathrm{Hess}(V)(x)\xi \ge \rho\, |\xi|^2$ for some $\rho > 0$, then $C_P(\nu) \le 1/\rho$. This result contains in particular the Gaussian case, and actually Gaussian measures achieve the Brascamp-Lieb bound as it is easily seen looking at linear functions f.

This result was extended to much more general "uniformly convex" situations through the celebrated Γ_2 theory introduced by Bakry and Emery (see the recent monograph [4] for an up to date state of the art of the theory) and the particular uniformly convex situation corresponds to the $\mathrm{CD}(\rho, \infty)$ curvature-dimension property in this theory. This theory has been recently revisited in [17] by using coupling techniques for the underlying stochastic process.

A particular property of the Poincaré inequality is the tensorization property [4, Proposition 4.3.1]

$$C_P(\nu_1 \otimes \ldots \otimes \nu_N) \le \max_{i=1,\ldots,N} C_P(\nu_i)\,.$$

It is of fundamental importance for the concentration of measure and for getting bounds for functionals of independent samples in statistics, due to its "dimension free" character. This "dimension free" character is captured by the Brascamp-Lieb inequality or the Bakry-Emery criterion, even for non-product measures.

Using a simple perturbation of V by adding a bounded (or a Lipschitz) term, one can show that uniform convexity "at infinity" is enough to get a Poincaré inequality [4, Proposition 4.2.7]. This result can also be proved by using reflection coupling (see [17, 26, 27]). However in this situation a "dimension free" bound for the optimal constant is hard to obtain, as it is well known for the double well potential $V(x) = |x|^4 - |x|^2$.

Uniform convexity (even at infinity) is not necessary as shown by the example of the symmetric exponential measure on the line, $v(dx) = \frac{1}{2} e^{-|x|}$ which satisfies $C_P(v) = 4$ (see [4] (4.4.3)). In 1999, Bobkov [10] has shown that any log-concave probability measure satisfies the Poincaré inequality. Here log-concave means that $v(dx) = e^{-V(x)} dx$ where V is a convex function with values in $\mathbb{R} \cup \{+\infty\}$. In particular uniform measures on convex bodies are log-concave. We refer to the recent book [16] for an overview on the topic of convex bodies, and to [49] for a survey of log-concavity in statistics. Another proof, applying to a larger class of measures, was given in [2] using Lyapunov functions as introduced in [3]. If it is now known that a Poincaré inequality is equivalent to the existence of a Lyapunov function (see [18, 22]), this approach is far to give good controls for the Poincaré constant.

Actually Bobkov's result is stronger since it deals with the $\mathbb{L}^1$ version of Poincaré inequality

$$v(|f - v(f)|) \leq C_C(v)\, v(|\nabla f|)\,, \tag{9.1.1}$$

which is often called Cheeger inequality. Another form of Cheeger inequality is

$$v(|f - m_v(f)|) \leq C'_C(v)\, v(|\nabla f|)\,. \tag{9.1.2}$$

Here again, C_C and C'_C have to be understood as the best constants satisfying the related inequalities. Using

$$\frac{1}{2}\, v(|f - v(f)|) \;\leq\; v(|f - m_v(f)|) \;\leq\; v(|f - v(f)|)\,,$$

it immediately follows that $\frac{1}{2} C_C \leq C'_C \leq C_C$.

It is well known that the Cheeger constant gives a natural control for the isoperimetric function Is_v of v. Recall that for $0 \leq u \leq \frac{1}{2}$,

$$\mathrm{Is}_v(u) = \inf_{A, v(A)=u} v_s(\partial A)$$

where

$$v_s(\partial A) = \liminf_{h \to 0} \frac{v(A + hB(0, 1)) - v(A)}{h}$$

denotes the surface measure of the boundary of A. It can be shown (see the introduction of [10]) that

$$\mathrm{Is}_v(u) = \frac{u}{C'_C(v)} \geq \frac{u}{C_C(v)}\,.$$

The Cheeger inequality is stronger than the Poincaré inequality and (see again [10])

$$C_P(\nu) \leq 4\,(C_C')^2(\nu)\,. \tag{9.1.3}$$

The first remarkable fact in the log-concave situation is that a converse inequality holds, namely if ν is log-concave,

$$(C_C')^2(\nu) \leq 36\,C_P(\nu)\,, \tag{9.1.4}$$

as shown by Ledoux [40, formula (5.8)].

Ledoux's approach is using the associated semi-group with generator $L = \Delta - \nabla V.\nabla$ for which the usual terminology corresponding to the convexity of V is zero curvature. Of course to define L one has to assume some smoothness of V. But if Z is a random variable with a log-concave distribution ν and G is an independent standard Gaussian random variable, Prekopa-Leindler theorem ensures that the distribution of $Z + \varepsilon\,G$ is still log concave for any $\varepsilon > 0$ and is associated to a smooth potential V_ε. Hence we may always assume that the potential V is smooth provided we may pass to the limit $\varepsilon \to 0$.

Log-concave measures deserve attention during the last 20 years in particular in Statistics. They are considered close to product measures in high dimension. It is thus important to get some tractable bound for their Poincaré constant, in particular to understand the role of the dimension.

Of course if $Z = (Z_1, \ldots, Z_n)$ is a random vector of $\mathbb{R}^n$,

$$C_P((\lambda_1\,Z_1, \ldots, \lambda_n\,Z_n)) \leq \max_i \lambda_i^2\,C_P((Z_1, \ldots, Z_n))\,,$$

and the Poincaré constant is unchanged if we perform a translation or an isometric change of coordinates. It follows that

$$C_P(Z) \leq \sigma^2(Z)\,C_P(Z')$$

where $\sigma^2(Z)$ denotes the largest eigenvalue of the covariance matrix $Cov_{i,j}(Z) = Cov(Z_i, Z_j)$, and Z' is an affine transformation of Z which is centered and with Covariance matrix equal to Identity. Such a random vector (or its distribution) is called isotropic (or in isotropic position for convex bodies and their uniform distribution). The reader has to take care about the use of the word isotropic, which has a different meaning in probability theory (for instance in Paul Lévy's work).

Applying the Poincaré inequality to linear functions show that $\sigma^2(Z) \leq C_P(Z)$. In particular, in the uniformly convex situation, $\sigma^2(Z) \leq 1/\rho$ with equality when Z is a Gaussian random vector. For the symmetric exponential measure on $\mathbb{R}$, we also have $\sigma^2(Z) = C_P(Z)$ while $\rho = 0$. It thus seems plausible that, even in positive curvature, $\sigma^2(Z)$ is the good parameter to control the Poincaré constant.

The following was conjectured by Kannan et al. [33]

Conjecture 9.1.1 (K-L-S Conjecture)   There exists a universal constant C such that for any log-concave probability measure ν on $\mathbb{R}^n$,

$$C_P(\nu) \leq C\,\sigma^2(\nu)\,,$$

where $\sigma^2(\nu)$ denotes the largest eigenvalue of the covariance matrix $Cov_{i,j}(\nu) = Cov_\nu(x_i, x_j)$, or if one prefers, there exists an universal constant C such that any isotropic log-concave probability measure ν, in any dimension n, satisfies $C_P(\nu) \leq C$.

During the last years a lot of works have been done on this conjecture. A recent book [1] is totally devoted to the description of the state of the art. We will thus mainly refer to this book for references, but of course apologize to all important contributors. We shall just mention part of these works we shall revisit and extend.

In this note we shall, on one hand investigate properties of log-concave measures that help to evaluate their Poincaré constant, on the other hand obtain explicit constants in many intermediate results. Let us explain on an example: in a remarkable paper [44], E. Milman has shown that one obtains an equivalent inequality if one replaces the energy of the gradient in the right hand side of the Poincaré inequality by the square of its Lipschitz norm, furnishing a much less demanding inequality $((2, +\infty)$ Poincaré inequality). The corresponding constant is sometimes called the spread constant. In other words the Poincaré constant of a log-concave measure ν is controlled by its spread constant. In the next section we shall give another proof of this result. Actually we shall extend it to weak forms of $(1, +\infty)$ inequalities. These weak forms allow us to directly compare the concentration profile of ν with the corresponding weak inequality. We shall also give explicit controls of the constants when one reduces the support of ν to an Euclidean ball as in [44] or a l^∞ ball, the latter being an explicit form of a result by Latala [38].

In Sect. 9.3 we shall describe several transference results using absolute continuity, concentration properties and distances between measures. There is an important overlap between this section and part of the results in [7] or [46]. In addition to different proofs, the main novelty here is that we compare a log-concave measure ν with another non necessarily log-concave measure μ and use the weak form of the $(1, +\infty)$ Poincaré inequality. For instance, we show that if the distance between ν and μ is small enough, then the Poincaré constant of μ controls the one of ν. This is shown for several distances: total variation, Wasserstein, Bounded Lipschitz.

Section 9.4 is concerned with mollification. The first part is a revisit of results by Klartag [36]. The second part studies convolution with a Gaussian kernel. It is shown that if γ is some Gaussian measure, $C_P(\nu * \gamma)$ controls $C_P(\nu)$. If the converse is well known, this implication is new and may have some applications elsewhere. The proof is based on stochastic calculus.

Finally in the last section and using what precedes we show that all the previous distances and the Lévy-Prokhorov distance define the same uniform structure on the set of log-concave measures independently of the dimension. We thus complete the transference results using distances. Some dimensional comparisons have been done in [43].

As said by the referee and the editors, this paper collects a (may be too) long list of results, some new, some already proven, presenting new proofs (in general simpler) for these last ones and trying to trace, in general non optimal, constants. We did our best to compare these results with the existing ones, but we apologize in advance for the omissions. We take the opportunity of these few lines to warmly thank an anonymous referee and a less anonymous editor for their suggestions and comments that improve the presentation of the paper at several places.

9.2 Revisiting E. Milman's Results

9.2.1 (p, q) Poincaré Inequalities

Following [44], the usual Poincaré inequality can be generalized in a (p, q) Poincaré inequality, for $1 \leq p \leq q \leq +\infty$,

$$B_{p,q} \, \nu^{1/p}(|f - \nu(f)|^p) \leq \nu^{1/q}(|\nabla f|^q). \tag{9.2.1}$$

For $p = q = 2$ we recognize the Poincaré inequality and $B_{2,2}^2 = 1/C_P(\nu)$, and for $p = q = 1$ the Cheeger inequality with $B_{1,1} = 1/C_C(\nu)$.

Among all (p, q) Poincaré inequalities, the weakest one is clearly the $(1, +\infty)$ one, the strongest the $(1, 1)$ one, we called Cheeger's inequality previously. Indeed for $1 \leq p \leq p' \leq q \leq +\infty$ except the case $p = q = +\infty$, one has the following schematic array between these Poincaré inequalities (see [44])

$$
\begin{array}{ccc}
(1,1) & \Rightarrow & (1,q) \\
\Downarrow & & \Uparrow \\
(p,p) & \Rightarrow & (p,q) \\
\Downarrow & & \Uparrow \\
(p',p') & \Rightarrow & (p',q)
\end{array}
$$

The meaning of all these inequalities is however quite unclear except some cases we shall describe below.

First remark that on $\mathbb{R}^n$,

$$|f(x) - f(a)| \leq \| \nabla f \|_\infty \, |x - a|$$

yielding

$$\nu(|f - m_\nu(f)|) = \inf_b \nu(|f - b|) \leq \inf_a \nu^{1/p}(|x - a|^p) \; \| \; |\nabla f| \; \|_\infty \, ,$$

so that since

$$\frac{1}{2} \nu(|f - \nu(f)|) \; \leq \; \nu(|f - m_\nu(f)|) \; \leq \; \nu(|f - \nu(f)|) \, , \tag{9.2.2}$$

the $(p, +\infty)$ Poincaré inequality is satisfied as soon as ν admits a p-moment. There is thus no hope for this inequality to be helpful unless we make some additional assumption.

Now look at the $(p, 2)$ Poincaré inequality ($1 \leq p \leq 2$). We may write, assuming that $\nu(f) = 0$,

$$\mathrm{Var}_\nu(f) \leq \| \; f \; \|_\infty^{2-p} \; \nu(|f|^p) \leq \frac{1}{B_{p,2}^p} \; \| \; f \; \|_\infty^{2-p} \; \nu^{p/2}(|\nabla f|^2)$$

which is equivalent to

$$\mathrm{Var}_\nu(f) \; \leq \; c \, s^{-\frac{2-p}{p}} \; \nu(|\nabla f|^2) + s \; \| \; f - \nu(f) \; \|_\infty^2 \qquad \text{for all } s > 0 \, ,$$

with

$$\frac{1}{B_{p,2}^2} = \frac{1}{p} \left(\frac{(c(2-p))^{2/p} + p^{2/p}}{(c(2-p))^{(2-p)/p}} \right) .$$

This kind of inequalities has been studied under the name of weak Poincaré inequalities (see [8, 20, 48]). They can be used to show the $\mathbb{L}^2 - \mathbb{L}^\infty$ convergence of the semi-group with a rate $t^{-p/(2-p)}$.

As shown in [48], any probability measure $\nu(dx) = e^{-V(x)} dx$ such that V is locally bounded, satisfies a weak Poincaré inequality. Indeed, using Holley-Stroock perturbation argument [4, Proposition 4.2.7] w.r.t. the (normalized) uniform measure on the Euclidean ball $B(0, R)$, it is easy to see that

$$\mathrm{Var}_\nu(f) \leq \frac{4R^2}{\pi^2} e^{Osc_R V} \nu(|\nabla f|^2) + 2 \nu(|x| > R) \; \| \; f - \nu(f) \; \|_\infty^2$$

where $Osc_R V = \sup_{|x| \leq R} V(x) - \inf_{|x| \leq R} V(x)$.

It is thus tempting to introduce weak versions of (p, q) Poincaré inequalities.

Definition 9.2.1 We shall say that ν satisfies a weak (p, q) Poincaré inequality if there exists some non increasing non-negative function β defined on $]0, +\infty[$ such that for all $s > 0$ and all smooth function f,

$$(\nu(|f - \nu(f)|^p))^{\frac{1}{p}} \leq \beta(s) \,\|\, |\nabla f| \,\|_q + s \operatorname{Osc}(f),$$

where Osc denotes the oscillation of f. We shall sometimes replace $\nu(f)$ by $m_\nu(f)$. In particular for $p = 1$ and $q = \infty$ we have

$$\beta^{med}(s) \leq \beta^{mean}(s) \leq 2\beta^{med}(s/2).$$

Of course if $\beta(0) < +\infty$ we recover the (p, q) Poincaré inequality. Any probability measure satisfies a particular weak $(1, +\infty)$ Poincaré inequality, namely:

Proposition 9.2.2 *Denote by α_ν the concentration profile of a probability measure ν, i.e.*

$$\alpha_\nu(r) := \sup\{1 - \nu(A + B(0, r)) \,;\, \nu(A) \geq \frac{1}{2}\}, \, r > 0,$$

where $B(y, r)$ denotes the Euclidean ball centered at y with radius r. Then for any probability measure ν and all $s > 0$,

$$\nu(|f - m_\nu(f)|) \leq \alpha_\nu^{-1}(s/2) \,\|\, |\nabla f| \,\|_\infty + s \operatorname{Osc} f.$$

and

$$\nu(|f - \nu(f)|) \leq 2\,\alpha_\nu^{-1}(s/4) \,\|\, |\nabla f| \,\|_\infty + s \operatorname{Osc}(f),$$

where α_ν^{-1} denotes the converse function of α_ν.

Proof Due to homogeneity we may assume that f is 1-Lipschitz. Hence $\nu(|f - m_\nu(f)| > r) \leq 2\alpha_\nu(r)$. Thus

$$\nu(|f - \nu(f)|) \leq 2\,\nu(|f - m_\nu(f)|) \leq 2\,r\,\nu(|f - m_\nu(f)| \leq r)$$
$$+ 2\operatorname{Osc}(f)\,\nu(|f - m_\nu(f)| > r) \leq 2r + 4\operatorname{Osc}(f)\,\alpha_\nu(r),$$

hence the result. $\square$

As asked to us by E. Milman, one should ask about a partial converse to the previous statement. Actually we do not know how to get something convincing.

These results will be surprisingly useful for the family of log-concave measures. Recall the following result is due to E. Milman (see Theorem 2.4 in [44]):

Theorem 9.2.3 (E. Milman's Theorem) *If $d\nu = e^{-V} dx$ is a log-concave probability measure in $\mathbb{R}^n$, there exists a universal constant C such that for all (p, q) and (p', q') (with $1 \leq p \leq q \leq +\infty$ and $1 \leq p' \leq q' \leq +\infty$)*

$$B_{p,q} \leq C\,p'\,B_{p',q'}.$$

Hence in the log-concave situation the $(1, +\infty)$ Poincaré inequality implies Cheeger's inequality, more precisely implies a control on the Cheeger's constant (that any log-concave probability satisfies a Cheeger's inequality is already well known). We shall revisit this last result in the next subsection.

9.2.2   Log-Concave Probability Measures

In order to prove Theorem 9.2.3 it is enough to show that the $(1, +\infty)$ Poincaré inequality implies the $(1, 1)$ one, and to use the previous array. We shall below reinforce E. Milman's result. The proof (as in [44]) lies on the concavity of the isoperimetric profile, namely the following proposition which was obtained by several authors (see [44] Theorem 1.8 for a list):

Proposition 9.2.4 *Let v be a (smooth) log-concave probability measure on $\mathbb{R}^n$. Then the isoperimetric profile $u \mapsto Is_v(u)$ is concave on $[0, \frac{1}{2}]$.*

The previous concavity assumption may be used to get some estimates on Poincaré and Cheeger constants.

Proposition 9.2.5 *Let v be a probability measure such that $u \mapsto Is_v(u)$ is concave on $[0, \frac{1}{2}]$. Assume that there exist some $0 \leq u \leq 1/2$ and some $C(u)$ such that for any Lipschitz function $f \geq 0$ it holds*

$$v(|f - v(f)|) \leq C(u)\, v(|\nabla f|) + u\, Osc(f). \tag{9.2.3}$$

Then for all measurable A such that $v(A) \leq 1/2$,

$$v_s(\partial A) \geq \frac{1 - 2u}{C(u)}\, v(A) \quad i.e. \quad C'_C(v) \leq \frac{C(u)}{1 - 2u}.$$

If we reinforce (9.2.3) as follows, for some $0 \leq u \leq 1$, $1 < p < \infty$ and some $C_p(u)$

$$v(|f - v(f)|) \leq C_p(u)\, v(|\nabla f|) + u \left(\int (f - v(f))^p dv \right)^{\frac{1}{p}}, \tag{9.2.4}$$

then for all measurable A such that $v(A) \leq 1/2$,

$$v_s(\partial A) \geq \frac{1 - u}{C_p(u)}\, v(A) \quad i.e. \quad C'_C(v) \leq \frac{C_p(u)}{1 - u}.$$

Proof Let A be some Borel subset with $v(A) = \frac{1}{2}$. According to Lemma 3.5 in [13] one can find a sequence f_n of Lipschitz functions with $0 \leq f_n \leq 1$, such that $f_n \to \mathbf{1}_{\bar{A}}$ pointwise ($\bar{A}$ being the closure of A) and $\limsup v(|\nabla f_n|) \leq v_s(\partial A)$. According to the proof of Lemma 3.5 in [13], we may assume that $v(\bar{A}) = v(A)$

(otherwise $v_s(\partial A) = +\infty$). Taking limits in the left hand side of (9.2.3) thanks to Lebesgue's bounded convergence theorem, we thus obtain

$$v(|\mathbf{1}_A - v(A)|) \leq C(u)\, v_s(\partial A) + u\,.$$

The left hand side is equal to $2v(A)(1-v(A)) = \frac{1}{2}$ so that we obtain $v_s(\partial A) \geq \frac{\frac{1}{2}-u}{C(u)}$. It remains to use the concavity of Is_v, which yields $Is_v(u) \geq 2\, Is_v(\frac{1}{2})\, u$.

If we replace (9.2.3) by (9.2.4), we similarly obtain, when $v(A) = \frac{1}{2}$, $\int(\mathbf{1}_A - v(A))^p dv = (1/2)^p$, so that $\frac{1}{2} \leq C_p(u)\, v_s(\partial A) + \frac{1}{2} u$ and the result follows similarly.   □

Remark 9.2.6 We may replace $v(f)$ by $m_v(f)$ in (9.2.3) without changing the proof, since the explicit form of the approximating f_n in [13] satisfies $m_v(f_n) \to v(A)$.   ◇

According to Proposition 9.2.4, the previous proposition applies to log-concave measures. But in this case one can weaken the required inequalities.

Theorem 9.2.7 *Let v a log-concave probability measure.*
  Assume that there exist some $0 \leq s < 1/2$ and some $\beta(s)$ such that for any Lipschitz function f it holds

$$v(|f - v(f)|) \leq \beta(s)\, \|\, |\nabla f|\, \|_\infty + s\, Osc(f)\,, \tag{9.2.5}$$

respectively, for some $0 \leq s < 1$ and some $\beta(s)$

$$v(|f - v(f)|) \leq \beta(s)\, \|\, |\nabla f|\, \|_\infty + s\, (Var_v(f))^{\frac{1}{2}}\,. \tag{9.2.6}$$

Then

$$C_C'(v) \leq \frac{4\beta(s)}{\pi\, (\frac{1}{2} - s)^2} \quad resp. \quad C_C'(v) \leq \frac{16\beta(s)}{\pi\, (1-s)^2}\,.$$

We may replace $v(f)$ by $m_v(f)$ in both cases.

Proof In the sequel P_t denotes the symmetric semi-group with infinitesimal generator $L = \Delta - \nabla V.\nabla$. Here we assume for simplicity that V is smooth on the interior of $D = \{V < +\infty\}$ (which is open and convex), so that the generator acts on functions whose normal derivative on ∂D is equal to 0. From the probabilistic point of view, P_t is associated to the diffusion process with generator L normally reflected at the boundary ∂D.

A first application of zero curvature is the following, that holds for all $t > 0$,

$$\|\, |\nabla P_t g|\, \|_\infty \leq \frac{1}{\sqrt{\pi t}}\, \|\, g\, \|_\infty\,. \tag{9.2.7}$$

This result is proved using reflection coupling in Proposition 17 of [17]. With the slightly worse constant $\sqrt{2t}$, it was previously obtained by Ledoux in [40]. According to Ledoux's duality argument (see (5.5) in [40]), we deduce, with $g = f - \nu(f)$,

$$\nu(|g|) \leq \sqrt{4t/\pi}\, \nu(|\nabla f|) + \nu(|P_t g|). \tag{9.2.8}$$

Note that $\nu(|P_t g|) = \nu(|P_t f - \nu(P_t f)|)$. Applying (9.2.5) with $P_t f$, we obtain

$$\nu(|f - \nu(f)|) \leq \sqrt{4t/\pi}\, \nu(|\nabla f|) + \beta(s)\, \|\, |\nabla P_t f|\, \|_\infty + s\, Osc(P_t f).$$

Applying (9.2.7) again, and the contraction property of the semi-group in $\mathbb{L}^\infty$, yielding $Osc(P_t f) \leq Osc(f)$, we get

$$\nu(|f - \nu(f)|) \leq \sqrt{4t/\pi}\, \nu(|\nabla f|) + \left(s + \frac{\beta(s)}{\sqrt{\pi t}}\right) Osc(f). \tag{9.2.9}$$

Choose $t = \frac{4\,\beta^2(s)}{\pi\,(\frac{1}{2}-s)^2}$. We may apply Proposition 9.2.3 (and Remark 9.2.6) with $u = (s + \frac{1}{2})/2$ which is less than $\frac{1}{2}$ and

$$C(u) = \frac{4\beta(s)}{\pi\,(\frac{1}{2} - s)}.$$

yielding the result.

If we want to deal with the case of $m_\nu(f)$ we have to slightly modify the proof. This time we choose $g = f - m_\nu(P_t f)$ so that, first $\nu(|f - m_\nu(f)|) \leq \nu(|g|)$, second $P_t g = P_t f - m_\nu(P_t f)$, so that we can apply (9.2.5) with the median. We are done by using Remark 9.2.6.

Next we apply (9.2.6) and the contraction property of the semi-group in $\mathbb{L}^2$. We get

$$\nu(|f - \nu(f)|) \leq \sqrt{4t/\pi}\, \nu(|\nabla f|) + \beta(s)\, \|\, |\nabla P_t f|\, \|_\infty + s\, (Var_\nu(f))^{\frac{1}{2}}.$$

But now, either $Var_\nu(f) \leq \frac{1}{4} Osc(f)$ or $Var_\nu(f) \geq \frac{1}{4} Osc(f)$.

In the first case we get

$$\nu(|f - \nu(f)|) \leq \sqrt{4t/\pi}\, \nu(|\nabla f|) + \beta(s)\, \|\, |\nabla P_t f|\, \|_\infty + \frac{s}{2} Osc(f),$$

and as we did before we finally get, for

$$\nu(|f - \nu(f)|) \leq \frac{8\,\beta(s)}{\pi\,(1 - s)}\, \nu(|\nabla f|) + \frac{s+1}{4} Osc(f). \tag{9.2.10}$$

One can notice that $\frac{s+1}{4} < \frac{1}{2}$.

In the second case, we first have

$$\| \, |\nabla P_t f| \, \|_\infty \leq \frac{2}{\sqrt{\pi t}} \, (\mathrm{Var}_\nu(f))^{\frac{1}{2}} \, ,$$

so that finally

$$\nu(|f - \nu(f)|) \leq \frac{8\,\beta(s)}{\pi\,(1-s)} \, \nu(|\nabla f|) + \frac{s+1}{2} \, (\mathrm{Var}_\nu(f))^{\frac{1}{2}} \, . \qquad (9.2.11)$$

Looking at the proof of Proposition 9.2.5 we see that both situations yield exactly the same bound for the surface measure of a subset of probability less than or equal to $1/2$ i.e. the desired result.

If V is not smooth we may approximate ν by convolving with tiny Gaussian mollifiers, so that the convolved measures are still log-concave according to Prekopa-Leindler theorem and with smooth potentials. If X has distribution ν and G is a standard Gaussian vector independent of X, ν_ε will denote the distribution of $X + \varepsilon G$.

It is immediate that for a Lipschitz function f,

$$\mathbb{E}(|f(X + \varepsilon G) - f(X)|) \leq \varepsilon \, \| \, |\nabla f| \, \|_\infty \, \mathbb{E}(|G|)$$
$$\leq \varepsilon \sqrt{n} \, \| \, |\nabla f| \, \|_\infty \, ,$$

so that if ν satisfies (9.2.5), ν_ε also satisfies (9.2.5) with $\beta_\varepsilon(s) = \beta(s) + 2\varepsilon\sqrt{n}$. We may thus use the result for ν_ε and let ε go to 0.

Assume now that ν satisfies (9.2.6). It holds

$$\nu_\varepsilon(|f - \nu_\varepsilon(f)|) \leq \nu(|f - \nu(f)|) + 2\sqrt{n}\,\varepsilon \, \| \, |\nabla f| \, \|_\infty$$
$$\leq (\beta(s) + 2\sqrt{n}\,\varepsilon) \, \| \, |\nabla f| \, \|_\infty + s \, (\mathrm{Var}_\nu(f))^{\frac{1}{2}} \, .$$

But, assuming that $\nu(f) = 0$ to simplify the notation,

$$\mathrm{Var}_\nu(f) = \mathbb{E}(f^2(X + \varepsilon G)) + \mathbb{E}((f(X + \varepsilon G) - f(X))^2) + 2\mathbb{E}(f(X + \varepsilon G)(f(X)$$
$$- f(X + \varepsilon G)))$$
$$\leq \mathrm{Var}_{\nu_\varepsilon}(f) + (\mathbb{E}(f(X + \varepsilon G)))^2 + n\varepsilon^2 \, \| \, |\nabla f| \, \|_\infty^2$$
$$+ 2\varepsilon\sqrt{n} \, \| \, |\nabla f| \, \|_\infty \, \mathbb{E}(|f(X + \varepsilon G)|)$$
$$\leq \mathrm{Var}_{\nu_\varepsilon}(f) + 4n\,\varepsilon^2 \, \| \, |\nabla f| \, \|_\infty^2 + 2\varepsilon\sqrt{n} \, \| \, |\nabla f| \, \|_\infty \, \mathbb{E}(|f(X)|) \, .$$

In particular if f is 1-Lipschitz and bounded by M, we get

$$\nu_\varepsilon(|f - \nu_\varepsilon(f)|) \leq \left(\beta(s) + 2\varepsilon\sqrt{n} + s(4n\varepsilon^2 + 2M\varepsilon\sqrt{n})^{\frac{1}{2}} \right) + s \, \mathrm{Var}_{\nu_\varepsilon}(f)$$

i.e. using homogeneity, ν_ε also satisfies (9.2.6) with some β_ε, and we may conclude as before.

For the median case just remark that $m_{\nu_\varepsilon}(f)$ goes to $m_\nu(f)$ as ε goes to 0.   □

Using (9.1.3) we get similar bounds for $C_P(\nu)$.

Remark 9.2.8 Of course if ν satisfies a weak $(1, +\infty)$ Poincaré inequality with function $\beta(u)$, we obtain

$$C'_C(\nu) \leq \inf_{0 \leq s < \frac{1}{2}} \frac{4\beta(s)}{\pi \left(\frac{1}{2} - s\right)^2} .$$

Using that β is non increasing, it follows that

$$C'_C(\nu) \leq \frac{4}{\pi \left(\frac{1}{2} - s_\nu\right)^4} \quad \text{where} \quad \beta(s_\nu) = \frac{1}{\left(\frac{1}{2} - s_\nu\right)^2} .$$

We should write similar statements replacing the Oscillation by the Variance. In a sense (9.2.5) looks more universal since the control quantities in the right hand side do not depend (except the constants of course) of ν. It is thus presumably more robust to perturbations. We shall see this later. Also notice that both (9.2.5) and (9.2.6) agree when $s = 0$, which corresponds to an explicit bound for the Cheeger constant in E. Milman's theorem.

The advantage of (9.2.6) is that it looks like a deficit in the Cauchy-Schwarz inequality, since we may take s close to 1.   ◇

Notice that in the non weak framework, a not too far proof (i.e. with a functional flavor) is given in [1] theorem 1.10.

Remark 9.2.9 Notice that (9.2.5) is unchanged if we replace f by $f + a$ for any constant a, hence we may assume that $\inf f = 0$. Similarly it is unchanged if we multiply f by any M, hence we may choose $0 \leq f \leq 1$ with $Osc(f) = 1$.   ◇

9.2.3 Some Variations and Some Immediate Consequences

9.2.3.1 Immediate Consequences

Now using our trivial Proposition 9.2.2 (more precisely the version with the median) we immediately deduce:

Corollary 9.2.10 *For any log-concave probability measure ν,*

$$C'_C(\nu) \leq \inf_{0 < s < \frac{1}{4}} \frac{16\,\alpha_\nu^{-1}(s)}{\pi \,(1 - 4s)^2} \quad \text{and} \quad C_P(\nu) \leq \inf_{0 < s < \frac{1}{4}} \left(\frac{32\,\alpha_\nu^{-1}(s)}{\pi \,(1 - 4s)^2} \right)^2 .$$

The fact that the concentration profile controls the Poincaré or the Cheeger constant of a log-concave probability measure was also discovered by E. Milman in [44]. Another (simpler) proof, based on the semi-group approach was proposed by Ledoux [41]. We shall not recall Ledoux's proof, but tuning the constants in this proof furnishes worse constants than ours. The introduction of the weak version of the $(1, +\infty)$ Poincaré inequality is what is important here, in order to deduce such a control without any effort.

A similar and even better result was obtained by E. Milman in Theorem 2.1 of [45], namely

$$C'_C(v) \leq \frac{\alpha_v^{-1}(s)}{1 - 2s}$$

that holds for all $s < \frac{1}{2}$. The proof of this result lies on deep geometric results (like the Heintze-Karcher theorem) while ours is elementary. In a sense it is the semi-group approach alternate proof mentioned by E. Milman after the statement of its result.

Also notice that the previous corollary gives a new proof of Ledoux's result (9.1.4) but with a desperately worse constant. Indeed if we combine the previous bound for C'_C and some explicit estimate in the Gromov-Milman theorem (see respectively [9, 50]) i.e.

$$\alpha_v(r) \leq 16\, e^{-r/\sqrt{2C_P}} \quad \text{or} \quad \alpha_v(r) \leq e^{-r/3\sqrt{C_P}},$$

we obtain $(C'_C)^2(v) \leq m\, C_P(v)$ for some m. The reader will check that m is much larger than 36.

But actually one can recover Ledoux's result in a much more simple way: indeed

$$v(|f - v(f)|) \leq \sqrt{v(|f - v(f)|^2)} \leq \sqrt{C_P(v)}\, v^{\frac{1}{2}}(|\nabla f|^2) \leq \sqrt{C_P(v)}\, \|\, |\nabla f|\, \|_\infty$$

furnishes, thanks to Theorem 9.2.7,

Proposition 9.2.11 *If v is a log-concave probability measure,*

$$C'_C(v) \leq C_C(v) \leq \frac{16}{\pi} \sqrt{C_P(v)}.$$

Since $16/\pi < 6$ this result is slightly better than Ledoux's result recalled in (9.1.4).

Remark 9.2.12 Another immediate consequence is the following: since

$$|f(x) - f(a)| \leq \|\, |\nabla f|\, \|_\infty\, |x - a|$$

we have for all a,

$$v(|f - m_v(f)|) = \inf_b \int (|f(x) - b|)v(dx) \leq \int |f(x) - f(a)|v(dx) \leq \| \, |\nabla f| \, \|_\infty$$
$$\int |x - a|v(dx) \, .$$

Taking the infimum with respect to a in the right hand side, we thus have

$$v(|f - m_v(f)|) \leq \| \, |\nabla f| \, \|_\infty \int |x - m_v(x)|v(dx)$$

$$\leq \| \, |\nabla f| \, \|_\infty \int |x - v(x)|v(dx) \, . \tag{9.2.12}$$

A stronger similar result (credited to Kannan et al. [33]) is mentioned in [1, p. 11], namely

$$\mathrm{Var}_v(f) \leq 4 \, \mathrm{Var}_v(x) \, v(|\nabla f|^2) \, , \tag{9.2.13}$$

where $\mathrm{Var}_v(x) = v(|x - v(x)|^2)$.
    According to (9.2.12)

$$C'_C(v) \leq \frac{16}{\pi} \int |x - m_v(x)| \, dv \leq \frac{16}{\pi} \mathrm{Var}_v^{1/2}(x) \, .$$

In particular Since $16/\pi < 5, 2$, a consequence is the bound $C_P(v) \leq 484 \, \mathrm{Var}_v(x)$.
    Notice that this result contains "diameter" bounds, i.e. if the support of v is compact with diameter D, $C'_C(v) \leq \frac{16 D}{\pi}$. In the isotropic situation one gets $C'_C(v) \leq \frac{16 \sqrt{n}}{\pi}$. $\diamond$

Remark 9.2.13 Consider an isotropic log-concave random vector X with distribution v. If f is a Lipschitz function we have for all a,

$$v(|f - a|) \leq \mathbb{E}\left[\left|f(X) - f\left(\sqrt{n} \, \frac{X}{|X|}\right)\right|\right] + \mathbb{E}\left[\left|a - f\left(\sqrt{n} \, \frac{X}{|X|}\right)\right|\right]$$

$$\leq \| \, |\nabla f| \, \|_\infty \, \mathbb{E}\left[\left||X| - \sqrt{n}\right|\right] + \mathbb{E}\left[\left|a - f\left(\sqrt{n} \, \frac{X}{|X|}\right)\right|\right] \, . \tag{9.2.14}$$

Hence, if we choose $a = \mathbb{E}\left[f\left(\sqrt{n} \, \frac{X}{|X|}\right)\right]$ and if we denote by v_{angle} the distribution of $X/|X|$ we obtain

$$v(|f - m_v(f)|) \leq 2 \, \| \, |\nabla f| \, \|_\infty \, \left(\mathbb{E}\left[\left||X| - \sqrt{n}\right|\right] + \sqrt{n} \, \sqrt{C_P(v_{angle})}\right) \, .$$
$$\tag{9.2.15}$$

This shows that the Cheeger constant of ν is completely determined by the concentration of the radial part of X around $\sqrt{n}$ (which is close to its mean), and the Poincaré constant (we should also use the Cheeger constant) of $X/|X|$.

In particular, if ν is spherically symmetric, the distribution of $X/|X|$ is the uniform measure on the sphere S^{n-1} which is known to satisfy (provided $n \geq 2$) a Poincaré inequality with Poincaré constant equal to $1/n$ for the usual Euclidean gradient (not the Riemannian gradient on the sphere). In addition, in this situation its known that

$$\mathbb{E}\left[\left||X| - \sqrt{n}\right|\right] \leq 1 \qquad \text{see [11] formula (6)}$$

so that we get that $C'_C(\nu) \leq \frac{64}{\pi}$ for an isotropic radial log-concave probability measure. Since $\frac{16}{\pi}\sqrt{12} < \frac{64}{\pi}$, this result is worse than the one proved in [11] telling that $C_P(\nu) \leq 12$ so that $C'_C(\nu) \leq \frac{16}{\pi}\sqrt{12}$ thanks to proposition 9.2.11 (12 may replace the original 13 thanks to a remark by Huet [32]). Actually, applying (9.2.13) in dimension 1, it seems that we may replace 12 by 4. $\qquad\qquad\diamond$

9.2.3.2   Variations: The $\mathbb{L}^2$ Framework

In some situations it is easier to deal with variances. We recalled that any (nice) absolutely continuous probability measure satisfies a weak $(2, 2)$ Poincaré inequality.

Theorem 9.2.14 *Let ν a log-concave probability measure satisfying the weak $(2, 2)$ Poincaré inequality, for some $s < \frac{1}{6}$,*

$$Var_\nu(f) \leq \beta(s)\nu(|\nabla f|^2) + s \, Osc^2(f) \,.$$

Then

$$C'_C(\nu) \leq \frac{4\sqrt{\beta(s)\ln 2}}{1 - 6s} \quad \text{and} \quad C_P(\nu) \leq 4\,(C'_C(\nu))^2 \,.$$

Proof We start with the following which is also due to Ledoux [39]: if ν is log-concave, then for any subset A,

$$\sqrt{t}\,\nu_s(\partial A) \geq \nu(A) - \nu\left((P_t\mathbf{1}_A)^2\right) \,.$$

But

$$\nu\left((P_t\mathbf{1}_A)^2\right) = Var_\nu(P_t\mathbf{1}_A) + (\nu(P_t\mathbf{1}_A))^2 = Var_\nu(P_t\mathbf{1}_A) + \nu^2(A) \,.$$

Define $u(t) = \mathrm{Var}_v(P_t \mathbf{1}_A)$. Using the semi-group property and the weak Poincaré inequality it holds

$$\frac{d}{dt}u(t) = -2\,v(|\nabla P_t \mathbf{1}_A|^2) \le \frac{-2}{\beta(s)}\,(u(t) - s)$$

since $Osc(P_t \mathbf{1}_A) \le 1$. Using Gronwall's lemma we thus obtain

$$\mathrm{Var}_v(P_t \mathbf{1}_A) \le e^{-2t/\beta(s)}\,v(A) + s\,\left(1 - e^{-2t/\beta(s)}\right)$$

so that finally, if $v(A) = 1/2$ we get

$$\sqrt{t}\,v_s(\partial A) \ge \left(1 - e^{-2t/\beta(s)}\right)\left(\frac{1}{2} - s\right) - \frac{1}{4}.$$

Choose $t = \beta(s)\,\ln 2$. The right hand side in the previous inequality becomes $\frac{1}{8}(1 - 6s)$, hence

$$\mathrm{Is}_v(1/2) \ge \frac{1 - 6s}{8\,\sqrt{\beta(s)\,\ln 2}}.$$

Hence the result arguing as in the previous proof. $\qquad\qquad\qquad\qquad\qquad\square$

9.2.3.3 Other Consequences: Reducing the Support

All the previous consequences are using either the weak or the strong $(1, +\infty)$ Poincaré inequality. The next consequence will use the full strength of what precedes.

Pick some Borel subset A. Let $a \in \mathbb{R}$ and f be a smooth function. Then, provided $\inf f \le a \le \sup f$,

$$v(|f - a|) \le \int_A |f - a| dv + (1 - v(A))\,\|\,f - a\,\|_\infty$$

$$\le \int_A |f - a| dv + (1 - v(A))\,Osc(f).$$

Denote $dv_A = \frac{\mathbf{1}_A}{v(A)}\,dv$ the restriction of v to A. Choosing $a = m_{v_A}(f)$ we have

$$v(|f - m_v(f)|) \le v(|f - a|) \le v(A)\,v_A(|f - a|) + (1 - v(A))\,Osc(f)$$

$$\le v(A)\left(\beta_{v_A}(u)\,\|\,|\nabla f|\,\|_\infty + u\,Osc(f)\right) + (1 - v(A))\,Osc(f)$$

$$\le v(A)\,\beta_{v_A}(u)\,\|\,|\nabla f|\,\|_\infty + (1 - v(A)(1 - u))\,Osc(f),$$

provided ν_A satisfies some $(1, +\infty)$ Poincaré inequality. We can improve the previous bound, if ν_A satisfies some Cheeger inequality and get

$$\nu(|f - m_\nu(f)|) \leq \nu(A)\, C'_C(\nu_A)\, \nu(|\nabla f|) + (1 - \nu(A))\, Osc(f).$$

Hence, applying Theorem 9.2.7 or Proposition 9.2.5 we have:

Proposition 9.2.15 *Let ν be a log-concave probability measure, A be any subset with $\nu(A) > \frac{1}{2}$ and $d\nu_A = \frac{1_A}{\nu(A)}\, d\nu$ be the (normalized) restriction of ν to A. Then*

(1)

$$C'_C(\nu) \leq \frac{\nu(A)\, C'_C(\nu_A)}{2\nu(A) - 1},$$

(2) *if ν_A satisfies a $(1, +\infty)$ weak Poincaré inequality with rate β_{ν_A}, then ν satisfies a $(1, +\infty)$ weak Poincaré inequality with rate β_ν satisfying for $u < 1 - \frac{1}{2\nu(A)}$,*

$$\beta_\nu(u) \leq \nu(A)\, \beta_{\nu_A}\left(1 - \frac{1 - u}{\nu(A)}\right)$$

so that

$$C'_C(\nu) \leq \frac{4\nu(A)\, \beta_{\nu_A}(u)}{\pi\, ((1 - u)\nu(A) - \frac{1}{2})^2}.$$

Remark 9.2.16 A similar result is contained in [44] namely if K is some convex body,

$$C'_C(\nu) \leq \frac{1}{\nu^2(K)}\, C'_C(\nu_K).$$

This result is similar when $\nu(K)$ is close to 1, but it requires the convexity of K. Convexity of K ensures that ν_K is still log-concave. We shall come back to this point later. Of course our result does not cover the situation of sets with small measure.

In addition a careful look at the proof of Lemma 5.2 in [44] indicates that when $\nu(A) > 1/2$, the convexity of A is not required. (5.2) in [44] thus furnishes the same result as (1) in the previous Proposition up to the pre-factor $\nu(A)$.   ◇

This result enables us to reduce the study of log-concave probability measures to the study of compactly supported distributions, arguing as follows. Let Z be a random variable (in $\mathbb{R}^n$) with log concave distribution ν. We may assume without loss of generality that Z is centered. Denote by $\sigma^2(Z)$ the largest eigenvalue of the covariance matrix of Z.

- l^2 *truncation.*

Thanks to Chebyshev inequality, for $a > 1$,

$$\mathbb{P}(|Z| > a \, \sigma(Z) \, \sqrt{n}) \leq \frac{1}{a^2} .$$

According to Proposition 9.2.15, if $a > \sqrt{2}$,

$$C'_C(Z) \leq \frac{a^2}{a^2 - 2} \, C'_C(Z(a))$$

where $Z(a)$ is the random variable $\mathbf{1}_{Z \in K_a} Z$ supported by $K_a = B(0, a \, \sigma(Z) \, \sqrt{n})$ with distribution $\frac{\mathbf{1}_{K_a}}{\nu(K_a)} \nu$. Of course the new variable $Z(a)$ is not necessarily centered, but we may, without changing the Poincaré constant(s), consider the variable $\bar{Z}(a) = Z(a) - \mathbb{E}(Z(a))$. It is easily seen that

$$|\mathbb{E}(Z_i(a))| \leq |\mathbb{E}(-Z_i \, \mathbf{1}_{K_a^c}(Z))| \leq \frac{\mathrm{Var}^{1/2}(Z_i)}{a} \leq \frac{\sigma(Z)}{a} ,$$

i.e.

$$\sum_{i=1}^{n} |\mathbb{E}(Z_i(a))|^2 \leq \frac{n\sigma^2(Z)}{a^2}$$

so that $\bar{Z}(a)$ is centered and supported by $B(0, \sqrt{a^2 + (1/a^2)} \, \sigma(Z) \, \sqrt{n})$.

Notice that for all i,

$$\mathrm{Var}(Z_i) \geq \mathrm{Var}(\bar{Z}_i(a)) \geq \mathrm{Var}(Z_i) \left(1 - \frac{\kappa}{a} - \frac{1}{a^2} \right) , \tag{9.2.16}$$

where κ is the universal Khinchine constant, i.e. satisfies

$$\mu(z^4) \leq \kappa^2 \, (\mu(z^2))^2$$

for all log concave probability measure on $\mathbb{R}$. According to the one dimensional estimate of Bobkov [10, corollary 4.3] we already recalled, we know that $\kappa \leq 7$.

The upper bound in (9.2.16) is immediate, while for the lower bound we use the following reasoning:

$$\begin{aligned}
\mathrm{Var}(\bar{Z}_i(a)) &= \mathbb{E}(Z_i^2 \, \mathbf{1}_{K_a}(Z)) - (\mathbb{E}(Z_i(a)))^2 \\
&\geq \mathrm{Var}(Z_i) - \mathbb{E}(Z_i^2 \, \mathbf{1}_{K_a^c}(Z)) - (\mathbb{E}(Z_i(a)))^2 \\
&\geq \mathrm{Var}(Z_i) - \frac{1}{a} \, (\mathbb{E}(Z_i^4))^{\frac{1}{2}} - \frac{\mathrm{Var}(Z_i)}{a^2}
\end{aligned}$$

according to the previous bound on the expectation. We conclude by using Khinchine inequality.

Similar bounds are thus available for $\sigma^2 \bar{Z}(a)$ in terms of $\sigma^2(Z)$.

Remark 9.2.17 Though we gave explicit forms for all the constants, they are obviously not sharp.

For instance we used the very poor Chebyshev inequality for reducing the support of ν to some Euclidean ball, while much more precise concentration estimates are known. For an isotropic log-concave random variable (vector) Z, it is known that its distribution satisfies some "concentration" property around the sphere of radius $\sqrt{n}$. We shall here recall the best known result, due to Lee and Vempala [42], improving on Guédon and Milman [31] (also see [16] chapter 13 for an almost complete overview of the state of the art):

Theorem 9.2.18 (Lee and Vempala) *Let Z be an isotropic log-concave random vector in $\mathbb{R}^n$. Then there exist some universal positive constants C and c such that for all $t > 0$,*

$$\mathbb{P}(|\,|Z| - \sqrt{n}\,| \geq t\sqrt{n}) \leq C \exp(-c\sqrt{n}\min\{t, t^2\}) . \qquad \diamond$$

- l^∞ truncation.

Instead of looking at Euclidean balls we shall look at hypercubes, i.e. l^∞ balls. We assume that Z is centered.

According to Prekopa -Leindler theorem again, we know that the distribution ν_1 of Z_1 is a log-concave distribution with variance $\lambda_1^2 = \text{Var}(Z_1)$. Hence ν_1 satisfies a Poincaré inequality with $C_P(\nu_1) \leq 12\lambda_1^2$ according to [10] proposition 4.1. Of course a worse bound was obtained in Remark 9.2.13.

Using Proposition 4.1 in [14] (see Lemma 9.3.7 in the next section), we have for all $1 \geq \epsilon > 0$

$$\nu_1(z_1 > a_1) \leq \frac{4-\epsilon}{\epsilon} e^{-\frac{(2-\epsilon)a_1}{\sqrt{C_P(\nu_1)}}} \leq \frac{4-\epsilon}{\epsilon} e^{-\frac{(2-\epsilon)a_1}{2\sqrt{3}\lambda_1}} . \qquad (9.2.17)$$

Of course changing Z in $-Z$ we get a similar bound for $\nu_1(z_1 < -a_1)$. Choosing $a_1 = \frac{2\sqrt{3}}{2-\epsilon}\lambda_1 a \ln n$ for some $a > 0$, we get

$$\nu_1(|z_1| > a_1) \leq 2\frac{4-\epsilon}{\epsilon n^a} .$$

Hence if

$$K_a = \left\{ \max_{i=1,\ldots,n} |z_i| < \frac{2\sqrt{3}}{2-\epsilon}\sigma(Z) a \ln n \right\} \qquad (9.2.18)$$

we have

$$\nu(K_a) \geq 1 - 2\frac{4 - \epsilon}{\epsilon n^{a-1}}.$$

Using Proposition 9.2.15 we have

Proposition 9.2.19 *For $n \geq 2$, K_a being defined by (9.2.18), we have*

$$C_C'(\nu) \leq \frac{n^{a-1}}{n^{a-1} - (8 - 2\epsilon)\epsilon^{-1}} \, C_C'(\nu_{K_a}).$$

Again we may center ν_{K_a}, introducing a random vector $\bar{Z}$ with distribution equal to the re-centered ν_{K_a}. This time it is easily seen that

$$(1 - \varepsilon(n)) \operatorname{Var}(Z_i) \leq \operatorname{Var}(\bar{Z}_i) \leq \operatorname{Var}(Z_i),$$

with $\varepsilon(n) \to 0$ as $n \to +\infty$.

Notice that we have written a more precise version of Latala's deviation result [38]

Theorem 9.2.20 *Let Z be an isotropic log-concave random vector in $\mathbb{R}^n$. Then for $n \geq 2$, $1 \geq \epsilon > 0$,*

$$\mathbb{P}\left(\max_{i=1,\dots,n} |Z_i| \geq t \ln n\right) \leq \frac{(8 - 2\epsilon)\epsilon^{-1}}{n^{\frac{(2-\epsilon)t}{2\sqrt{3}} - 1}} \quad \text{for } t \geq \frac{2\sqrt{3}}{2 - \epsilon}.$$

Let us give another direct application. ν is said to be unconditional if it is invariant under the transformation $(x_1, \dots, x_n) \mapsto (\varepsilon_1 x_1, \dots, \varepsilon_n x_n)$ for any n-uple $(\varepsilon_1, \dots, \varepsilon_n) \in \{-1, +1\}^n$. Defining K_a as in (9.2.18) (with $\sigma(Z) = 1$ here), the restricted measure ν_{K_a} is still unconditional.

So we may apply Theorem 1.1 in [23] (and its proof in order to find an explicit expression of the constant) saying that ν_{K_a} satisfies some weighted Poincaré inequality

$$\operatorname{Var}_{\nu_{K_a}}(f) \leq (4\sqrt{3} + 1)^2 \sum_{i=1}^{n} \nu_{K_a}\left(\nu_{K_a}^{i-1}(x_i^2) (\partial_i f)^2\right),$$

where $\nu_{K_a}^j$ denotes the conditional distribution of ν_{K_a} w.r.t. the sigma field generated by $(x_1, \dots, x_j)$. Actually, up to the pre-constant, we shall replace the weight $\nu_{K_a}^{i-1}(x_i^2)$ by the simpler one $x_i^2 + \nu_{K_a}(x_i^2)$ according to [35].

In all cases, since x_i is ν_{K_a} almost surely bounded by a constant times $\ln(n)$, we have obtained thanks to Proposition 9.2.19 the following result originally due to Klartag [34]

Proposition 9.2.21 *There exists an universal constant c such that, if ν is an isotropic and unconditional log-concave probability measure,*

$$C_P(\nu) \leq c \, \max(1, \ln^2 n) \,.$$

Of course what is needed here is the invariance of ν_{K_a} with respect to some symmetries (see [6, 23]). But it is not easy to see how ν_{K_a} inherits such a property satisfied by the original ν, except in the unconditional case.

9.3 Transference of the Poincaré Inequality

9.3.1 Transference via Absolute Continuity

If μ and ν are probability measures it is well known (Holley-Stroock argument) that

$$C_P(\nu) \leq \| \frac{d\nu}{d\mu} \|_\infty \| \frac{d\mu}{d\nu} \|_\infty \, C_P(\mu) \,,$$

the same being true with the same pre-factor if we replace C_P by C_C or C'_C. Similar results are known for weak $(2, 2)$ Poincaré inequalities too. In this section we shall give several transference principles allowing us to reinforce this result, at least under some curvature assumption.

Such transference results have been obtained in [7] using transference results for the concentration profile.

We first recall the statement of Proposition 2.2 in [7], in a simplified form:

Proposition 9.3.1 (Barthe-Milman) *Recall that the concentration profile α_ν is defined in Proposition 9.2.2.*
Assume that for some $1 < p \leq +\infty$,

$$\int \left| \frac{d\nu}{d\mu} \right|^p d\mu = M_p^p < +\infty \,.$$

Then if $q = \frac{p}{p-1}$, for all $r > 0$,

$$\alpha_\nu(r) \leq 2M_p \, \alpha_\mu^{1/q}(r/2) \,.$$

We may use this result together with Corollary 9.2.10 to deduce

Corollary 9.3.2 *Under the assumptions of Proposition 9.3.1, if ν is log-concave,*

$$C'_C(\nu) \leq \inf_{0 < s < \frac{1}{4}} \frac{32 \, \alpha_\mu^{-1}((s/2M_p)^q)}{\pi (1 - 4s)^2} \,.$$

Notice that, using the equivalence results between concentration and Poincaré inequality of [44], Barthe and E. Milman have shown a comparison result in [7] Theorem 2.7, namely:

Corollary 9.3.3 (Barthe-Milman) *If ν is log-concave and satisfies the assumptions of Proposition 9.3.1, there exists some universal constant c such that*

$$C_P(\nu) \le c \, \frac{p}{p-1} \, (1 + \ln(M_p)) \, C_P(\mu) \,.$$

The goal of the remaining part of this subsection is to derive similar results using our approach. This allows us to obtain more general results since we shall consider only weak Poincaré inequalities. We can also trace the constants. However, in the log-concave situation, despite the explicit constants, our estimates have a worse dependence in M_p. Notice that we will also consider the relative entropy instead of the $\mathbb{L}^p$ norm. The latter has also been done in [46, Theorem 5.7]. Comparison of this Theorem and the results below seems difficult.

Theorem 9.3.4 *Let ν and μ be two probability measures.*

(1) *If for some $1 < p$,*

$$\int \left| \frac{d\nu}{d\mu} \right|^p d\mu = M_p^p < +\infty \,,$$

then

$$\beta_\nu(s) \le M_p^{p/(p-1)} s^{-1/(p-1)} C(\mu) \,,$$

where $C(\mu)$ can be chosen as $C'_C(\mu)$, $1/B_{1,\infty}(\mu)$, $C_C(\mu)$ or $\sqrt{C_P(\mu)}$.
In particular if ν is log-concave

$$C'_C(\nu) \le D \, C(\mu) \, M_p^{p/(p-1)}$$

where $D = \dfrac{16 \, (p+1)^{1/(p-1)}}{\pi \, (\frac{1}{2} - \frac{1}{p+1})^2}$.

(2) *Let ν be log-concave. If the relative entropy $D(\nu\|\mu) := \int \ln(d\nu/d\mu) \, d\nu$ is finite, then with the same $C(\mu)$ and any $u < \frac{1}{2}$,*

$$C'_C(\nu) \le \frac{4 \left(e^{2 \max(1, D(\nu\|\mu))/u} - 1 \right)}{\pi \, (\frac{1}{2} - u)^2} \, C(\mu) \,.$$

Proof Let f be a smooth function. It holds, provided $\inf f \leq a \leq \sup f$,

$$\nu(|f - m_\nu(f)|) \leq \nu(|f - a|)$$

$$\leq \int |f - a| \frac{d\nu}{d\mu} d\mu$$

$$\leq K \int \mathbf{1}_{d\nu/d\mu \leq K} |f - a| d\mu + Osc(f) \int \mathbf{1}_{d\nu/d\mu > K} d\nu$$

$$\leq K \int |f - a| d\mu + Osc(f) \int \mathbf{1}_{d\nu/d\mu > K} d\nu. \qquad (9.3.1)$$

In order to control the last term we may use Hölder (or Orlicz-Hölder) inequality. In the usual $\mathbb{L}^p$ case we have, using Markov inequality

$$\int \mathbf{1}_{d\nu/d\mu > K} d\nu \leq M_p \, \mu^{\frac{1}{q}} \left(\frac{d\nu}{d\mu} > K \right)$$

$$\leq M_p \frac{M_p^{p/q}}{K^{p/q}} = \frac{M_p^{1+(p/q)}}{K^{p/q}} = \frac{M_p^p}{K^{p-1}},$$

so that choosing $K = M_p^{p/(p-1)} u^{-1/(p-1)}$ we have obtained

$$\nu(|f - m_\nu(f)|) \leq M_p^{p/(p-1)} u^{-1/(p-1)} \int |f - a| d\mu + u \, Osc(f).$$

Then using either $a = m_\mu(f)$ or $a = \mu(f)$ we get the desired result, with $C(\mu) = C'_C(\mu)$ or $1/B_{1,\infty}(\mu)$ or $C_C(\mu)$ or $\sqrt{C_P(\mu)}$. If in addition ν is log-concave it follows

$$C'_C(\nu) \leq \frac{16}{\pi \left(\frac{1}{2} - u \right)^2 u^{1/(p-1)}} M_p^{\frac{p}{p-1}} C(\mu).$$

It remains to optimize in u for $0 < u < \frac{1}{2}$ and elementary calculations show that the maximum is attained for $u = 1/(p+1)$.

Starting with (9.3.1) we may replace Hölder's inequality by Orlicz-Hölder's inequality for a pair of conjugate Young functions θ and θ^*, that is

$$\int \mathbf{1}_{d\nu/d\mu > K} d\nu \leq 2 \, \| \frac{d\nu}{d\mu} \|_{\mathbb{L}_\theta(d\mu)} \| \mathbf{1}_{d\nu/d\mu > K} \|_{\mathbb{L}_{\theta^*}(d\mu)}.$$

Here the chosen Orlicz norm is the usual gauge (Luxemburg) norm, i.e.

$$\| h \|_{L_\theta(\mu)} = \inf\{b \geq 0 \quad s.t. \quad \int \theta(|h|/b) \, d\mu \leq 1\},$$

and recall that for any $\lambda > 0$,

$$\| h \|_{L_\theta(\mu)} \leq \frac{1}{\lambda} \max\left(1 , \int \theta(\lambda|h|)d\mu\right). \tag{9.3.2}$$

For simplicity we will perform the calculations only for the pair of conjugate Young functions

$$\theta^*(u) = e^{|u|} - 1 \quad , \quad \theta(u) = (|u| \ln(|u|) + 1 - |u|).$$

According to what precedes

$$\| \frac{dv}{d\mu} \|_{\mathbb{L}_\theta(d\mu)} \leq \max(1 , D(v||\mu))$$

and

$$\int \theta(\mathbf{1}_{dv/d\mu > K}/b) \, d\mu = (e^{1/b} - 1) \, \mu\left(\frac{dv}{d\mu} > K\right)$$

$$\leq (e^{1/b} - 1) \frac{1}{K},$$

the final bound being non optimal since we only use $\mathbf{1}_{dv/d\mu > K} \leq \frac{1}{K} \frac{dv}{d\mu}$ and not the better integrability of the density. Using the best integrability does not substantially improve the bound. We thus obtain

$$\| \mathbf{1}_{dv/d\mu > K} \|_{\mathbb{L}_{\theta^*(d\mu)}} = \frac{1}{\ln(1 + K)}$$

and we may conclude as before.   □

The method we used in the previous proof is quite rough. One can expect (in particular in the entropic case) to improve upon the constants, using more sophisticated tools. This is the goal of the next result.

Theorem 9.3.5 *Let v and μ be two probability measures.*

(1) *If for some $1 < p \leq 2$,*

$$\int \left|\frac{dv}{d\mu}\right|^p d\mu = M_p^p < +\infty,$$

then

$$\frac{1}{B_{1,\infty}(v)} \leq \left(\frac{p}{p-1} 8^{\frac{p}{(p-1)}} C_P(\mu)\right)^{\frac{1}{2}} M_p.$$

If in addition v is log-concave

$$C_C'(v) \leq \frac{16\sqrt{p}}{\pi\sqrt{p-1}} 8^{\frac{p}{2(p-1)}} \sqrt{C_P(\mu)}\, M_p \,.$$

(2) *If the relative entropy $D(v\|\mu) := \int \ln(dv/d\mu)\, dv$ is finite, then*

$$\frac{1}{B_{1,\infty}(v)} \leq 2\sqrt{C_P(\mu)} \max\left(1, 3e^{\sqrt{C_P(\mu)}}\right) \max\left(1, D(v\|\mu)\right).$$

If in addition v is log-concave

$$C_C'(v) \leq \frac{32}{\pi} \sqrt{C_P(\mu)} \max\left(1, 3e^{\sqrt{C_P(\mu)}}\right) \max\left(1, D(v\|\mu)\right).$$

Proof Let f be a smooth function. We have

$$\frac{1}{2} v(|f - v(f)|) \leq v(|f - m_v(f)|) \leq v(|f - \mu(f)|)$$

$$\leq \left(\int |f - \mu(f)|^q \, d\mu\right)^{1/q} \left(\int \left|\frac{dv}{d\mu}\right|^p \, d\mu\right)^{1/p}. \quad (9.3.3)$$

Now we can use the results in [21], in particular the proof of Theorem 1.5 (see formulae 2.3, 2.7 and 2.8 therein) where the following is proved:

Lemma 9.3.6 *For all $q \geq 2$, it holds*

$$\int |f - \mu(f)|^q \, d\mu \leq q\, 8^q\, C_P(\mu) \int |f - \mu(f)|^{q-2} |\nabla f|^2 \, d\mu \,.$$

It is at this point that we need $p \leq 2$. Using Hölder inequality we deduce

$$\int |f - \mu(f)|^q \, d\mu \leq (q\, 8^q\, C_P(\mu))^{\frac{q}{2}} \int |\nabla f|^q \, d\mu \,. \quad (9.3.4)$$

It follows

$$v(|f - m_v(f)|) \leq (q\, 8^q\, C_P(\mu))^{\frac{1}{2}} M_p \, \| |\nabla f| \|_\infty \,.$$

Since v is log-concave, we get the desired result, using Theorem 9.2.7.

Now we turn to the second part of the Theorem which is based on Proposition 4.1 in [14] we recall now:

Lemma 9.3.7 (Bobkov-Ledoux) *If g is Lipschitz,*

$$\mu(e^{\lambda g}) \leq \left(\frac{2 + \lambda C_P^{\frac{1}{2}}(\mu) \; \| \, |\nabla g| \, \|_\infty}{2 - \lambda C_P^{\frac{1}{2}}(\mu) \; \| \, |\nabla g| \, \|_\infty} \right) e^{\lambda\mu(g)} ,$$

provided $2 > \lambda C_P^{\frac{1}{2}}(\mu) \; \| \, |\nabla g| \, \|_\infty > 0$.

Hence, in (9.3.3), we may replace the use of Hölder inequality by the one of the Orlicz-Hölder inequality

$$\nu(|f - m_\nu(f)|) \leq 2 \; \| \, f - \mu(f) \, \|_{L_\theta(\mu)} \| \frac{d\nu}{d\mu} \|_{L_{\theta^*}(\mu)} \; .$$

Again we are using the pair of conjugate Young functions

$$\theta(u) = e^{|u|} - 1 \quad , \quad \theta^*(u) = |u| \ln(|u|) + 1 - |u| .$$

Without loss of generality we can first assume that f (hence $f - \mu(f)$) is 1-Lipschitz.

We then apply (9.3.2) and Lemma 9.3.7 with $g = |f - \mu(f)|$ and $\lambda = 1/\sqrt{C_P(\mu)}$. Since

$$\mu(|g|) \leq \mu^{\frac{1}{2}}(|g|^2) \leq C_P^{\frac{1}{2}}(\mu) \, \mu^{\frac{1}{2}}(|\nabla g|^2) \leq C_P^{\frac{1}{2}}(\mu) \; \| \, |\nabla g| \, \|_\infty ,$$

we obtain

$$\| \, f - \mu(f) \, \|_{L_\theta(\mu)} \leq \sqrt{C_P(\mu)} \, \max \left(1 , 3 e^{\sqrt{C_P(\mu)}} \right) \; \| \, |\nabla f| \, \|_\infty \; .$$

Similarly

$$\| \frac{d\nu}{d\mu} \|_{L_{\theta^*}(\mu)} \leq \max \left(1 , D(\nu\|\mu) \right) .$$

Again we conclude thanks to Theorem 9.2.7. □

As usual we should try to optimize in p for p going to 1 depending on the rate of convergence of M_p to 1.

We cannot really compare both theorems, since the constant $C(\mu)$ in the first theorem can take various values, while it is the usual Poincaré constant in the second theorem. In the $\mathbb{L}^p$ case, the first theorem seems to be better for large p's and the second one for small p's. In the entropic case, the second one looks better.

Finally, let us recall the following beautiful transference result between two log-concave probability measures proved in [44] and which is a partial converse of the previous results

Proposition 9.3.8 (E. Milman) *Let μ and v be two log-concave probability measures. Then*

$$C_C'(v) \leq \left\|\frac{d\mu}{dv}\right\|_\infty^2 C_C'(\mu).$$

9.3.2 Transference Using Distances

As shown in [44, Theorem 5.5], the ratio of the Cheeger constants of two log-concave probability measures is controlled by their total variation distance (which is the half of the W_0 Wasserstein distance). Recall the equivalent definitions of the total variation distance

Definition 9.3.9 If μ and v are probability measures the total variation distance $d_{TV}(\mu, v)$ is defined by one of the following equivalent expressions

$$\begin{aligned}
d_{TV}(\mu, v) &:= \frac{1}{2} \sup_{\|f\|_\infty \leq 1} |\mu(f) - v(f)| \\
&= \sup_{0 \leq f \leq 1} |\mu(f) - v(f)| \\
&= \inf\{\mathbb{P}(X \neq Y) \, ; \, \mathcal{L}(X) = \mu, \, \mathcal{L}(Y) = v\}
\end{aligned}$$

which is still equal to

$$\frac{1}{2} \int \left|\frac{d\mu}{dx} - \frac{dv}{dx}\right| dx$$

when μ and v are absolutely continuous w.r.t. Lebesgue measure.

The second equality is immediate just noticing that for $0 \leq f \leq 1$, $\mu(f - \frac{1}{2}) - v(f - \frac{1}{2}) = \mu(f) - v(f)$ and that $\|f - \frac{1}{2}\|_\infty \leq \frac{1}{2}$.

More precisely one can show the following explicit result, due to Milman [44, Theorem 5.5], except the bound for the constant κ:

Theorem 9.3.10 (E. Milman) *Let $\mu(dx) = e^{-V(x)}dx$ and $v(x) = e^{-W(x)}dx$ be two log-concave probability measures. If $d_{TV}(\mu, v) = 1 - \varepsilon$, for some $\varepsilon > 0$, then*

$$C_C'(v) \leq \frac{\kappa}{\varepsilon^2} \left(1 \vee \ln\left(\frac{1}{\varepsilon}\right)\right) C_C'(\mu),$$

for some universal constant κ one can choose equal to $(192 \, e/\pi)$.

Proof We give a short proof (adapted from [44]) but that does not use concentration results.

First if Z_μ and Z_ν are random variables with respective distribution μ and ν, and $\lambda > 0$ the total variation distance between the distributions of λZ_μ and λZ_ν is unchanged, hence still equal to $1 - \varepsilon$. Choosing $\lambda = 1/\sqrt{C_P(\mu)}$ we may thus assume that $C_P(\mu) = 1$.

Introduce the probability measure $\theta(dx) = \frac{1}{\varepsilon} \min(e^{-V(x)}, e^{-W(x)}) \, dx$ which is still log-concave and such that $d\theta/d\mu$ and $d\theta/d\nu$ are bounded by $1/\varepsilon$. Using Proposition 9.3.8 we first have $C'_C(\nu) \leq \frac{1}{\varepsilon^2} C'_C(\theta)$. Next $D(\theta\|\mu) \leq \ln(1/\varepsilon)$ so that using Theorem 9.3.5 (2) we have $C'_C(\theta) \leq \frac{96\,e}{\pi} \max(1, \ln(1/\varepsilon))$. It follows that

$$C'_C(\nu) \leq \frac{96\,e}{\pi\,\varepsilon^2} \max(1, \ln(1/\varepsilon))$$

provided $C_P(\mu) = 1$. It remains to use $C'_C(\lambda Z_\nu) = \lambda\, C'_C(\nu) = (1/\sqrt{C_P(\mu)})\, C'_C(\nu)$ and $\sqrt{C_P(\mu)} \leq 2 C'_C(\mu)$ to get the result. $\qquad\square$

But since the $(1, +\infty)$ Poincaré inequality deals with Lipschitz functions, it is presumably more natural to consider the W_1 Wasserstein distance

$$W_1(\nu, \mu) := \sup_{f\ 1-Lipschitz} \int f\,(d\mu - d\nu) = \inf_{\mathcal{L}(X)=\mu, \mathcal{L}(Y)=\nu} \mathbb{E}(|X - Y|)\,.$$

Actually we have the following:

Proposition 9.3.11 *Assume that μ and ν satisfy weak $(1, +\infty)$ Poincaré inequalities with respective rates β_μ and β_ν. Then for all $s > 0$,*

$$\beta_\nu(s) \leq \beta_\mu(s) + 2\,W_1(\nu, \mu)\,.$$

Proof Let f be 1-Lipschitz. We have

$$\nu(|f - \nu(f)|) \leq \nu(|f - \mu(f)|) + |\nu(f) - \mu(f)|$$
$$\leq \mu(|f - \mu(f)|) + W_1(\nu, \mu) + |\nu(f) - \mu(f)|$$
$$\leq \beta_\mu(s) + 2\,W_1(\nu, \mu) + s\,\mathrm{Osc}(f)\,.$$

Here we used that $|\mu(|f - \mu(f)|) - \nu(|f - \mu(f)|)| \leq W_1(\mu, \nu)$ since $|f - \mu(f)|$ is still 1-Lipschitz and that $|\nu(f) - \mu(f)| \leq W_1(\nu, \mu)$ too, i.e. the W_1 control for two different functions. $\qquad\square$

We immediately deduce:

Corollary 9.3.12 *Let ν be a log-concave probability measure. Then for all μ,*

$$C'_C(\nu) \leq \frac{16}{\pi}\,(C_C(\mu) + 2\,W_1(\nu, \mu))\,.$$

Remark 9.3.13 Comparison of Poincaré constants using Wasserstein distance is not new. In [46] Theorem 5.5, E. Milman already derived a result in this direction. However the formulation of the previous Corollary looks simpler and the constant is explicit.   ◇

The proof of Proposition 9.3.11 can be modified in order to give another approach of Theorem 9.3.10. Consider a Lipschitz function f satisfying $0 \leq f \leq 1 = Osc(f) = \| f \|_\infty$ (recall Remark 9.2.9), then

$$\nu(|f - \nu(f)|) \leq \mu(|f - \mu(f)|) + |\mu(|f - \mu(f)|) - \nu(|f - \mu(f)|)| + |\nu(f) - \mu(f)| .$$

Since

$$\mu(|f - \mu(f)|) \leq \beta_\mu(s) \ \| |\nabla f| \|_\infty + s \, Osc(f) ,$$

while for $0 \leq g \leq 1$,

$$|\nu(g) - \mu(g)| = |\nu(g - \inf(g)) - \mu(g - \inf(g))| \leq d_{TV}(\mu, \nu) \ \| g - \inf g \|_\infty$$
$$= d_{TV}(\mu, \nu) \, Osc(g) ,$$

we get applying the previous bound with $g = f$ and $g = |f - \mu(f)|$,

$$\nu(|f - \nu(f)|) \leq \beta_\mu(s) \ \| |\nabla f| \|_\infty + (s + 2 \, d_{TV}(\mu, \nu)) \, Osc(f) .$$

Hence, for any μ and ν, for all s' such that $s' > 2 \, d_{TV}(\mu, \nu)$, we have

$$\beta_\nu(s') \leq \beta_\mu \left(s' - 2 \, d_{TV}(\nu, \mu) \right) . \tag{9.3.5}$$

We thus have shown:

Proposition 9.3.14 *Let ν be a log-concave probability measure. Then for all μ such that $d_{TV}(\nu, \mu) \leq 1/4$, we have for all $s < \frac{1}{2} - 2d_{TV}(\mu, \nu)$,*

$$C_C'(\nu) \leq \frac{16 \, \beta_\mu(s)}{\pi \, (1 - 2s - 4 \, d_{TV}(\mu, \nu))^2} .$$

In particular for $s = 0$ we get

$$C_C'(\nu) \leq \frac{16}{\pi \, (1 - 4 \, d_{TV}(\nu, \mu))^2} \, C_C(\mu) .$$

Of course the disappointing part of the previous result is that, even if the distance between ν and μ goes to 0, we cannot improve on the pre factor. In comparison with Theorem 9.3.10 we do not require μ to be log-concave, but the previous proposition is restricted to the case of not too big distance between μ and ν while we may take

ε close to 0 in Theorem 9.3.10. Notice that we really need the weak form of the $(1, \infty)$ inequality here (the non weak form of E. Milman is not sufficient), since we only get $\beta_\nu(s')$ for s' large enough.

Remark 9.3.15 The previous result with $\mu = \nu_A$, see Proposition 9.2.15, furnishes a worse bound than in this proposition.

The fact that we have to use the total variation bounds for two different functions, prevents us to localize the method, i.e. to build an appropriate μ for each f as in Eldan's localization method (see e.g. [1, 28]). Let us explain the previous sentence.

Pick some function β. Let f be a given function satisfying $0 \le f \le 1$. Assume that one can find a measure μ_f such that $\beta_{\mu_f} \le \beta$ and

$$|\mu_f(|f - \mu_f(f)|) - \nu(|f - \mu_f(f)|)| + |\nu(f) - \mu_f(f)| \le \varepsilon \le \frac{1}{2}.$$

Then

$$\nu(|f - \nu(f)|) \le \beta(s) \, \| \, |\nabla f| \, \|_\infty + (s + \varepsilon) \, Osc(f)$$

so that one can conclude as in Proposition 9.3.14. Eldan's localization method is close to this approach, at least by controlling $|\nu(f) - \mu_f(f)|$, but not $|\mu_f(|f - \mu_f(f)|) - \nu(|f - \mu_f(f)|)|$. We shall come back to this approach later. $\diamond$

Remark 9.3.16 In the proof of Proposition 9.3.14 we may replace the total variation distance by the Bounded Lipschitz distance

$$d_{BL}(\mu, \nu) = \sup\{\mu(f) - \nu(f) \ \text{for} f \, 1\text{-Lipschitz and bounded by } 1\},$$

or the Dudley distance (also called the Fortet-Mourier distance)

$$d_{Dud}(\mu, \nu) = \sup\{\mu(f) - \nu(f) \ \text{for} \ \| \, f \, \|_\infty + \| \, |\nabla f| \, \|_\infty \le 1\}.$$

Recall that

$$d_{Dud}(\nu, \mu) \le d_{BL}(\nu, \mu) \le 2 \, d_{Dud}(\nu, \mu).$$

Provided one replaces $\| \, g \, \|_\infty$ by $\| \, g \, \|_\infty + \| \, |\nabla g| \, \|_\infty$, (9.3.5) is replaced by

$$\beta_\nu(s') \le \beta_\mu(s' - 2d_{Dud}(\nu, \mu)) + 2d_{Dud}(\nu, \mu), \tag{9.3.6}$$

so that, when ν is log-concave, we get

$$C'_C(\nu) \le \frac{16 \, (\beta_\mu(s) + 2d_{Dud}(\nu, \mu))}{\pi \, (1 - 2s - 4 \, d_{Dud}(\mu, \nu))^2}. \tag{9.3.7}$$

One can of course replace d_{Dud} by the larger d_{BL} in these inequalities.

When μ and ν are isotropic log-concave probability measures, it is known that

$$d_{TV}(\mu, \nu) \leq C \sqrt{n \, d_{BL}(\mu, \nu)}$$

according to proposition 1 in [43]. Combined with Corollary 9.3.10 this bound is far to furnish the previous result since it gives a dimension dependent result. In addition we do not assume that μ is log concave. ◇

Remark 9.3.17 Remark that with our definitions $d_{Dud} \leq d_{BL} \leq 2 \, d_{TV} \leq 2$. ◇

9.4 Mollifying the Measure

In this section we shall study inequalities for mollified measures. If Z is a random variable we will call mollified variable the sum $Z + X$ where X is some independent random variable, i.e. the law ν_Z is replaced by the convolution product $\nu_Z * \mu_X$. In this situation it is very well known that

$$C_P(\sqrt{\lambda} Z + \sqrt{1 - \lambda} X) \leq \lambda C_P(Z) + (1 - \lambda) C_P(X)$$

for $0 \leq \lambda \leq 1$ (see [5]). Taking $\lambda = 1/2$ it follows that

$$C_P(Z + X) \leq C_P(Z) + C_P(X). \tag{9.4.1}$$

It is well known that mollifying the measure can improve on functional inequalities. For instance if ν is a compactly supported probability measure, the convolution product of ν with a Gaussian measure will satisfy a logarithmic Sobolev inequality as soon as the variance of the Gaussian is large enough (see e.g. [51]), even if ν does not satisfy any "interesting" functional inequality (for instance if ν has disconnected support); but the constant is desperately dimension dependent. We shall see that adding the log-concavity assumption for ν will help to improve on similar results.

9.4.1 Mollifying Using Transportation

A first attempt to look at mollified measures was done by Klartag who obtained the following transportation inequality on the hypercube in [36]:

Theorem 9.4.1 (Klartag) *Let $R \geq 1$ and let Q be some cube in $\mathbb{R}^n$ of side length 1 parallel to the axes. Let $\mu = p(x)dx$ be a log-concave probability measure on Q satisfying in addition*

$$p(\lambda x + (1 - \lambda)y) \leq R \, (\lambda p(x) + (1 - \lambda)p(y)) \tag{9.4.2}$$

for any $0 \leq \lambda \leq 1$ and any pair $(x, y) \in Q^2$ such that all Cartesian coordinates of $x - y$ are vanishing except one ($x - y$ is proportional to some e_j where e_j is the canonical orthonormal basis).

Then, μ satisfies a T_2 Talagrand inequality, i.e. there exists some C (satisfying $C \leq 40/9$) such that for any μ',

$$W_2^2(\mu', \mu) \leq C R^2 D(\mu'||\mu),$$

where W_2 denotes the Wasserstein distance and $D(.||.)$ the relative entropy.

In particular μ satisfies a Poincaré inequality with $C_P(\mu) \leq \frac{C R^2}{2}$.

The final statement is an easy and well known consequence of the T_2 transportation inequality, as remarked in [36, corollary 4.6].

In the sequel Q will denote the usual unit cube $[-\frac{1}{2}, \frac{1}{2}]^n$, and for $\theta > 0$, θQ will denote its homothetic image of ratio θ.

For $\theta > 0$, let Z_θ be a log-concave random vector whose distribution $\mu_\theta = p_\theta(x) dx$ is supported by θQ and satisfies (9.4.2) for any pair $(x, y) \in (\theta Q)^2$ and some given R. Then the distribution μ of $Z = Z_\theta/\theta$ satisfies the assumptions in Theorem 9.4.1 with the same R since its probability density is given for $x \in Q$ by

$$p(x) = \theta^n p_\theta(\theta x).$$

In particular $C_P(Z) \leq \frac{1}{2} C R^2$ so that $C_P(Z_\theta) \leq \frac{1}{2} C R^2 \theta^2$.

In the sequel we can thus replace Q by θQ. We shall mainly look at two examples of such p's: convolution products with the uniform density and the Gaussian density.

9.4.1.1 Convolution with the Uniform Distribution

Consider U_θ a uniform random variable on θQ, $\theta > 1$. Its density $p(x) = \theta^{-n} \mathbf{1}_{\theta Q}(x)$ satisfies (9.4.2) in θQ with $R = 1$. It is immediate that

$$p(\lambda x + (1 - \lambda)x' - y) \leq \lambda p(x - y) + (1 - \lambda) p(x' - y)$$

for all x, x', y such that $x - y$ and $x' - y$ belong to θQ.

Let Z be a log-concave random variable whose law μ is supported by Q. The law

$$v_\theta(dx) = \left(\int p(x - y) v(dy) \right) dx$$

of $Z + U_\theta$ is still log-concave according to Prekopa-Leindler and satisfies (9.4.2) with $R = 1$ on $(\theta - 1)Q$. According to what precedes, its restriction

$$v_{\theta,1}(dx) = \frac{\mathbf{1}_{(\theta-1)Q}}{v_\theta((\theta - 1)Q)} v_\theta(dx)$$

to $(\theta - 1)Q$ satisfies

$$C_P(\nu_{\theta,1}) \le \frac{1}{2} C (\theta - 1)^2 .$$

Thus

$$C'_C(\nu_{\theta,1}) \le \frac{6}{\sqrt{2}} \sqrt{C} (\theta - 1) ,$$

according to Ledoux's comparison result.

9.4.1.2 Convolution with the Gaussian Distribution

Let $\gamma_\beta(x) = (2\pi\beta^2)^{-n/2} \exp\left(-\frac{|x|^2}{2\beta^2}\right)$ be the density of a centered Gaussian variable βG (as before G is the standard Gaussian).

It is elementary to show that γ_β satisfies the following convexity type property close to (9.4.2): for all pair (x, x') and all $\lambda \in [0, 1]$,

$$\gamma_\beta(\lambda x + (1 - \lambda)x') \le e^{\frac{|x-x'|^2}{8\beta^2}} \left(\lambda \gamma_\beta(x) + (1 - \lambda) \gamma_\beta(x')\right) , \qquad (9.4.3)$$

this inequality being optimal and attained for pairs $(x, x') = (x, -x)$.

It immediately follows that

$$\gamma_\beta(\lambda x + (1 - \lambda)x' - y) \le e^{\frac{|x-x'|^2}{8\beta^2}} \left(\lambda \gamma_\beta(x - y) + (1 - \lambda) \gamma_\beta(x' - y)\right) ,$$

for all (x, x', y). It follows that for all log-concave random vector Z, the distribution of $Z + \beta G$ is still satisfying (9.4.3). We have thus obtained:

Proposition 9.4.2 *Let β and θ be positive real numbers. Let Z be some log-concave random vector and denote by ν_β the distribution of $Z + \beta G$. Then the restriction*

$$\nu_{\beta,\theta} = \frac{\mathbb{1}_{\theta Q}}{\nu_\beta(\theta Q)} \nu_\beta$$

satisfies

$$C_P(\nu_{\beta,\theta}) \le \frac{20}{9} \theta^2 e^{\frac{\theta^2}{8\beta^2}} .$$

Notice that if we let β go to infinity, $\nu_{\beta,\theta}$ converges to the uniform measure on θQ so that the order θ^2 is the good one (if not the constant $\frac{20}{9}$). Also notice that if ν is supported by αQ, we may replace βG by a random variable whose distribution is γ_β restricted to $(\alpha + \theta)Q$.

The control obtained in Proposition 9.4.2 (or in the uniform case) can be very interesting for practical uses, in particular for applied statistical purposes using censored random variables. But if we want to use it in order to get some information on the original v, the result will depend dramatically on the dimension, since $v_\beta(\theta Q)$ is very small.

9.4.2 Mollifying with Gaussian Convolution: A Stochastic Approach

In this section we shall introduce our approach for controlling the Poincaré constant using some appropriate stochastic process. To this end, we first consider a standard Ornstein-Uhlenbeck process $X_{\cdot}$, i.e. the solution of

$$dX_t = dB_t - \frac{1}{2} X_t \, dt \, . \tag{9.4.4}$$

The law of X_t^x (i.e. the process starting from point x) will be denoted by $G(t, x, .) = \gamma(t, x, .) \, dx$, $\gamma(t, x, .)$ being thus the density of a Gaussian random variable with mean $e^{-t/2} x$ and covariance matrix $(1 - e^{-t}) \, Id$. The standard Gaussian measure γ is thus the unique invariant (reversible) probability measure for the process. The law of the O-U process starting from μ will be denoted by $\mathbb{G}_\mu$.

One way to recover (9.4.1) is to remark that, if Z denotes the initial random variable X_0,

$$X_T = e^{-(T-s)/2} Z + \sqrt{1 - e^{-(T-s)}} G$$

where G is an independent random variables with distribution γ, and next to apply the "local" Poincaré inequalities for the Ornstein-Uhlenbeck semi-group (see [4, Theorem 4.7.2]). An alternate proof using synchronous coupling is given in [17].

One may thus ask whether a converse inequality can be obtained by simply reversing time.

We shall see that in the log-concave situation, it is actually possible.

Let $v(dx) = e^{-V(x)} \, dx$ be a probability measure, V being smooth. We assume that v is log-concave. Let

$$h(x) = (dv/d\gamma)(x) = (2\pi)^{n/2} \, e^{((|x|^2/2) - V(x))} \, .$$

The relative entropy $D(v||\gamma)$ satisfies

$$D(v||\gamma) = \frac{n}{2} \log(2\pi) + \int ((|x|^2/2) - V(x)) \, v(dx) < +\infty \, ,$$

since V is non-negative outside some compact subset. We may define for all t (Mehler formula),

$$\mathbb{E}(h(X_t^x)) = G_t h(x) = \int h(e^{-t/2}x + \sqrt{1 - e^{-t}}y)\, \gamma(dy),$$

which is well defined, positive, smooth (C^∞), and satisfies for all $t > 0$,

$$\partial_t G_t h(x) = \frac{1}{2}\Delta G_t h(x) - \frac{1}{2}\langle x, \nabla G_t h(x)\rangle.$$

In particular we may consider the solution of

$$dY_t = dB_t - \frac{1}{2}Y_t\, dt + \nabla \log G_{T-t}h(Y_t) = dB_t + b(t, Y_t)\, dt, \qquad (9.4.5)$$

for $0 \leq t \leq T$. This stochastic differential equation has a strongly unique solution (since the coefficients are smooth) up to some explosion time $\xi \geq T$.

Since $D(\nu\|\gamma) < +\infty$, this explosion time is almost surely infinite when the initial distribution is given by $G_T h\, \gamma$. This well known result follows from the discussion below and Proposition 2.3 in [19].

In addition the law $\mathbb{Q}$ of the solution satisfies

$$\frac{d\mathbb{Q}}{d\mathbb{G}_\gamma}|_{\mathcal{F}_T} = h(\omega_T).$$

Here of course $\mathcal{F}_T$ denotes the natural filtration on the path space. In particular the law $\nu(X_s)$ of the process satisfies,

$$\nu(X_s) = \mathbb{Q} \circ \omega_s^{-1} = G_{T-s}h\, \gamma \qquad \text{for all } 0 \leq s \leq T,$$

thanks to the stationarity of $\mathbb{G}_\gamma$. Of course this is nothing else but the h-process corresponding to h and the O-U process, but with a non bounded h.

If one prefers the solution $\mathbb{Q}$ of (9.4.5) is defined up to and including time T and is simply the law of the time reversal, at time T, of an Ornstein Uhlenbeck process with initial law ν. For more details see e.g. [17].

A specific feature of the O-U process is that, according to Prekopa-Leindler theorem, $\nu(X_s)$ is still a log-concave measure as the law of $e^{-(T-s)/2}Z + \sqrt{1 - e^{-(T-s)}}G$ where Z and G are two independent random variables with respective distribution ν and γ (so that the pair (Z, G) has a log-concave distribution). This property of conservation of log-concave distributions is typical to the "linear" diffusion processes of O-U type (see [37]). Hence $\nu(X_s)(dx) = e^{-V_s(x)}\, dx$ for some potential V_s which is smooth thanks to the Mehler formula we recalled before, and convex.

It follows that $G_{T-s}h\,\gamma$ is log-concave, and finally that $b(t, .)$ (defined in (9.4.5)) satisfies the curvature condition

$$2\,\langle b(t, x) - b(t, y)\,,\, x - y \rangle \;\leq\; |x - y|^2\,, \tag{9.4.6}$$

for all t, x and y. (9.4.6) is called condition (H.C.-1) in [17]. It follows that for all $T > 0$,

$$C_P(v) \;\leq\; e^T\,C_P(G_T h\,\gamma) + (e^T - 1)\,. \tag{9.4.7}$$

For time homogeneous drifts this is nothing else but the so called "local" Poincaré inequality in [4] (Theorem 4.7.2). The extension to time dependent drift is done in [17] Theorem 5. Actually since the time non-homogeneous semi-group associated to (9.4.5) is defined on $[0, T - \varepsilon]$ for any $\varepsilon > 0$, we have first to use Theorem 5 of [17] up to time $T - \varepsilon$ and then to pass to the limit $\varepsilon \to 0$.

We have thus obtained:

Theorem 9.4.3 *Let Z be a random variable with log-concave distribution v and G be a standard Gaussian random variable independent of Z. Then for all $0 < \lambda \leq 1$ it holds*

$$C_P(Z) \;\leq\; \frac{1}{\lambda}\,C_P(\sqrt{\lambda}\,Z + \sqrt{1-\lambda}\,G) + \left(\frac{1}{\lambda} - 1\right)\,.$$

Of course since $C_P(aY) = a^2\,C_P(Y)$ for any random variable and any $a \in \mathbb{R}$, we obtain that for all real numbers α and β, defining $\lambda = \frac{\alpha^2}{\alpha^2 + \beta^2}$,

$$C_P(Z) \;\leq\; \frac{1}{\alpha^2}\,C_P(\alpha\,Z + \beta\,G) + \frac{\beta^2}{\alpha^2}\,. \tag{9.4.8}$$

Using all the comparison results we already quoted, it follows

$$C_C'(Z) \;\leq\; \frac{12}{\alpha}\,C_C'(\alpha Z + \beta G) + \frac{6\beta}{\alpha}\,. \tag{9.4.9}$$

Remark 9.4.4 In the proof we may replace the standard Ornstein-Uhlenbeck process by any O-U process

$$dX_t = dB_t - \frac{1}{2}\,A\,X_t\,dt\,,$$

where A is some non-negative symmetric matrix. Up to an orthogonal transformation we may assume that A is diagonal with non-negative diagonal terms a_i. Assume that

$$\max_{i=1,\dots,n}\, a_i \leq 1\,.$$

Then (9.4.6) is still satisfied, and no better inequality is. Similarly the Poincaré constant of the corresponding Gaussian variable G' is unchanged so that (9.4.7) is still satisfied, and no better inequality is. We may thus in (9.4.8) replace the standard Gaussian vector G by any centered Gaussian vector, still called G, with independent entries G_i such that $\mathrm{Var}(G_i) \leq 1$ for all i; in particular a degenerate Gaussian vector. $\diamond$

Remark 9.4.5 Equation (9.4.8) with $\alpha = 1$ and $\beta = \sqrt{T}$, can also be derived using a similar time reversal argument for the Brownian motion (with reversible measure Lebesgue) instead of the O-U process. The corresponding drift b satisfies $\langle b(t, x) - b(t, y), x - y \rangle \leq 0$ which directly yields the result. Unfortunately the invariant measure is no more a probability measure. $\diamond$

Remark 9.4.6 Theorem 9.4.3 can be viewed as a complement of previous similar results comparing the behavior of log-concave distributions and their Gaussian mollification. Indeed, if Z is an isotropic log-concave probability measure and G a standard Gaussian vector, it is elementary to show that for all $t \geq 0$,

$$\mathrm{Var}(|Z + \sqrt{t}\,G|^2) = \mathrm{Var}(|Z|^2) + 2nt\,(2+t)\,, \tag{9.4.10}$$

so that if for some t_0, $\mathrm{Var}(|Z + \sqrt{t_0}\,G|^2) \leq Cn$ then the same bound is true for $\mathrm{Var}(|Z|^2)$ i.e. the variance conjecture is true. Similarly, it is recalled in [31] (just before formula (4.5) therein), that a Fourier argument due to Klartag furnishes a control for the deviations

$$\mathbb{P}(|Z| > (1+t)\sqrt{n}) \leq C\,\mathbb{P}(|Z+G| > (1+(1+t)^2)\sqrt{n})$$

and

$$\mathbb{P}(|Z| < (1-t)\sqrt{n}) \leq C\,\mathbb{P}(|Z+G| < (1+(1-t)^2)\sqrt{n})$$

for some universal constant C. $\diamond$

9.5 Probability Metrics and Log-Concavity

9.5.1 *Comparing Bounded Lipschitz and Total Variation Distances*

Let μ and ν be two probability measures. It is immediate to show the analogue of (9.4.1), i.e. if $\mu * \nu$ denotes the convolution product of both measures

$$\beta_{\mu*\nu}(s) \leq \beta_\mu(s/2) + \beta_\nu(s/2)\,, \tag{9.5.1}$$

here β corresponds to the usual centering with the mean (not the median).

Let $0 < t$. Denote by γ_t the distribution of tG, that is the Gaussian distribution with covariance $t^2 Id$ (whose density is $\tilde{\gamma}_t$). For $0 \leq g \leq 1$, one has

$$(v * \gamma_t)(g) = v(g * \tilde{\gamma}_t)$$

and $g_t = g * \tilde{\gamma}_t$ is still bounded by 1 and $1/t$-Lipschitz (actually $\sqrt{2}/t\sqrt{\pi}$) according to (9.2.7) applied to the Brownian motion semi-group at time t^2. It follows that

$$(v * \gamma_t)(g) - (\mu * \gamma_t)(g) = v(g_t) - \mu(g_t)$$

so that

$$d_{TV}(\mu * \gamma_t, v * \gamma_t) \leq \left(1 \vee \frac{\sqrt{2}}{t\sqrt{\pi}}\right) d_{BL}(\mu, v)$$

and   (9.5.2)

$$d_{TV}(\mu * \gamma_t, v * \gamma_t) \leq \left(1 + \frac{\sqrt{2}}{t\sqrt{\pi}}\right) d_{Dud}(\mu, v).$$

We may thus apply (9.3.5), and the fact that $C'_C(\gamma_t) = t$ in order to get, provided respectively

$$s > 2(1 \vee \sqrt{2}/t\sqrt{\pi}) d_{BL}(v, \mu) \text{ or } s > 2(1 + \sqrt{2}/t\sqrt{\pi}) d_{Dud}(\mu, v),$$

the following

$$\beta_{v*\gamma_t}(s) \leq \beta_{\mu*\gamma_t}\left(s - 2(1 \vee \sqrt{2}/t\sqrt{\pi}) d_{BL}(\mu, v)\right)$$

$$\leq \beta_\mu\left(\frac{s}{2} - (1 \vee \sqrt{2}/t\sqrt{\pi}) d_{BL}(\mu, v)\right) + t,   (9.5.3)$$

or

$$\beta_{v*\gamma_t}(s) \leq \beta_\mu\left(\frac{s}{2} - (1 + \sqrt{2}/t\sqrt{\pi}) d_{Dud}(\mu, v)\right) + t.$$

Gathering the previous inequality, Theorem 9.2.7 and (9.4.8), we get new versions of Proposition 9.3.14, slightly different from the one we gave in Remark 9.3.16.

For example, for a log-concave v and for all μ such that $d_{BL}(v, \mu) \leq 1/4$, if we choose $t = \sqrt{2}/\sqrt{\pi}$ it holds

$$C'_C(v * \gamma_t) \leq \frac{16\left(C'_C(\mu) + \sqrt{2/\pi}\right)}{\pi(1 - 4d_{BL}(\mu, v))^2},$$

so that

$$C_P(\nu) \leq \frac{2}{\pi} + 4 \left(\frac{16 \left(C_C'(\mu) + \sqrt{2/\pi} \right)}{\pi (1 - 4 \, d_{BL}(\mu, \nu))^2} \right)^2 .$$

But we may use (9.5.2) in a potentially more interesting direction. Indeed, using Theorem 9.3.10 and provided ν and μ (hence $\nu * \gamma$ and $\mu * \gamma$) are log-concave

$$C_C'(\nu * \gamma) \leq \frac{\kappa}{(1 - d_{BL}(\nu, \mu))^2} \left(1 \vee \ln(1/(1 - d_{BL}(\nu, \mu))) \right) C_C'(\mu * \gamma),$$

$$(9.5.4)$$

for some $\kappa \leq 192e/\pi$, provided $d_{BL}(\nu, \mu) \leq 1$ (we have skipped the $\sqrt{2/\pi}$ for simplicity). Hence using (9.4.9), the comparison between C_P and C_C' and (9.4.1) we have obtained a partial analogue of Theorem 9.3.10 with the weaker bounded Lipschitz distance:

Theorem 9.5.1 *Let ν and μ be two log-concave probability measures on $\mathbb{R}^n$, γ be the standard Gaussian distribution on $\mathbb{R}^n$.*

(1) *Then if $d_{BL}(\nu, \gamma) = 1 - \varepsilon$,*

$$C_C'(\nu) \leq \frac{C}{\varepsilon^2} \left(1 \vee \ln(1/\varepsilon) \right) + 6,$$

where the universal constant C can be chosen less than $\frac{13824 \, e}{\pi}$.

(2) *If $d_{BL}(\nu, \mu) = 1 - \varepsilon$ then*

$$C_C'(\nu) \leq \frac{D}{\varepsilon^2} \left(1 \vee \ln(1/\varepsilon) \right) \left(2 C_C'(\mu) + 1 \right) + 6,$$

where the universal constant D can be chosen less than $6C$.

9.5.2 Comparing Total Variation and W_1

We still use the notation G_T for the Ornstein-Uhlenbeck semi-group we introduced in (9.4.4), in particular we write $G_T \nu$ for the law at time T of the O-U process with initial distribution ν.

It is well known that $W_1(G_T \nu, G_T \mu) \leq e^{-T/2} W_1(\nu, \mu)$. Recall that this contraction property is an immediate consequence of synchronous coupling, i.e. if we build two solutions $X_.$ and $Y_.$ of (9.4.4) with the same Brownian motion it holds

$$|X_t - Y_t| = |X_0 - Y_0| - \frac{1}{2} \int_0^t |X_s - Y_s| ds$$

implying the result by choosing an optimal coupling (X_0, Y_0).

If we want to replace the W_1 distance by the total variation distance (or the bounded Lipschitz distance) one has to replace the synchronous coupling by a coupling by reflection following the idea of Eberle [27] we already used in [17]. This yields (see [17, subsection 7.4]) the following inequality

$$d_{TV}(G_T\nu, G_T\mu) \leq \frac{e^{-T/2}}{\sqrt{2\pi\,(1 - e^{-T})}}\, W_1(\nu, \mu)\,. \tag{9.5.5}$$

What we did before allows us to state a negative result:

Proposition 9.5.2 *The Ornstein-Uhlenbeck semi-group is not a contraction in total variation distance, nor in bounded Lipschitz distance.*

Proof Since $d_{TV} \leq 1$ and applying the semi-group property, any uniform decay $d_{TV}(G_T\nu, \gamma) \leq h(T)$ with h going to 0 implies an exponential decay and the contraction property, for some $T > 0$, $d_{TV}(G_T\nu, \gamma) \leq \frac{1}{2}$. If ν_λ is the log-concave distribution of some random vector λX, it follows that for universal constants C and C',

$$C_P(\nu_\lambda) \leq C\,(C_P(G_T\nu_\lambda) + 1) \leq C'\,.$$

But $C_P(\nu_\lambda) = \lambda^2\, C_P(\nu_1)$ yielding a contradiction.

Similarly, if $d_{BL}(G_T\nu, \gamma) \leq 1/2$ then $d_{TV}(G_{T+1}\nu, \gamma) \leq \frac{1}{2}$ which is impossible. □

9.5.3 Comparison with Other Metrics on Probability Measures

The weakest distance we introduced is the Bounded Lipschitz distance. It is known that this distance metrizes weak convergence. We may thus compare d_{BL} with the Lévy-Prokhorov distance.

Definition 9.5.3 If μ and ν are two probability measures the Lévy-Prokhorov distance $d_{LP}(\mu, \nu)$ is defined as

$$d_{LP}(\mu, \nu) = \inf\{\varepsilon \geq 0\;;\; \mu(A) \leq \nu(A + B(0, \varepsilon)) + \varepsilon\;;\; \text{for all Borel set } A.\}.$$

It is well known that d_{LP} is a metric (in particular $d_{LP}(\mu, \nu) = d_{LP}(\nu, \mu)$ despite the apparent non symmetric definition, just taking complements), clearly bounded by 1, that actually metrizes the convergence in distribution (weak convergence). We may also replace Borel sets A by closed sets A, hence defining $\rho(\mu, \nu)$ and symmetrizing the definition i.e. $d_{LP}(\mu, \nu) = max(\rho(\mu, \nu), \rho(\nu, \mu))$.

The following properties can be found in [24, Corollaries 2 and 3] (and using that d_{LP} is less than 1), or [25, problem 5 p. 398] or in [47] (with a worse constant)

Proposition 9.5.4

(1) *It holds*

$$\frac{1}{4} d_{BL}(\nu, \mu) \leq \frac{1}{2} d_{Dud}(\nu, \mu) \leq d_{LP}(\nu, \mu) \leq \sqrt{\frac{3}{2} d_{Dud}(\nu, \mu)}$$
$$\leq \sqrt{\frac{3}{2} d_{BL}(\nu, \mu)},$$

(2) $d_{LP}(\nu, \mu) = \inf\{K(X, Y) ; \mathcal{L}(X) = \nu, \mathcal{L}(Y) = \mu\}$ *where*

$$K(X, Y) = \inf\{\varepsilon \geq 0 ; \mathbb{P}(|X - Y| > \varepsilon) \leq \varepsilon\}$$

is the Ky Fan distance between X and Y, and $\mathcal{L}(X)$ denotes the probability distribution of X.

Assume that μ and ν are log-concave. Then $\mu \otimes \nu$ is also log-concave according to Prekopa-Leindler theorem, so that $(x, y) \mapsto x - y$ is a polynomial of degree 1 on $\mathbb{R}^n \otimes \mathbb{R}^n$. For such a polynomial, moment controls have been obtained by several authors. We shall use the version in [29] Corollary 4 of a result by Guédon [30]

Theorem 9.5.5 (Guédon) *If η is a log-concave probability measure and P is a polynomial of degree 1, then for all $c > 0$ and all $t \geq 1$,*

$$\eta(x ; |P(x)| > ct) \leq \eta(x ; |P(x)| > c)^{\frac{1+t}{2}},$$

provided the left hand side is (strictly) positive.

If X and Y are independent log-concave random variables, we deduce that for $t \geq 1$,

$$\mathbb{P}(|X - Y| > t K(X, Y)) \leq (K(X, Y))^{\frac{1+t}{2}}. \tag{9.5.6}$$

Using

$$\mathbb{E}(|X - Y|) = \int_0^{+\infty} \mathbb{P}(|X - Y| > t) \, dt$$

we have thus obtained

$$\mathbb{E}(|X - Y|) \leq K(X, Y) \left(1 + \frac{2 K(X, Y)}{\ln(1/K(X, Y))}\right), \tag{9.5.7}$$

so that taking an optimal coupling on the right hand side we have obtained

Corollary 9.5.6 *Let μ and ν be two log-concave probability measures. Then*

$$d_{LP}^2(\mu, \nu) \leq W_1(\mu, \nu) \leq d_{LP}(\mu, \nu) \left(1 + \frac{2\, d_{LP}(\mu, \nu)}{\ln(1/d_{LP}(\mu, \nu))}\right),$$

so that,

$$C_C'(\nu) \leq \frac{16}{\pi} \left(C_C(\mu) + 2\, d_{LP}(\mu, \nu) \left(1 + \frac{2\, d_{LP}(\mu, \nu)}{\ln(1/d_{LP}(\mu, \nu))}\right)\right).$$

Recall that the left hand side of the inequality between distances is always true (see e.g. [12, (10.1) p. 1045]).

Combining all what precedes we have also obtained

Corollary 9.5.7 *The Ornstein-Uhlenbeck semi-group is not a contraction in Lévy-Prokhorov distance. However if ν is log-concave*

$$d_{LP}(G_T\nu, \gamma) \leq e^{-T/4} \left[d_{LP}(\gamma, \nu) \left(1 + \frac{2\, d_{LP}(\gamma, \nu)}{\ln(1/d_{LP}(\gamma, \nu))}\right)\right]^{\frac{1}{2}}.$$

Provided $d_{BL}(\mu, \nu) \leq 2/3$ we also have,

$$W_1(\mu, \nu) \leq \sqrt{\frac{3}{2}\, d_{BL}(\mu, \nu)} \left(1 + \frac{\sqrt{6d_{BL}(\mu, \nu)}}{\ln(\sqrt{2}/\sqrt{3d_{BL}(\mu, \nu)})}\right),$$

so that we get a new bound for two log-concave measures

$$C_C'(\nu) \leq \frac{16}{\pi} \left(C_C(\mu) + \sqrt{6d_{BL}(\mu, \nu)} \left(1 + \frac{\sqrt{6d_{BL}(\mu, \nu)}}{\ln(\sqrt{2}/\sqrt{3d_{BL}(\mu, \nu)})}\right)\right).$$

For small values of $d_{BL}(\mu, \nu)$, this bound is better than (9.3.7) (but here we need both measures to be log-concave) and Theorem 9.5.1, which is true for large values of $d_{BL}(\mu, \nu)$.

One can also compare Corollary 9.5.6 with proposition 4 in [43] giving a dimensional inequality

$$W_1(\mu, \nu) \leq C \left(\sqrt{n} \vee \ln\left(\frac{\sqrt{n}}{d_{BL}(\mu, \nu)}\right)\right) d_{BL}(\mu, \nu),$$

for isotropic log-concave probability measures.

Remark 9.5.8 The previous results give some hints on the (somehow bad) structure of isotropic log-concave measures. Indeed look, on one hand at the uniform measure μ^n on $A = [-\sqrt{3}, \sqrt{3}]^n$ associated to a random variable U, on the other hand at the standard Gaussian distribution γ^n associated to G. $\mu^n(A) = 1$ when

$$\gamma^n(A + B(0, \varepsilon)) \leq \gamma^n([-\sqrt{3} - \varepsilon, \sqrt{3} + \varepsilon]^n) = (\gamma^1([-\sqrt{3} - \varepsilon, \sqrt{3} + \varepsilon]))^n,$$

so that

$$d_{LP}(\mu^n, \gamma^n) \geq 1 - u^n$$

with $u = \gamma^1([-\sqrt{3}, \sqrt{3}])$, hence, for large n, $d_{LP}(\mu^n, \gamma^n)$ is as close to 1 as we want. Consequently we cannot expect to get a dimension free nice upper bound for the Lévy-Prokhorov distance. The question is then whether such a bound is true if we consider the set of $\nu * \gamma_t$ where ν describes the set of *isotropic* log-concave distributions, or not. We know that such a bound does not exist for all log-concave distributions, according to Corollary 9.5.7.   ◇

If the Lévy-Prokhorov distance seems difficult to estimate, one can relate it to a Wasserstein distance for a new distance. Introduce

$$W_{LP}(\nu, \mu) = \inf\left\{ \int \frac{|x - y|}{1 + |x - y|} \pi(dx, dy) \; ; \; \pi \circ x^{-1} = \nu \; , \; \pi \circ y^{-1} = \mu \right\}.$$
$$(9.5.8)$$

Proposition 9.5.9 *Let* $K^*(X, Y) = \mathbb{E}\left(\frac{|X-Y|}{1+|X-Y|}\right)$. *It holds*

$$\frac{1}{2} K^*(X, Y) \leq K(X, Y) \leq \sqrt{2K^*(X, Y)}.$$

Consequently

$$\frac{1}{2} W_{LP}(\mu, \nu) \leq d_{LP}(\mu, \nu) \leq \sqrt{2\, W_{LP}(\mu, \nu)}.$$

Proof Denote $Z = |X - Y|$ and $Z^* = \frac{Z}{1+Z}$, so that $Z = \frac{Z^*}{1-Z^*}$.
   On one hand, since $Z^* \leq 1$,

$$\mathbb{E}(Z^*) \leq \mathbb{P}(Z^* > \eta) + \eta.$$

But $\mathbb{P}(Z^* > \eta) = \mathbb{P}(Z > \varepsilon)$ for $\eta = \frac{\varepsilon}{1+\varepsilon}$, so that

$$\mathbb{E}(Z^*) \leq K(X, Y) + \frac{K(X, Y)}{1 + K(X, Y)} \leq 2\, K(X, Y).$$

Conversely, using what precedes and Markov inequality,

$$\mathbb{P}(Z > \varepsilon) \leq \frac{\mathbb{E}(Z^*)}{\frac{\varepsilon}{1+\varepsilon}}$$

so that $\mathbb{P}(Z > \varepsilon) \leq \varepsilon$ provided $\mathbb{E}(Z^*) \leq \frac{\varepsilon^2}{1+\varepsilon}$ in particular if $\mathbb{E}(Z^*) \leq \frac{\varepsilon^2}{2}$ (because we only have to consider $\varepsilon \leq 1$) yielding the result.   □

Since the cost $c(x, y) = \frac{|x-y|}{1+|x-y|}$ is concave we also have

$$K^*(X, Y) = \mathbb{E}(c(|X - Y|)) \leq c(\mathbb{E}(|X - Y|)) = \frac{E(|X - Y|)}{1 + E(|X - Y|)},$$

so that

$$W_{LP}(\nu, \mu) \leq \frac{W_1(\nu, \mu)}{1 + W_1(\nu, \mu)}. \tag{9.5.9}$$

References

1. D. Alonso-Gutierrez, J. Bastero, Approaching the Kannan-Lovasz-Simonovits and variance conjectures, in *LNM*, vol 2131 (Springer, Berlin, 2015)
2. D. Bakry, F. Barthe, P. Cattiaux, A. Guillin, A simple proof of the Poincaré inequality for a large class of probability measures. Electron. Commun. Probab. **13**, 60–66 (2008)
3. D. Bakry, P. Cattiaux, A. Guillin, Rate of convergence for ergodic continuous Markov processes: Lyapunov versus Poincaré. J. Funct. Anal. **254**, 727–759 (2008)
4. D. Bakry, I. Gentil, M. Ledoux, Analysis and Geometry of Markov diffusion operators, in *Grundlehren der Mathematischen Wissenchaften*, vol 348 (Springer, Berlin, 2014)
5. K. Ball, F. Barthe, A. Naor, Entropy jumps in the presence of a spectral gap. Duke Math. J. **119**, 41–63 (2003)
6. F. Barthe, D. Cordero-Erausquin, Invariances in variance estimates. Proc. Lond. Math. Soc. **106**(1), 33–64 (2013)
7. F. Barthe, E. Milman, Transference principles for Log-Sobolev and Spectral-Gap with applications to conservative spin systems. Commun. Math. Phys. **323**, 575–625 (2013)
8. F. Barthe, P. Cattiaux, C. Roberto, Concentration for independent random variables with heavy tails. AMRX **2005**(2), 39–60 (2005)
9. N. Berestycki, R. Nickl, *Concentration of Measure* (2009). http://www.statslab.cam.ac.uk/~beresty/teach/cm10.pdf
10. S.G. Bobkov, Isoperimetric and analytic inequalities for log-concave probability measures. Ann. Probab. **27**(4), 1903–1921 (1999)
11. S.G. Bobkov, Spectral gap and concentration for some spherically symmetric probability measures, in *Geometric Aspects of Functional Analysis, Israel Seminar 2000–2001*. Lecture Notes in Mathematics, vol 1807, pp. 37–43 (Springer, Berlin, 2003)
12. S.G. Bobkov, Proximity of probability distributions in terms of Fourier-Stieltjes transforms. Russ. Math. Surv. **71**(6), 1021–1079 (2016)
13. S.G. Bobkov, C. Houdré, Isoperimetric constants for product probability measures. Ann. Probab. **25**(1), 184–205 (1997)

14. S.G. Bobkov, M. Ledoux, Poincaré's inequalities and Talagrand's concentration phenomenon for the exponential distribution. Probab. Theory Relat. Fields **107**(3), 383–400 (1997)
15. H.J. Brascamp, E.H. Lieb, On extensions of the Brunn-Minkowski and Prékopa-Leindler theorems, including inequalities for log concave functions, and with an application to the diffusion equation. J. Funct. Anal. **22**, 366–389 (1976)
16. S. Brazitikos, A. Giannopoulos, P. Valettas, B.H. Vritsiou, Geometry of isotropic convex bodies, in *Mathematics Surveys and Monographs*, vol 196 (AMS, Providence, 2014)
17. P. Cattiaux, A. Guillin, Semi log-concave Markov diffusions, in *Séminaire de Probabilités XLVI*. Lecture Notes in Mathematics, vol 2014, pp. 231–292 (2015)
18. P. Cattiaux, A. Guillin, Hitting times, functional inequalities, Lyapunov conditions and uniform ergodicity. J. Funct. Anal. **272**(6), 2361–2391 (2017)
19. P. Cattiaux, C. Léonard, Minimization of the Kullback information of diffusion processes Ann. Inst. Henri Poincaré. Prob. Stat. **30**(1), 83–132 (1994). and correction in *Ann. Inst. Henri Poincaré* **31**, 705–707 (1995)
20. P. Cattiaux, N. Gozlan, A. Guillin, C. Roberto, Functional inequalities for heavy tails distributions and applications to isoperimetry. Electron. J. Probab. **15**, 346–385 (2010)
21. P. Cattiaux, A. Guillin, C. Roberto, Poincaré inequality and the L^p convergence of semi-groups. Electron. Commun. Probab. **15**, 270–280 (2010)
22. P. Cattiaux, A. Guillin, P.A. Zitt, Poincaré inequalities and hitting times. Ann. Inst. Henri Poincaré. Prob. Stat. **49**(1), 95–118 (2013)
23. D. Cordero-Erausquin, N. Gozlan, Transport proofs of weighted Poincaré inequalities for log-concave distributions. Bernoulli **23**(1), 134–158 (2017)
24. R.M. Dudley, Distances of probability measures and random variables. Ann. Math. Stat. **39**, 1563–1572 (1968)
25. R.M. Dudley, *Real Analysis and Probability* (Cambridge University Press, Cambridge, 2002)
26. A. Eberle, Reflection coupling and Wasserstein contractivity without convexity. *C. R. Acad. Sci. Paris Ser. I Math.* **349**, 1101–1104 (2011)
27. A. Eberle, Reflection couplings and contraction rates for diffusions. Probab. Theory Relat. Fields **166**(3), 851–886 (2016)
28. R. Eldan, Thin shell implies spectral gap up to polylog via a stochastic localization scheme. Geom. Funct. Anal. **23**(2), 532–569 (2013)
29. M. Fradelizi, Concentration inequalities for s -concave measures of dilations of Borel sets and applications. Electron. J. Probab. **14**(71), 2068–2090 (2009)
30. O. Guédon, Kahane-Khinchine type inequalities for negative exponent. Mathematika **46**(1), 165–173 (1999)
31. O. Guédon, E. Milman, Interpolating thin shell and sharp large deviation estimates for isotropic log-concave measures. Geom. Funct. Anal. **21**, 1043–1068 (2011)
32. N. Huet, Isoperimetry for spherically symmetric log-concave probability measures. Rev. Mat. Iberoamericana **27**(1), 93–122 (2011)
33. R. Kannan, L. Lovasz, M. Simonovits, Isoperimetric problems for convex bodies and a localization lemma. Discret. Comput. Geom. **13**(3-4), 541–559 (1995)
34. B. Klartag, A Berry-Esseen type inequality for convex bodies with an unconditional basis. Probab. Theory Relat. Fields **145**(1-2), 1–33 (2009)
35. B. Klartag, Poincaré inequalities and moment maps. Ann. Fac. Sci. Toulouse Math. **22**(1), 1–41 (2013)
36. B. Klartag, Concentration of measures supported on the cube. Isr. J. Math. **203**(1), 59–80 (2014)
37. A.V. Kolesnikov, On diffusion semigroups preserving the log-concavity. J. Funct. Anal. **186**(1), 196–205 (2001)
38. R. Latala, Order statistics and concentration of L^r norms for log-concave vectors. J. Funct. Anal. **261**, 681–696 (2011)
39. M. Ledoux, A simple analytic proof of an inequality by P. Buser. Proc. Am. Math. Soc. **121**, 951–959 (1994)

40. M. Ledoux, Spectral gap, logarithmic Sobolev constant, and geometric bounds, in *Surveys in Differential Geometry*, vol IX, pp. 219–240 (International Press, Somerville, 2004)
41. M. Ledoux, From concentration to isoperimetry: semigroup proofs, in *Concentration, Functional Inequality and Isoperimetry*. Contemporary Mathematics, vol 545, pp. 155–166 (American Mathematical Society, New York, 2011)
42. Y.T. Lee, S.S. Vempala, *Stochastic localization + Stieltjes barrier = tight bound for Log-Sobolev* (2017). Available on Math. ArXiv. 1712.01791 [math.PR]
43. E. Meckes, M.W. Meckes, On the equivalence of modes of convergence for log-concave measures, in *Geometric Aspects of Functional Analysis*. Lecture Notes in Mathematics, vol 2116, pp. 385–394 (Springer, New York, 2014)
44. E. Milman, On the role of convexity in isoperimetry, spectral-gap and concentration. Invent. Math. **177**, 1–43 (2009)
45. E. Milman, Isoperimetric bounds on convex manifolds, in *Concentration, Functional Inequality and Isoperimetry*. Contemporary Mathematics, vol 545, pp. 195–208 (American Mathematical Society, New York, 2011)
46. E. Milman, Properties of isoperimetric, functional and transport-entropy inequalities via concentration. Probab. Theory Relat. Fields **152**(3-4), 475–507 (2012)
47. S.T. Rachev, L.B. Klebanov, S.V. Stoyanov, F.J. Fabozzi, *The Methods of Distances in the Theory of Probability and Statistics* (Springer, New York, 2013)
48. M. Röckner, F.Y. Wang, Weak Poincaré inequalities and L^2-convergence rates of Markov semigroups. J. Funct. Anal. **185**(2), 564–603 (2001)
49. A. Saumard, J.A. Wellner, Log-concavity and strong log-concavity: a review. Stat. Surv. **8**, 45–114 (2014)
50. M. Troyanov, *Concentration et inégalité de Poincaré* (2001). https://infoscience.epfl.ch/record/118471/files/concentration2001.pdf
51. D. Zimmermann, Logarithmic Sobolev inequalities for mollified compactly supported measures. J. Funct. Anal. **265**, 1064–1083 (2013)

Chapter 10
On Poincaré and Logarithmic Sobolev Inequalities for a Class of Singular Gibbs Measures

Djalil Chafaï and Joseph Lehec

Abstract This note, mostly expository, is devoted to Poincaré and log-Sobolev inequalities for a class of Boltzmann–Gibbs measures with singular interaction. Such measures allow to model one-dimensional particles with confinement and singular pair interaction. The functional inequalities come from convexity. We prove and characterize optimality in the case of quadratic confinement via a factorization of the measure. This optimality phenomenon holds for all beta Hermite ensembles including the Gaussian unitary ensemble, a famous exactly solvable model of random matrix theory. We further explore exact solvability by reviewing the relation to Dyson–Ornstein–Uhlenbeck diffusion dynamics admitting the Hermite–Lassalle orthogonal polynomials as a complete set of eigenfunctions. We also discuss the consequence of the log-Sobolev inequality in terms of concentration of measure for Lipschitz functions such as maxima and linear statistics.

10.1 Introduction

The aim of this note is first to provide synthetic exposition gathering material from several distant sources, and second to provide extensions and novelty about optimality.

D. Chafaï
CEREMADE, CNRS UMR 7534, Université Paris-Dauphine, PSL, Paris, France
e-mail: djalil@chafai.net; http://djalil.chafai.net/

J. Lehec (✉)
CEREMADE, CNRS UMR 7534, Université Paris-Dauphine, PSL, Paris, France

DMA, CNRS UMR 8553, École Normale Supérieure, Paris, France
e-mail: lehec@ceremade.dauphine.fr; https://www.ceremade.dauphine.fr/~lehec/

© Springer Nature Switzerland AG 2020
B. Klartag, E. Milman (eds.), *Geometric Aspects of Functional Analysis*,
Lecture Notes in Mathematics 2256,
https://doi.org/10.1007/978-3-030-36020-7_10

Let $n \in \{1, 2, \ldots\}$. For a given $\rho \in \mathbb{R}$, we say that a function $\phi \colon \mathbb{R}^n \to \mathbb{R} \cup \{+\infty\}$ is ρ-convex when $x \mapsto \phi(x) - \rho |x|^2/2$ is convex, where $|x| := \sqrt{x_1^2 + \cdots + x_n^2}$ is the Euclidean norm. In particular a 0-convex function is just a convex function. An equivalent condition is that for all $x, y \in \mathbb{R}^n$ and $\lambda \in [0, 1]$,

$$\phi((1 - \lambda)x + \lambda y) \leq (1 - \lambda)\phi(x) + \lambda \phi(y) - \frac{\rho \lambda(1 - \lambda)}{2} |y - x|^2.$$

If ϕ is C^2-smooth on its domain then this is yet equivalent to $\mathrm{Hess}(\phi) \geq \rho \, I_n$ as quadratic forms, pointwise, where I_n is the identity: $\langle \mathrm{Hess}(\varphi)(x)y, y \rangle \geq \rho |y|^2$ for all $x, y \in \mathbb{R}^n$.

Let $V \colon \mathbb{R}^n \to \mathbb{R}$ and $W \colon \mathbb{R} \to \mathbb{R} \cup \{+\infty\}$ be two functions, called "confinement potential" and "interaction potential" respectively. We assume that

- V is ρ-convex for some $\rho > 0$;
- W is convex with domain $(0; +\infty)$. In particular $W \equiv +\infty$ on $(-\infty; 0]$.

The energy of a configuration $x = (x_1, \ldots, x_n) \in \mathbb{R}^n$ is

$$U(x) = V(x_1, \ldots, x_n) + \sum_{i < j} W(x_i - x_j) = V(x) + U_W(x) \in \mathbb{R} \cup \{+\infty\}.$$

The nature of W gives that $U(x)$ is finite if and only if x belongs to the "Weyl chamber"

$$D = \{x \in \mathbb{R}^n : x_1 > \cdots > x_n\}.$$

Assuming that

$$Z_\mu = \int_{\mathbb{R}^n} e^{-U(x_1, \ldots, x_n)} dx_1 \cdots dx_n < \infty$$

we define a probability measure μ on $\mathbb{R}^n$ by

$$\mu(dx) = \frac{e^{-U(x_1, \ldots, x_n)}}{Z_\mu} dx. \tag{10.1.1}$$

The support of μ is $\overline{D} = \{x \in \mathbb{R}^n : x_1 \geq \cdots \geq x_n\}$. Note that if

$$W(u) = \begin{cases} -\beta \log u, & \text{if } u > 0 \\ +\infty & \text{otherwise} \end{cases} \tag{10.1.2}$$

where β is a positive parameter, and if X is a random vector of $\mathbb{R}^n$ distributed according to μ, then for every $\sigma > 0$, the scaled random vector σX follows the law μ with same W but with V replaced by $V(\cdot/\sigma)$.

Following Edelman [34], the beta Hermite ensemble corresponds to the case

$$V(x) = \frac{n}{2}|x|^2 = \frac{n}{2}(x_1^2 + \cdots + x_n^2),$$

and W given by (10.1.2). In this case μ rewrites using a Vandermonde determinant as

$$d\mu(x) = \frac{e^{-\frac{n}{2}|x|^2}}{Z_\mu} \prod_{i<j}(x_i - x_j)^\beta \, \mathbb{1}_{\{x_1 \geq \cdots \geq x_n\}} dx. \tag{10.1.3}$$

The normalizing constant Z_μ can be explicitly computed in terms of Gamma functions by reduction to a classical Selberg integral, but this is useless for our purposes in this work. The Gaussian unitary ensemble (GUE) of Dyson [37] corresponds to $\beta = 2$, namely

$$d\mu(x) = \frac{e^{-\frac{n}{2}|x|^2}}{Z_\mu} \prod_{i<j}(x_i - x_j)^2 \mathbb{1}_{\{x_1 \geq \cdots \geq x_n\}} dx. \tag{10.1.4}$$

Note that on $\mathbb{R}^n$ the density of the beta Hermite ensemble (10.1.3) with respect to the Gaussian law $\mathcal{N}(0, \frac{1}{n}I_n)$ is equal up to a multiplicative constant to $\prod_{i<j}(x_i - x_j)^\beta$ times the indicator function of the Weyl chamber. The cases $\beta = 1$ and $\beta = 4$ are known as the Gaussian orthogonal ensemble (GOE) and the Gaussian symplectic ensemble (GSE).

Let $L^2(\mu)$ be the Lebesgue space of measurable functions from $\mathbb{R}^n$ to $\mathbb{R}$ which are square integrable with respect to μ. Let $H^1(\mu)$ be the Sobolev space of functions in $L^2(\mu)$ with weak derivative in $L^2(\mu)$ in the sense of Schwartz distributions.

We provide in Sect. 10.2 some useful or beautiful facts about (10.1.1), (10.1.3), and (10.1.4).

10.1.1   Functional Inequalities and Concentration of Measure

Given $f \in L^2(\mu)$ we define the variance of f with respect to μ by

$$\mathrm{var}_\mu(f) = \int_{\mathbb{R}^n} f^2 \, d\mu - \left(\int_{\mathbb{R}^n} f \, d\mu \right)^2.$$

If additionally $f \geq 0$, then we define similarly the entropy of f with respect to μ by

$$\mathrm{ent}_\mu(f) = \int_{\mathbb{R}^n} f \log f \, d\mu - \left(\int_{\mathbb{R}^n} f \, d\mu \right) \log \left(\int_{\mathbb{R}^n} f \, d\mu \right).$$

Theorem 10.1.1 (Poincaré Inequality) *Let μ be as in* (10.1.1). *For all $f \in H^1(\mu)$,*

$$\mathrm{var}_\mu(f) \le \frac{1}{\rho} \int_{\mathbb{R}^n} |\nabla f|^2 \,\mathrm{d}\mu.$$

This holds in particular with $\rho = n$ for the beta Hermite ensemble (10.1.3) *for all $\beta > 0$.*

Theorem 10.1.2 (Log-Sobolev Inequality) *Let μ be as in* (10.1.1). *For all $f \in H^1(\mu)$,*

$$\mathrm{ent}_\mu(f^2) \le \frac{2}{\rho} \int_{\mathbb{R}^n} |\nabla f|^2 \,\mathrm{d}\mu.$$

This holds in particular with $\rho = n$ for the beta Hermite ensemble (10.1.3) *for all $\beta > 0$.*

Theorem 10.1.3 (Optimality for Poincaré and Log-Sobolev Inequalities) *Let μ be as in* (10.1.1). *Assume that V is quadratic: $V(x) = \rho|x|^2/2$ for some $\rho > 0$. This is in particular the case for the beta Hermite ensemble* (10.1.3) *for all $\beta > 0$. Then equality is achieved in the Poincaré inequality of Theorem 10.1.1 for*

$$f : x \in \mathbb{R}^n \mapsto \lambda(x_1 + \cdots + x_n) + c, \quad \lambda, c \in \mathbb{R}.$$

Moreover equality is achieved in the logarithmic Sobolev inequality of Theorem 10.1.2 for

$$f : x \in \mathbb{R}^n \mapsto \mathrm{e}^{\lambda(x_1 + \cdots + x_n) + c}, \quad \lambda, c \in \mathbb{R}.$$

Lastly, in both cases these are the only extremal functions.

Theorems 10.1.1 and 10.1.2 are proved in Sect. 10.3.1 and Theorem 10.1.3 in Sect. 10.3.2.

Poincaré and logarithmic Sobolev inequalities for beta ensembles are already known in the literature about random matrix theory, see for instance [1, 38] and references therein. However the optimality that we point out here seems to be new.

The following corollary of Theorem 10.1.2 provides concentration of measure around the mean for Lipschitz functions, including linear statistics and maximum.

Corollary 10.1.4 (Gaussian Concentration Inequality for Lipschitz Functions) *Let μ be as in* (10.1.1). *For every Lipschitz function $F : \mathbb{R}^n \to \mathbb{R}$ and for all real parameter $r > 0$,*

$$\mu\left(\left|F - \int F \,\mathrm{d}\mu\right| \ge r\right) \le 2\exp\left(-\frac{\rho}{\|F\|_{\mathrm{Lip}}^2} \frac{r^2}{2}\right). \tag{10.1.5}$$

In particular for any measurable $f : \mathbb{R} \to \mathbb{R}$ and all $r > 0$, with $L_n(f)(x) = \frac{1}{n} \sum_{i=1}^{n} f(x_i)$,

$$\mu\left(\left|L_n(f) - \int L_n(f)\mathrm{d}\mu\right| \geq r\right) \leq 2\exp\left(-n\frac{\rho}{\|f\|_{\mathrm{Lip}}^2}\frac{r^2}{2}\right). \tag{10.1.6}$$

Additionally, for all $r > 0$,

$$\mu\left(\left|x_1 - \int x_1\mu(\mathrm{d}x)\right| \geq r\right) \leq 2\exp\left(-\rho\frac{r^2}{2}\right). \tag{10.1.7}$$

This holds in particular with $\rho = n$ for the beta Hermite ensemble (10.1.3) for all $\beta > 0$.

The proof of Corollary 10.1.4 and some additional comments are given in Sect. 10.3.3.

The scale in (10.1.7) is not optimal for the beta Hermite ensemble, the largest particle is actually more concentrated than what is predicted by Corollary 10.1.4. Indeed, it is proved for instance in [63] that $n^{2/3}(\lambda_1 - 2)$ converges in law as n tends to infinity to a Tracy–Widom distribution of parameter β. In particular fluctuations of λ_1 are of order $n^{-2/3}$, whereas (10.1.7) only predicts an upper bound of order $n^{-1/2}$. See also [4, 56, 57] for a concentration property that matches the correct order of fluctuations.

Note also that (10.1.6) allows to get concentration for the Cauchy–Stieltjes transform of the empirical distribution by taking f equal to the real or imaginary part of $x \mapsto 1/(x - z)$ where $z = a + ib$ is a fixed complex parameter with $b > 0$.

The function $(x_1, \ldots, x_n) \mapsto L_n(f)(x) = \frac{1}{n} \sum_{i=1}^{n} f(x_i)$ is called a linear statistics. The inequality (10.1.6) appears for the spectrum of random matrix models in the work of Guionnet and Zeitouni [45] via the logarithmic Sobolev inequality, see also [44] and [1, Section 4.4.2], and [1, Exercise 4.4.33] for beta ensembles. The exponential speed n^2 in (10.1.6) is optimal according to the large deviation principle satisfied by L_n under μ established by Ben Arous and Guionnet [10] for the GUE, see [24] and references therein for the general case (10.1.1). Concentration inequalities and logarithmic Sobolev inequalities for spectra of some random matrix models at the correct scale can also be obtained using coupling methods or exact decompositions, see for instance [60, 61] and references therein.

Many proofs involve the following simple transportation facts:

$$\mathcal{N}(0, n^{-1}I_n) \xrightarrow{\text{Caffarelli}} \mu \xrightarrow{x_1+\cdots+x_n} \mathcal{N}(0, 1)$$

and

$$\mathcal{N}(0, n^{-1}I_n) \xrightarrow{x_1+\cdots+x_n} \mathcal{N}(0, 1)$$

and

$$\mathrm{Law}(H) \xrightarrow{\quad\text{Spectrum}\quad} \mu \xrightarrow{\quad x_1 + \cdots + x_n \quad} \mathcal{N}(0, 1)$$

and

$$\mathrm{Law}(H) \xrightarrow{\quad\text{Trace}\quad} \mathcal{N}(0, 1)$$

where H is a random Hermitian matrix as in Theorem 10.2.1 or 10.2.2.

10.1.2 Dynamics

Let us assume in this section that the functions V and W are smooth on $\mathbb{R}^n$ and $(0, +\infty)$ respectively. Then the energy U is smooth on its domain D. Fix $X_0 \in D$ and consider the overdamped Langevin diffusion associated to the potential U starting from X_0, solving the stochastic differential equation

$$X_t = X_0 + \sqrt{2}B_t - \int_0^t \nabla U(X_s)\,\mathrm{d}s + \Phi_t, \ t \geq 0, \tag{10.1.8}$$

where $(B_t)_{t \geq 0}$ is a standard Brownian Motion of $\mathbb{R}^n$, and where Φ_t is a reflection at the boundary of D which constrains the process X to stay in D. More precisely

$$\Phi_t = - \int_0^t \mathrm{n}_s\, L(\mathrm{d}s)$$

where L is a random measure depending on X and supported on $\{t \geq 0 : X_t \in \partial D\}$ and where n_t is an outer unit normal to the boundary of D at X_t for every t in the support of L. The process L is called the "local time" at the boundary of D. The stochastic differential equation (10.1.8) writes equivalently

$$\mathrm{d}X_t = \sqrt{2}\,\mathrm{d}B_t - \nabla U(X_t)\,\mathrm{d}t - \mathrm{n}_t\, L(\mathrm{d}t).$$

It is not obvious that Eq. (10.1.8) admits a solution. Such diffusions with reflecting boundary conditions were first considered by Tanaka. He proved in [65] that if ∇U is globally Lipschitz on D and grows at most linearly at infinity then (10.1.8) does admit a unique strong solution.

If it exists, the solution is a Markov process. Its generator is the operator G where

$$\mathrm{G} = \Delta - \langle \nabla U, \nabla \rangle = \sum_{i=1}^n \partial_{x_i}^2 - \sum_{i=1}^n (\partial_{x_i} V)(x)\partial_{x_i} - \sum_{i \neq j} W'(x_i - x_j)\partial_{x_i} \tag{10.1.9}$$

with Neumann boundary conditions at the boundary of D. Stokes formula then shows that G is symmetric in $L^2(\mu)$. As a result the measure μ is reversible for the process (X_t). By integration by parts the density f_t of X_t with respect to the Lebesgue measure satisfies the Fokker–Planck equation $\partial_t f_t = \Delta f_t + \text{div}(f_t \nabla U)$.

It is common to denote $X_t = (X_t^1, \ldots, X_t^n)$ and to interpret $X_t^1, \ldots, X_t^n$ as interacting particles on the real line experiencing confinement and pairwise interactions. Let us discuss now the particular case of the beta Hermite ensemble (10.1.3), for which (10.1.8) rewrites

$$dX_t^i = \sqrt{2}\, dB_t^i - nX_t^i dt + \beta \sum_{j:\, j\neq i} \frac{1}{X_t^i - X_t^j}\, dt, \quad 1 \leq i \leq n \qquad (10.1.10)$$

as long as the particles have not collided. We call this diffusion the Dyson–Ornstein–Uhlenbeck process. Without the confinement term $-X_t^i dt$ this diffusion is known in the literature as the Dyson Brownian motion. Indeed Dyson proved in [36] the following remarkable fact: if (M_t) is an Ornstein–Uhlenbeck process taking values in the space of complex Hermitian matrices then the eigenvalues of (M_t) follow the diffusion (10.1.10) with parameter $\beta = 2$, while if (M_t) is an Ornstein–Uhlenbeck process taking values in the space of real symmetric matrices then the same holds true with $\beta = 1$. Dyson also proved an analogue result for the eigenvalues of a Brownian motion on the unitary group. It is natural to ask whether the repulsion term $1/(X_t^i - X_t^j)$ is strong enough to actually prevent the collision of particles. This was investigated by Rogers and Shi in [28], see also [1]. They proved that if $\beta \geq 1$ then there are no collisions: (10.1.10) admits a unique strong solution and with probability 1, the process (X_t) stays in the Weyl chamber D for all time. This means that in that case, Tanaka's equation (10.1.8) does admit a unique strong solution, but the reflection at the boundary Φ_t is actually identically 0. This critical phenomenon was also observed 25 years ago by Calogero in [17]. Besides, although it is not explicitly written in Rogers and Shi's article, when $\beta < 1$ collisions do occur in finite time, so that the reflection Φ_t enters the picture. In that case though, the existence of a process (X_t) satisfying (10.1.8) does not follow from Tanaka's theorem [65], as the potential U is singular at the boundary of D. Still (10.1.8) does admit a unique strong solution. Indeed, this was established by Cépa and Lépingle in [20] using an existence result for multivalued stochastic differential equations due to Cépa [19]. See also the work of Demni [31, 32].

Long Time Behavior of the Dynamics Let us assume that the process (10.1.8) is well defined. We denote by (P_t) the associated semigroup: For every test function f

$$P_t f(x) = \mathbb{E}(f(X_t) \mid X_0 = x).$$

Given a probability measure ν on $\mathbb{R}^n$ we denote νP_t the law of the process at time t when initiated from ν. Recall that the measure μ is stationary: $\mu P_t = \mu$ for all time.

For all real number $p \geq 1$, the L^p Kantorovich or Wasserstein distance between μ and ν is

$$W_p(\nu, \mu) = \inf_{\substack{(X,Y) \\ X \sim \nu \\ Y \sim \mu}} \mathbb{E}(|X - Y|^p)^{1/p}. \qquad (10.1.11)$$

Note that $W_p(\nu, \mu) < \infty$ if $|\cdot|^p \in L^1(\nu) \cap L^1(\mu)$. It can be shown that the convergence for W_p is equivalent to weak convergence together with convergence of p-th moment. If ν has density f with respect to μ, the relative entropy of ν with respect to μ is

$$H(\nu \mid \mu) = \int_{\mathbb{R}^n} \log f \, d\nu. \qquad (10.1.12)$$

If ν is not absolutely continuous we set $H(\nu \mid \mu) = +\infty$ by convention.

Theorem 10.1.5 (Convergence to Equilibrium) *For any two probability measures ν_0, ν_1 on $\mathbb{R}^n$ we have, for all $p \geq 1$ and $t \geq 0$, in $[0, +\infty]$,*

$$W_p(\nu_0 P_t, \nu_1 P_t) \leq e^{-\rho t} W_p(\nu_0, \nu_1).$$

In particular, choosing $\nu_1 = \mu$ yields

$$W_p(\nu_0 P_t, \mu) \leq e^{-\rho t} W_p(\nu_0, \mu).$$

Moreover we also have, for all $t \geq 0$,

$$H(\nu_0 P_t \mid \mu) \leq e^{-\rho t} H(\nu_0 \mid \mu). \qquad (10.1.13)$$

A proof of Theorem 10.1.5 is given in Sect. 10.3.4.

10.1.3   Hermite–Lassalle Orthogonal Polynomials

Recall that for all $n \geq 1$, the classical Hermite polynomials $(H_{k_1,...,k_n})_{k_1 \geq 0,...,k_n \geq 0}$ are the orthogonal polynomials for the standard Gaussian distribution γ_n on $\mathbb{R}^n$. The tensor product structure $\gamma_n = \gamma_1^{\otimes n}$ gives $H_{k_1,...,k_n}(x_1, \ldots, x_n) = H_{k_1}(x_1) \cdots H_{k_n}(x_n)$ where $(H_k)_{k \geq 0}$ are the orthogonal polynomials for the one-dimensional Gaussian distribution γ_1. Among several remarkable characteristic properties, these polynomials satisfy a differential equation which writes

$$LH_{k_1,...,k_n} = -(k_1 + \cdots + k_n)H_{k_1,...,k_n} \quad \text{where} \quad L = \Delta - \langle x, \nabla \rangle \qquad (10.1.14)$$

is the infinitesimal generator of the Ornstein–Uhlenbeck process, which admits γ_n as a reversible invariant measure. In other words these orthogonal polynomials form a complete set of eigenfunctions of this operator. Such a structure is relatively rare, see [59] for a complete classification when $n = 1$.

Lassalle discovered in the 1990s that a very similar phenomenon takes place for beta Hermite ensembles and the Dyson–Ornstein–Uhlenbeck process, provided that we restrict to symmetric polynomials. Observe first that this cannot hold for all polynomials, simply because the infinitesimal generator

$$G = \sum_{i=1}^{n} \partial_{x_i}^2 - n \sum_{i=1}^{n} x_i \partial_{x_i} + \beta \sum_{i \neq j} \frac{1}{x_i - x_j} \partial_{x_i}, \tag{10.1.15}$$

of the Dyson–Ornstein–Uhlenbeck process, which is a special case of (10.1.9), does not preserve polynomials, for instance we have $Gx_1 = -nx_1 + \beta \sum_{j \neq 1} \frac{1}{x_1 - x_j}$. However, rewriting this operator by symmetrization as

$$G = \sum_{i=1}^{n} \partial_{x_i}^2 - n \sum_{i=1}^{n} x_i \partial_{x_i} + \frac{\beta}{2} \sum_{i \neq j} \frac{1}{x_i - x_j} (\partial_{x_i} - \partial_{x_j}), \tag{10.1.16}$$

it is easily seen that the set of symmetric polynomials in n variables is left invariant by G.

Let μ be the beta Hermite ensemble defined in (10.1.3). Lassalle studied in [51] multivariate symmetric polynomials $(P_{k_1,\ldots,k_n})_{k_1 \geq \cdots \geq k_n \geq 0}$ which are orthogonal with respect to μ. He called them "generalized Hermite" but we decide to call them "Hermite–Lassalle". For all $k_1 \geq \cdots \geq k_1 \geq 0$ and $k_1' \geq \cdots \geq k_n' \geq 0$,

$$\int P_{k_1,\ldots,k_n}(x_1,\ldots,x_n) P_{k_1',\ldots,k_n'}(x_1,\ldots,x_n) \mu(dx) = \mathbb{1}_{(k_1,\ldots,k_n)=(k_1',\ldots,k_n')}. \tag{10.1.17}$$

They can be obtained from the standard basis of symmetric polynomials by using the Gram–Schmidt algorithm in the Hilbert space $L^2_{\text{sym}}(\mu)$ of square integrable symmetric functions. The total degree of $P_{k_1,\ldots,k_n}$ is $k_1 + \cdots + k_n$, in particular $P_{0,\ldots,0}$ is a constant polynomial. The numbering in terms of $k_1,\ldots,k_n$ used in [51] is related to Jack polynomials. Beware that [51] comes without proofs. We refer to [5] for proofs, and to [35] for symbolic computation via Jack polynomials.

The Hermite–Lassalle symmetric polynomials form an orthogonal basis in $L^2_{\text{sym}}(\mu)$ of eigenfunctions of the Dyson–Ornstein–Uhlenbeck operator G. Restricted to symmetric functions, this operator is thus exactly solvable, just like the classical Ornstein–Uhlenbeck operator. Here is the result of Lassalle in [51], see [5] for a proof.

Theorem 10.1.6 (Eigenfunctions and Eigenvalues) *For all $n \geq 2$ and $k_1 \geq \cdots \geq k_n \geq 0$,*

$$\mathrm{G} P_{k_1,\ldots,k_n} = -n(k_1 + \cdots + k_n) P_{k_1,\ldots,k_n}. \tag{10.1.18}$$

where G *is the operator* (10.1.15).

When $\beta = 0$ then G becomes the Ornstein–Uhlenbeck operator. For all $\beta > 0$, the spectrum of G is identical to the one of the Ornstein–Uhlenbeck operator. This can be guessed from the fact that the eigenfunctions are polynomials together with the fact that the interaction term (the non O.-U. part) lowers the degree of polynomials.

The spectral gap of G in $L^2_{\mathrm{sym}}(\mu)$ is n: if $f \in L^2_{\mathrm{sym}}(\mu)$ is orthogonal to constants then

$$n \int f^2 \, d\mu \leq - \int f \, \mathrm{G} f \, d\mu = \int |\nabla f|^2 \, d\mu.$$

Theorem 10.1.1 shows that this inequality holds actually for all f, not only symmetric ones.

Hermite–Lassalle polynomials can be decomposed in terms of Jack polynomials, and this decomposition generalizes the hypergeometric expansion of classical Hermite polynomials.

Remark 10.1.7 (Examples and Formulas) It is not difficult to check that up to normalization

$$x_1 + \cdots + x_n \quad \text{and} \quad x_1^2 + \cdots + x_n^2 - 1 - \beta \frac{n-1}{2}.$$

are Hermite–Lassalle polynomials. In the GUE case, $\beta = 2$, Lassalle gave in [51], using Jack polynomials and Schur functions, a formula for $P_{k_1,\ldots,k_n}$ in terms of a ratio of a determinant involving classical Hermite polynomials and a Vandermonde determinant.

10.1.4   Comments and Open Questions

Regarding functional inequalities, one can probably extend the results to the class of Gaussian φ-Sobolev inequalities such as the Beckner inequality [9], see also [22]. Lassalle has studied not only the beta Hermite ensemble in [51], but also the beta Laguerre ensemble in [53] with density proportional to

$$x \in D \mapsto \prod_{k=1}^{n} x_k^a e^{-bnx_k} \prod_{i<j} (x_i - x_j)^\beta \mathbb{1}_{x_1 \geq \cdots \geq x_n \geq 0},$$

and the beta Jacobi ensemble in [52] with density proportional to

$$x \in D \mapsto \prod_{k=1}^{n} x_k^{a-1}(1 - x_k)^{b-1} \prod_{i<j}(x_i - x_j)^{\beta} \mathbb{1}_{1 \geq x_1 \geq \cdots \geq x_n \geq 0}.$$

It is tempting to study functional inequalities and concentration of measure for these ensembles. The proofs of Lassalle, based on Jack polynomials, are not in [51–53] but can be found in the work [5] by Baker and Forrester. We refer to [66] for the link with Macdonald polynomials. It is natural (maybe naive) to ask about direct proofs of these results without using Jack polynomials. The study of beta ensembles can be connected to H-transforms and to the work [42] on Brownian motion in a Weyl chamber, see also [33]. The analogue of the Dyson Brownian motion for the Laguerre ensemble is studied in [14], see also [33, 43, 50, 67]. Tridiagonal matrix models for Dyson Brownian motion are studied in [48].

The natural isometry between $L^2(\gamma_n)$ and $L^2(\mathrm{d}x)$ leads to associate to the Ornstein–Uhlenbeck operator a real Schrödinger operator which turns out to be the quantum harmonic oscillator. Similarly, the natural isometry between $L^2_{\mathrm{sym}}(\mu)$ and $L^2_{\mathrm{sym}}(\mathrm{d}x)$ leads to associate to the Dyson–Ornstein–Uhlenbeck operator a real Schödinger operator known as the Calogero–Moser–Sutherland operator, which is related to radial Dunkl operators, see for instance [64, 66]. The fact that the eigenfunctions of such operators are explicit and involve polynomials goes back at least to Calogero [17], more than 25 years before Lassalle!

The factorization phenomenon captured by Lemma 10.2.6, which is behind the optimality provided by Theorem 10.1.3, reminds some kind of concentration-compactness related to continuous spins systems as in [21] and [58] for instance. The factorization Lemma 10.2.6 remains valid for other ensembles such as the Beta-Ginibre ensemble with density proportional to

$$z \in \mathbb{C}^n \mapsto \mathrm{e}^{-n \sum_{k=1}^{n} |z_i|^2} \prod_{j<k} |z_j - z_k|^{\beta}, \tag{10.1.19}$$

see [25, Remark 5.4] for the case $n = \beta = 2$. However, in contrast with the Beta-Hermite ensemble, the interaction term is not convex in the complex case, and it is not clear at all what are the Poincaré and log-Sobolev constants of the Ginibre ensemble. See [25] for an upper bound and further discussions on the associated dynamics.

10.2 Useful or Beautiful Facts

10.2.1 Random Matrices, GUE, and Beta Hermite Ensemble

The following result from random matrix theory goes back to Dyson, see [1, 37, 40, 62].

Theorem 10.2.1 (Gaussian Random Matrices and GUE) *The Gaussian unitary ensemble* μ *defined by* (10.1.4) *is the law of the ordered eigenvalues of a random* $n \times n$ *Hermitian matrix* H *with density proportional to* $h \mapsto$ $e^{-\frac{n}{2}\mathrm{Trace}(h^2)} = e^{-\frac{n}{2}\sum_{i=1}^{n} h_{ii}^2 - n\sum_{i<j} |h_{ij}|^2}$ *in other words the* n^2 *real random variables* $\{H_{ii}, \Re H_{ij}, \Im H_{ij}\}_{1 \leq i < j \leq n}$ *are independent, with* $\Re H_{ij}$ *and* $\Im H_{ij} \sim \mathcal{N}(0, 1/(2n))$ *for any* $i < j$ *and* $H_{ii} \sim \mathcal{N}(0, 1/n)$ *for any* $1 \leq i \leq n$.

There is an analogue theorem for the GOE case $\beta = 1$ with random Gaussian real symmetric matrices, and for the GSE case $\beta = 4$ with random Gaussian quaternion selfdual matrices. The following result holds for all beta Hermite ensemble (10.1.3), see [34].

Theorem 10.2.2 (Tridiagonal Random Matrix Model for Beta Hermite Ensemble) *The beta Hermite ensemble* μ *defined by* (10.1.3) *is the distribution of the ordered eigenvalues of the random tridiagonal symmetric* $n \times n$ *matrix*

$$
H = \frac{1}{\sqrt{2n}}
\begin{pmatrix}
\mathcal{N}(0, 2) & \chi_{(n-1)\beta} & & & \\
\chi_{(n-1)\beta} & \mathcal{N}(0, 2) & \chi_{(n-2)\beta} & & \\
& \ddots & \ddots & \ddots & \\
& & \chi_{2\beta} & \mathcal{N}(0, 2) & \chi_\beta \\
& & & \chi_\beta & \mathcal{N}(0, 2)
\end{pmatrix}
$$

where, up to the scaling prefactor $1/\sqrt{2n}$, *the entries in the upper triangle including the diagonal are independent, follow a Gaussian law* $\mathcal{N}(0, 2)$ *on the diagonal, and* χ-*laws just above the diagonal with a decreasing parameter with step* β *from* $(n-1)\beta$ *to* β.

In particular the trace follows the Gaussian law $\mathcal{N}(0, 1)$. Such random matrix models with independent entries allow notably to compute moments of (10.1.3) via traces of powers.

10.2.2 Isotropy of Beta Hermite Ensembles

This helps to understand the structure. Let μ be the beta Hermite ensemble (10.1.3), and let $\widetilde{\mu}$ be the probability measure obtained from μ by symmetrizing coordinates: For every test function $f : \mathbb{R}^n \to \mathbb{R}$ we have

$$
\int f \, d\widetilde{\mu} = \int f_* \, d\mu
$$

where f_* is the symmetrization of f, defined by

$$
f_*(x_1, \ldots, x_n) = \frac{1}{n!} \sum_{\sigma \in \Sigma_n} f(x_{\sigma(1)}, \ldots, x_{\sigma(n)})
$$

where Σ_n is the symmetric group of permutations of $\{1, \ldots, n\}$. Of course the probability measures μ and $\widetilde{\mu}$ coincide on symmetric test functions. The probability measure $\widetilde{\mu}$ is by definition invariant by permutation of the coordinates, and its density with respect to the Lebesgue measure is

$$\frac{\mathrm{d}\widetilde{\mu}}{\mathrm{d}x} = \frac{\mathrm{e}^{-\frac{n}{2}|x|^2}}{n! Z_\mu} \prod_{i < j} |x_i - x_j|^\beta$$

Note that the support of $\widetilde{\mu}$ is the whole space and that $\widetilde{\mu}$ is not log-concave, even though μ is.

Corollary 10.2.3 (Isotropy of Beta Hermite Ensemble) *For every $1 \leq i \neq j \leq n$,*

$$\int x_i \, \mathrm{d}\widetilde{\mu} = 0, \quad \int x_i^2 \, \mathrm{d}\widetilde{\mu} = \frac{\beta}{2} + \frac{2 - \beta}{2n}, \quad \int x_i x_j \, \mathrm{d}\widetilde{\mu} = -\frac{\beta}{2n}.$$

In particular, the law $\widetilde{\mu}$ is asymptotically isotropic.

Recall that isotropy means zero mean and covariance matrix multiple of the identity.

In the extremal case $\beta = 0$, the measure $\widetilde{\mu}$ is the Gaussian law $\mathcal{N}(0, \frac{1}{n} I_n)$.

Proof of Corollary 10.2.3 Observe first that if $X \sim \mu$ then $\sum X_i$ is a standard Gaussian. This can be seen using Theorem 10.2.2, and observing that $\sum X_i$ coincides with the trace of the matrix H. Actually this is true regardless of the interaction potential W, see Lemma 10.2.6 below. In particular

$$\int (x_1 + \cdots + x_n) \, \mu(\mathrm{d}x) = 0,$$

hence, by definition $\widetilde{\mu}$,

$$\int x_i \, \widetilde{\mu}(\mathrm{d}x) = \frac{1}{n} \int (x_1 + \cdots + x_n) \, \mu(\mathrm{d}x) = 0,$$

for every $i \leq n$. Since $\sum X_i$ is a standard Gaussian we also have

$$\int (x_1 + \cdots + x_n)^2 \, \mu(\mathrm{d}x) = 1. \tag{10.2.1}$$

Next we compute $\int |x|^2 \, \mathrm{d}\mu$. This can be done using Theorem 10.2.2, namely

$$\int |x|^2 \, \mu(\mathrm{d}x) = \mathbb{E}(\mathrm{Trace}(H^2)) = 1 + \frac{\beta}{n} \sum_{k=1}^{n-1} k = 1 + \frac{(n-1)\beta}{2}. \tag{10.2.2}$$

Note that the matrix model gives more: indeed, using the algebra of the Gamma laws,

$$\text{Trace}(H^2) \sim \text{Gamma}\left(\frac{n}{2} + \frac{\beta n(n-1)}{4}, \frac{n}{2}\right).$$

Alternatively one can use the fact that the square of the norm $|\cdot|^2$ is, up to an additive constant, an eigenvector of G, see Remark 10.1.7. Namely, recall the definition (10.1.15) of the operator G and note that

$$G(|\cdot|^2)(x) = 2n - 2n|x|^2 + 2\beta \sum_{i \neq j} \frac{x_i}{x_i - x_j} = 2n - 2n|x|^2 + n(n-1)\beta.$$

In particular $G(|\cdot|^2) \in L^2(\mu)$. Since μ is stationary, we then have $\int G|x|^2 \, d\mu = 0$, and we thus recover (10.2.2).

Combining (10.2.1) and (10.2.2) we get

$$\int x_i^2 \, \widetilde{\mu}(\mathrm{d}x) = \frac{1}{n} \int |x|^2 \, \mu(\mathrm{d}x) = \frac{\beta}{2} + \frac{2-\beta}{2n},$$

and

$$\int x_i x_j \, \widetilde{\mu}(\mathrm{d}x) = \frac{1}{n(n-1)} \int (x_1 + \cdots + x_n)^2 - (x_1^2 + \cdots + x_n^2) \, \widetilde{\mu}(\mathrm{d}x) = -\frac{\beta}{2n}.$$

$\square$

Remark 10.2.4 (Mean and Covariance of Beta Hermite Ensembles) Let μ and $\widetilde{\mu}$ be as in Corollary 10.2.3. In contrast with the probability measure $\widetilde{\mu}$, the probability measure μ is log-concave but is not centered, even asymptotically as $n \to \infty$, and this is easily seen from $0 \notin D$. Moreover, if $X_n = (X_{n,1}, \ldots, X_{n,n}) \sim \mu$ then the famous Wigner theorem for the beta Hermite ensemble, see for instance [45], states that almost surely and in L^1, regardless of the way we choose the common probability space,

$$\frac{1}{n} \sum_{i=1}^{n} \delta_{X_{n,i}} \xrightarrow[n \to \infty]{\text{weak}} \nu_\beta \tag{10.2.3}$$

where

$$\nu_\beta = \arg\inf_{\mu} \left(\int \frac{x^2}{2} \mathrm{d}\mu(x) - \beta \iint \log(x-y) \mathrm{d}\mu(x)\mathrm{d}\mu(y) \right)$$

$$= \frac{\sqrt{2\beta - x^2}}{\beta\pi} \mathbb{1}_{[-\sqrt{2\beta}, \sqrt{2\beta}]}(x)\mathrm{d}x. \tag{10.2.4}$$

This follows for instance from a large deviation principle. Moreover it can be shown that $X_{n,1} \xrightarrow[n \to \infty]{} \sqrt{2\beta}$ and $X_{n,n} \xrightarrow[n \to \infty]{} -\sqrt{2\beta}$. This suggests in a sense that asymptotically, as $n \to \infty$, the mean is supported by the whole interval $[-\sqrt{2\beta}, \sqrt{2\beta}]$. It is quite natural to ask about the asymptotic shape of the covariance matrix of μ. Elements of answer can be found in the work of Gustavsson [46].

10.2.3 Log-Concavity and Curvature

The following Lemma is essentially the key of the proof of Theorems 10.1.1 and 10.1.2.

Lemma 10.2.5 (Log-Concavity and Curvature) *Let μ be as in* (10.1.1). *Then U is ρ-convex. In particular, for the beta Hermite ensemble* (10.1.3), *the potential U is n-convex, for all $\beta > 0$.*

Proof Recall from (10.1) that $U(x) = V(x) + U_W(x)$. Observe that U_W is convex as a sum of linear maps composed with the convex function W. Thus, if V is ρ-convex then so is U. □

10.2.4 Factorization by Projection

The following factorization lemma is the key of the proof of Theorem 10.1.3. Let u be the unit vector of $\mathbb{R}^n$ given by the diagonal direction:

$$u = \frac{1}{\sqrt{n}}(1, \ldots, 1)$$

and let π and $\pi^\perp$ be the orthogonal projection onto $\mathbb{R}u$ and $(\mathbb{R}u)^\perp = \{v \in \mathbb{R}^n : \langle v, u \rangle = 0\}$.

Lemma 10.2.6 (Factorization by Projection) *Let μ be as in* (10.1.1) *and let X be a random vector distributed according to μ. Assume that the confinement potential V is quadratic: $V = \rho |\cdot|^2 / 2$ for some $\rho > 0$. Then μ has a Gaussian factor in the direction u in the sense that $\pi(X) = \langle X, u \rangle u$ and $\pi^\perp(X)$ are independent and*

$$\langle X, u \rangle \sim \mathcal{N}\left(0, \frac{1}{\rho}\right).$$

Moreover $\pi^\perp(X)$ has density proportional to e^{-U} with respect to the Lebesgue measure on $(\mathbb{R}u)^\perp$.

In the special case of the beta Hermite ensemble (10.1.3), the law of $\langle X, u \rangle = \text{Trace}(H)/\sqrt{n}$ is easily seen on the random matrix model H provided by Theorems 10.2.1 and 10.2.2.

An extension of Lemma 10.2.6 to higher dimensional gases in considered is [26].

Proof of Lemma 10.2.6 Since $x = \pi(x) + \pi^\perp(x)$ and $\pi(x) = \langle x, u \rangle u$, we have

$$|x|^2 = \langle x, u \rangle^2 + |\pi^\perp(x)|^2.$$

Besides it is easily seen that $U_W(x) = U_W(\pi^\perp(x))$ for all x, a property which comes from the shift invariance of the interaction energy U_W along $\mathbb{R}u$. Therefore

$$\mathrm{e}^{-U(x)} = \mathrm{e}^{-\rho\langle x,u \rangle^2/2} \times \mathrm{e}^{-\rho|\pi^\perp(x)|^2/2 - U_W(\pi^\perp(x))} = \mathrm{e}^{-\rho\langle x,u \rangle^2/2} \times \mathrm{e}^{-U(\pi^\perp(x))}.$$

So the density of X is the product of a function of $\langle x, u \rangle$ by a function of $\pi^\perp(x)$. □

The result extends naturally by the same proof to the more general quadratic case $V = \langle Ax, x \rangle$ where A is a symmetric positive definite $n \times n$ matrix, provided that the diagonal direction u is an eigenvector of A.

Remark 10.2.7 (Gaussian Factor and Orthogonal Polynomials) Let μ be as in Lemma 10.2.6. Let H_i and H_j be two distinct univariate (Hermite) orthogonal polynomials with respect to the standard Gaussian law $\mathcal{N}(0, I_n)$. Then it follows from Lemma 10.2.6 that the symmetric multivariate polynomials $H_i(\sqrt{\rho/n}(x_1 + \cdots + x_n))$ and $H_j(\sqrt{\rho/n}(x_1 + \cdots + x_n))$ are orthogonal with respect to μ. In particular, when $\rho = n$ and with $H_i(x) = x$ and $H_j(x) = x^2 - 1$, we get that $x_1 + \cdots + x_n$ and $(x_1 + \cdots + x_n)^2 - 1$ are orthogonal for μ.

10.3 Proofs

10.3.1 Proof of Theorems 10.1.1 and 10.1.2

Proof of Theorems 10.1.1 and 10.1.2 Let us first mention that Theorem 10.1.2 actually implies Theorem 10.1.1. Indeed it is well-known that applying log-Sobolev to a function f of the form $f = 1 + \epsilon h$ and letting ϵ tend to 0 yields the Poincaré inequality for h, with half the constant if the log-Sobolev inequality. See for instance [2] or [7] for details.

In the discussion below, we call *potential* of a probability measure μ the function $-\log \rho$, where ρ is the density of μ with respect to the Lebesgue measure. In view of Lemma 10.2.5 it is enough to prove that a probability measure μ on $\mathbb{R}^n$ whose potential U is ρ-convex for some positive ρ satisfies the logarithmic Sobolev inequality with constant $2/\rho$. This is actually a well-known fact. It can be seen in various ways which we briefly spell out now. Some of these arguments require extra assumptions on U, namely that the domain of U equals $\mathbb{R}^n$ (equivalently μ has full support) and that U is C^2-smooth on $\mathbb{R}^n$. For this reason we first explain a regularization procedure showing that these hypothesis can be added without loss of generality.

Regularization Procedure Let γ be the Gaussian measure whose density is proportional to $e^{-\rho|x|^2/2}$ and let f be the density of μ with respect to γ. Clearly U is ρ-convex if and only if $\log f$ is concave. Next let (Q_t) be the Ornstein–Uhlenbeck semigroup having γ as a stationary measure, namely for every test function g

$$Q_t g(x) = \mathbb{E}\left[g\left(e^{-t}x + \sqrt{1 - e^{-2t}}G\right)\right]$$

where $G \sim \gamma$. Since γ is reversible for (Q_t) the measure μQ_t has density $Q_t f$ with respect to γ. Moreover the semigroup (Q_t) satisfies the following property

$$f \text{ log-concave} \quad \Rightarrow \quad Q_t f \text{ log-concave}.$$

This is indeed an easy consequence of the Prékopa–Leindler inequality, see (10.3.3) below. As a result the potential U_t of μQ_t is also ρ-convex. Besides U_t is clearly C^∞ smooth on the whole $\mathbb{R}^n$. Lastly since $\lim_{t \to 0} Q_t f(x) = f(x)$ for almost every x, we have $\mu P_t \to \mu$ weakly as t tends to 0. As a result, if μP_t satisfies log-Sobolev with constant $2/\rho$ for every t, then so does μ.

First Proof: The Brascamp–Lieb Inequality A theorem due to Brascamp and Lieb [13] states that if the potential of μ is smooth and satisfies $\text{Hess}(U)(x) > 0$ for all $x \in \mathbb{R}^n$, then for any C^∞ compactly supported test function $f : \mathbb{R}^n \to \mathbb{R}$, we have the inequality

$$\text{var}_\mu(f) \leq \int_{\mathbb{R}^n} \left\langle \text{Hess}(U)^{-1}\nabla f, \nabla f \right\rangle d\mu.$$

If U is ρ-convex then $\text{Hess}(U)^{-1} \leq (1/\rho)I_n$ and we obtain

$$\text{var}_\mu(f) \leq \frac{1}{\rho} \int_{\mathbb{R}^n} |\nabla f|^2 \, d\mu.$$

The extension of this inequality to all $f \in H^1(\mu)$ follows by truncation and regularization. Note that this method only works for Poincaré. The Brascamp–Lieb inequality does not seem to admit a logarithmic Sobolev inequality counterpart, see [12] for a discussion.

Second Proof: Caffarelli's Contraction Theorem Again let γ be the Gaussian measure on $\mathbb{R}^n$ whose density is proportional to $e^{-\rho|x|^2/2}$. The theorem of Caffarelli [15, 16] states that if the potential of μ is ρ-convex then the Brenier map from γ to μ is 1-Lipschitz. This easily implies that the Poincaré constant of μ is at least as good as that of γ, namely $1/\rho$. Let us sketch the argument briefly. Let T

be the Brenier map from γ to μ and let f be a smooth function on $\mathbb{R}^n$. Using the fact that T pushes forward γ to μ, the Poincaré inequality for γ and the Lipschitz property of T we get

$$
\begin{aligned}
\mathrm{var}_\mu(f) = \mathrm{var}_\gamma(f \circ T) &\leq \frac{1}{\rho} \int_{\mathbb{R}^n} |\nabla(f \circ T)|^2 \, \mathrm{d}\gamma \\
&\leq \frac{1}{\rho} \int_{\mathbb{R}^n} |\nabla f|^2 \circ T \, \mathrm{d}\gamma \qquad (10.3.1) \\
&= \frac{1}{\rho} \int_{\mathbb{R}^n} |\nabla f|^2 \, \mathrm{d}\mu.
\end{aligned}
$$

This contraction principle works just the same for log-Sobolev.

Third Proof: The Bakry–Émery Criterion Assume that U is finite and smooth on the whole $\mathbb{R}^n$ and consider the Langevin diffusion

$$
\mathrm{d}X_t = \sqrt{2} \, \mathrm{d}B_t - \nabla U(X_t) \, \mathrm{d}t.
$$

The generator of the diffusion is the operator $\mathrm{G} = \Delta - \langle \nabla U, \nabla \rangle$. The carré du champ Γ and its iterated version Γ_2 are easily computed:

$$
\begin{aligned}
\Gamma(f, g) &= \frac{1}{2}(\mathrm{G}(fg) - f\mathrm{G}(g) - g\mathrm{G}(f)) = \langle \nabla f, \nabla g \rangle \\
\Gamma_2(f, g) &= \frac{1}{2}(\mathrm{G}\Gamma(f, g) - \Gamma(f, \mathrm{G}g) - \Gamma(\mathrm{G}f, g)) \qquad (10.3.2) \\
&= \mathrm{Tr}\,(\mathrm{Hess}(f)\mathrm{Hess}(g)) + \langle \mathrm{Hess}(U)\nabla f, \nabla g \rangle.
\end{aligned}
$$

We also set $\Gamma(f) = \Gamma(f, f)$ and similarly for Γ_2. The hypothesis that U is ρ-convex thus implies that

$$
\Gamma_2(f) \geq \rho\Gamma(f),
$$

for every suitable f. Actually this inequality is equivalent to the condition that U is ρ-convex, as can be seen by plugging in linear functions. In the language of Bakry–Émery, see [2, 6, 7], the diffusion satisfies the curvature dimension criterion $\mathrm{CD}(\rho, \infty)$. This criterion implies that the stationary measure μ satisfies the following logarithmic Sobolev inequality

$$
\mathrm{ent}_\mu(f^2) \leq \frac{2}{\rho} \int_{\mathbb{R}^n} \Gamma(f) \, d\mu,
$$

see [7, Proposition 5.7.1]. Formally this proof also works if μ does not have full support by adding a reflection at the boundary, just as in Sect. 10.1.2. However this poses some technical issues which are not always easy to overcome. As a matter of fact, diffusions with reflecting boundary conditions are not treated in the book [7].

Fourth Proof: An Argument of Bobkov and Ledoux This fourth proof is the one that requires the least background. Another nice feature is that the regularization procedure is not needed for this proof. It is based on the Prékopa-Leindler inequality. The latter, which is a functional form of the Brunn–Minkowski inequality, states that if f, g, h are functions on $\mathbb{R}^n$ satisfying

$$(1 - t)f(x) + tg(y) \leq h((1 - t)x + ty)$$

for every $x, y \in \mathbb{R}^n$ and for some $t \in [0, 1]$, then

$$\left(\int e^f \, dx \right)^{1-t} \left(\int e^g \, dx \right)^t \leq \int e^h \, dx. \tag{10.3.3}$$

We refer to [8] for a nice presentation of this inequality. Let $F : \mathbb{R}^n \to \mathbb{R}$ be a smooth function with compact support, and for $s > 0$ let $R_s F$ be the infimum convolution

$$R_s F(x) = \inf_{y \in \mathbb{R}^n} \left\{ F(x + y) + \frac{1}{2s} |y|^2 \right\}.$$

Fix $t \in (0, 1)$. Using the ρ-convexity of U:

$$(1 - t)U(x) + tU(y) \leq U((1 - t)x + ty) - \frac{\rho t(1 - t)}{2} |x - y|^2,$$

it is easily seen that the functions $f = R_{t/\rho}F - U$, $g = -U$ and $h = (1 - t)F - U$ satisfy the hypothesis of the Prékopa-Leindler inequality. The conclusion (10.3.3) rewrites in this case

$$\left(\int_{\mathbb{R}^n} e^{R_{t/\rho}F} \, d\mu \right)^{1-t} \leq \int_{\mathbb{R}^n} e^{(1-t)F} \, d\mu. \tag{10.3.4}$$

It is well-known that (R_s) solves the Hamilton–Jacobi equation

$$\partial R_s F + \frac{1}{2} |\nabla R_s F|^2 = 0,$$

see for instance [39]. Using this and differentiating the inequality (10.3.4) at $t = 0$ yields

$$\mathrm{ent}_\mu(e^F) \leq \frac{1}{2\rho} \int_{\mathbb{R}^n} |\nabla F|^2 e^F \, d\mu$$

which is equivalent to the desired log-Sobolev inequality. We refer to the article [12] for more details.   □

10.3.2 Proof of Theorem 10.1.3 and Comments on Optimality

In the case of the beta ensemble, Theorem 10.1.6 shows that $x \in \mathbb{R}^n \mapsto x_1 + \cdots + x_n$ is the only symmetric function optimal in the Poincaré inequality, up to additive and multiplicative constants. Our goal now is to study the optimality far beyond this special case.

We have just seen that if a measure μ has density $\mathrm{e}^{-\phi}$ where ϕ is ρ-convex for some $\rho > 0$ then it satisfies Poincaré with constant $1/\rho$. This constant is sharp in the case where μ is the Gaussian measure whose density is proportional to $\mathrm{e}^{-\rho|x|^2/2}$. Indeed the Poincaré constant of that Gaussian measure is equal to $1/\rho$ and extremal functions are affine functions, see for instance [2, 7]. Similarly, its log-Sobolev constant is $2/\rho$ and extremal functions are log-affine functions, see [18]. The next lemma asserts that conversely, if the Poincaré constant of μ or its log-Sobolev constant matches the bound predicted by the strict convexity of its potential, then μ must be Gaussian in some direction.

Recall the notion of having a Gaussian factor in a given direction, used in Lemma 10.2.6.

Lemma 10.3.1 *Let μ be a probability measure on $\mathbb{R}^n$ with density $\mathrm{e}^{-\phi}$ where ϕ is ρ-convex for some $\rho > 0$, and assume that there exists a non constant function f such that*

$$\mathrm{var}_\mu(f) = \frac{1}{\rho} \int_{\mathbb{R}^n} |\nabla f|^2 \, \mathrm{d}\mu. \tag{10.3.5}$$

Then the following properties hold true:

(i) *The function f is affine: there exists a vector u and a constant b such that*

$$f(x) = \langle u, x \rangle + b.$$

(ii) *The measure μ has a Gaussian factor of variance $1/\rho$ in the direction u.*

Besides, there is a similar statement for the log-Sobolev inequality: if there exists a non constant function f such that

$$\mathrm{ent}_\mu(f^2) = \frac{2}{\rho} \int_{\mathbb{R}^n} |\nabla f|^2 \, \mathrm{d}\mu,$$

then $\log f$ is affine and μ has a Gaussian factor in the corresponding direction.

The Poincaré case is contained in the main result of Cheng and Zhou's article [27]. The general result is a consequence of the works of de Philippis and Figalli [30], and also Courtade and Fathi [29]. These authors actually establish a stability estimate for this lemma: if there exists a function f which is near optimal in Poincaré then μ nearly has a Gaussian factor. We sketch a proof of Lemma 10.3.1 based on their ideas.

Proof of Lemma 10.3.1 We analyze the equality case in the second proof of the main theorem, the one based on Caffarelli's contraction theorem. Recall that the Brenier map T from γ to μ is the gradient of a convex function and that it pushes forward γ to μ. Recall also that Caffarelli's theorem asserts that under the hypothesis of the lemma T is 1-Lipschitz. Therefore T is differentiable almost everywhere, and its differential is a symmetric matrix satisfying

$$0 \leq (\mathrm{d}T)_x \leq I_n \qquad (10.3.6)$$

as quadratic forms. Now, observe that if f satisfies (10.3.5) then every inequality in (10.3.1) must actually be an equality. In particular $f \circ T$ must be optimal in the Poincaré inequality for γ. This implies that $f \circ T$ is affine. Also there is equality in the inequality

$$|\nabla(f \circ T)(x))|^2 \leq |\nabla f(Tx)|^2$$

for almost every x. Because of (10.3.6) this actually implies that

$$(\mathrm{d}T)_x(\nabla f(Tx)) = \nabla f(Tx),$$

for almost every x. Since $f \circ T$ is affine the left hand side is constant, and we obtain that f itself must be affine. Thus, there exists a vector u and a constant b such that $f(x) = \langle u, x \rangle + b$, and moreover $(\mathrm{d}T)_x(u) = u$ for almost every x. By a change of variable, we can assume that u is a multiple of the first coordinate vector. The differential of T at x thus has the form

$$(\mathrm{d}T)_x = \begin{pmatrix} 1 & 0 \\ 0 & * \end{pmatrix}$$

for almost every x. Therefore

$$T(x_1, \ldots, x_n) = (x_1 + a, S(x_2, \ldots, x_n))$$

for some constant a and some map S from $\mathbb{R}^{n-1}$ to itself. This implies that the image μ of γ by T is a product measure, and that the first factor is the Gaussian measure with mean a and variance $1/\rho$. This finishes the proof of the first part of the lemma. The log-Sobolev version can be obtained very similarly and we omit the details. $\square$

Remark 10.3.2 (Alternate Proof Based on the Bakry–Émery Calculus) We spell out briefly an alternative proof of Lemma 10.3.1 based on the Bakry–Émery calculus starting from a work by Ledoux [54]. Let G, Γ, and Γ_2 be as in (10.3.2), and let $(P_t)_{t \geq 0}$ be the Markov semigroup generated by G. The usual Bakry–Émery method gives, up to regularity considerations,

$$\mathrm{var}_\mu(f) = \frac{1}{\rho} \int \Gamma f \mathrm{d}\mu - \frac{2}{\rho} \int_0^\infty \left(\int (\Gamma_2 - \rho\Gamma)(P_t f)\mathrm{d}\mu \right) \mathrm{d}t.$$

This is a Taylor–Lagrange formula expressing the "deficit" in the Poincaré inequality. It shows that if $\Gamma_2 \geq \rho\Gamma$ and $\text{var}_\mu(f) = \frac{1}{\rho} \int \Gamma f \, d\mu$ then $(\Gamma_2 - \rho\Gamma)(P_t f)(x) = 0$ almost everywhere in t and x. Up to regularity issues, we get in particular

$$(\Gamma_2 - \rho\Gamma)(f) = 0.$$

In other words, denoting $\|\cdot\|_{\text{HS}}$ the Hilbert–Schmidt norm,

$$\|\text{Hess}(f)\|_{\text{HS}}^2 + \langle(\text{Hess}(U) - \rho I_n)\nabla f, \nabla f\rangle = 0.$$

Since both term are non negative this actually implies that

$$\text{Hess}(f) = 0 \quad \text{and} \quad \langle(\text{Hess}(U) - \rho I_n)\nabla f, \nabla f\rangle = 0.$$

Thus f is affine: there exists a unit vector u and two constants λ and c such that

$$f(x) = \lambda\langle u, x\rangle + c,$$

and moreover $\langle\text{Hess}(U)u, u\rangle = \rho$. Since $\text{Hess}(U) \geq \rho I_n$ this actually implies that

$$\text{Hess}(U)u = \rho u$$

pointwise. Proceeding as in the proof of Lemma 10.3.1, we then see that μ has a Gaussian factor of variance $1/\rho$ in the direction u. There is a similar argument for the log-Sobolev inequality using

$$\text{ent}_\mu(f) = \frac{1}{2\rho} \int \Gamma(\log f) \, f \, d\mu - \frac{1}{\rho} \int_0^\infty \left(\int (\Gamma_2 - \rho\Gamma)(\log P_t f) \, P_t f \, d\mu \right) dt.$$

This leads to the fact that if f is optimal in the logarithmic Sobolev inequality then f is of the form $f(x) = e^{\lambda\langle u, x\rangle + c}$ and μ has a Gaussian factor in the direction u. As usual, this seductive approach requires to justify rigorously the computations via delicate handling of regularity and smoothing, see for instance [7], [3], and [1, Section 4.4.2].

Proof of Theorem 10.1.3 According to Lemma 10.2.6, if $V = \rho |\cdot|^2 /2$ for some $\rho > 0$ then μ has a Gaussian factor in the diagonal direction $u = (1, \ldots, 1)/\sqrt{n}$. As we have seen, this Gaussian satisfies Poincaré with constant $1/\rho$ and log-Sobolev with constant $2/\rho$. Moreover, affine functions are optimal in Poincaré and log-affine functions are optimal in log Sobolev. This shows that we have equality in the Poincaré inequality of Theorem 10.1.1 for functions f of the form $f(x) = \lambda(x_1 + \cdots + x_n) + c$ for some constants λ and c, and equality in the logarithmic Sobolev inequality of Theorem 10.1.2 for functions whose logarithm is of the preceding type.

Let us now prove that these are the only optimal functions. Assume that f is non constant and extremal in the Poincaré inequality. Then by Lemma 10.3.1, there exists a vector v and a constant b such that $f(x) = \langle v, x \rangle + b$, and moreover μ has a Gaussian factor in the direction v. Since the support of μ is the set $\{x_1 \geq \cdots \geq x_n\}$ this can only happen if v is proportional to the diagonal direction u, which is the result. The proof for log-Sobolev is similar.   □

10.3.3   Proof of Corollary 10.1.4 and Comments on Concentration

Proof of Corollary 10.1.4 The Gaussian concentration can be deduced from the log-Sobolev inequality via an argument due to Herbst, see for instance [55], which consists in using log-Sobolev with $f = e^F$ to get the Gaussian upper bound on the Laplace transform

$$\int e^F \, d\mu \leq \exp\left(\int F \, d\mu + \frac{\|F\|_{\mathrm{Lip}}^2}{2\rho} \right), \tag{10.3.7}$$

which leads in turn to the concentration inequality (10.1.5) via the Markov inequality. Alternatively we can use the intermediate inequality (10.3.4) obtained in the course of the fourth proof of Theorem 10.1.2. Indeed applying Jensen's inequality to the right-hand side of (10.3.4) and letting $t \to 1$, we obtain

$$\int e^{R_{1/\rho}F} \, d\mu \leq \exp\left(\int F \, d\mu \right). \tag{10.3.8}$$

Moreover, if F is Lipschitz it is easily seen that

$$R_{1/\rho}F \geq F - \frac{1}{2\rho} \|F\|_{\mathrm{Lip}}^2.$$

Plugging this into the previous inequality yields (10.3.7). Note that a result due to Bobkov and Götze states that (10.3.8) is equivalent to a Talagrand W_2 transportation inequality for μ, see for instance [55] and references therein.

In the case $F(x_1, \ldots, x_n) = \frac{1}{n} \sum_{i=1}^n f(x_i) = L_n(f)(x)$ we have

$$\|F\|_{\mathrm{Lip}} \leq \frac{\|f\|_{\mathrm{Lip}}}{\sqrt{n}}$$

so that (10.1.6) follows from (10.1.5).

Finally taking $F(x_1, \ldots, x_n) = \max(x_1, \ldots, x_n) \; (= x_1 \text{ on } D)$ in (10.1.5) gives (10.1.7).   □

Remark 10.3.3 (Concentration of Measure in Transportation Distance) Following Gozlan (personal communication), it is possible to obtain concentration of measure inequalities in Kantorovich–Wasserstein distance W_2 from the Hoffman–Wielandt inequality. Namely, given a Hermitian matrix A, we let $x_1(A) \geq \cdots \geq x_n(A)$ be the eigenvalues of A, arranged in decreasing order, and

$$L_A = \frac{1}{n} \sum_{i=1}^{n} \delta_{x_i(A)}$$

be the corresponding empirical measure. If B is another Hermitian matrix, we get from the Hoffman–Wielandt inequality

$$nW_2(L_A, L_B)^2 = \sum_{i=1}^{n}(x_i(A) - x_i(B))^2 \leq \mathrm{Trace}((A - B)^2) = \|A - B\|_{\mathrm{HS}}^2.$$

$$(10.3.9)$$

Thanks to the triangle inequality for W_2, this implies that for every probability measure v on $\mathbb{R}$ with finite second moment, the Lipschitz constant of the map $A \mapsto W_2(L_A, \mu)$ with respect to the Hilbert–Schmidt norm is at most $1/\sqrt{n}$. If G is a Gaussian matrix with density proportional to $\mathrm{e}^{-n\mathrm{Trace}(X^2)}$, the Gaussian concentration inequality then yields

$$\mathbb{P}\left(|W_2(L_G, v) - \mathbb{E}W_2(L_G, v)| > r\right) \leq 2\mathrm{e}^{-\frac{n^2}{2}r^2}.$$

Note that v is arbitrary. This inequality can be reformulated as follows: If μ is the Gaussian unitary ensemble in $\mathbb{R}^n$ and L is the map $x \in \mathbb{R}^n \mapsto \frac{1}{n}\sum_{i \leq n} \delta_{x_i}$ then for any probability measure v on $\mathbb{R}$ we have

$$\mu\left(\left|W_2(L, v) - \int W_2(L, v)\, d\mu\right| > r\right) \leq 2\mathrm{e}^{-\frac{n^2}{2}r^2}.$$

More generally this inequality remains valid when μ is the law of the eigenvalues of a random matrix satisfying Gaussian concentration with rate n. This is the case for instance if the matrix has independent entries satisfying a logarithmic Sobolev inequality with constant $1/n$.

Remark 10.3.4 (Proof for GUE/GOE via Hoffmann–Wielandt Inequality) For the GUE and the GOE one can give a fifth proof, based on the contraction principle, like the proof using Caffarelli's theorem above. The Hoffman–Wielandt inequality [11, 47, 49], states that for all $n \times n$ Hermitian matrices A and B with ordered eigenvalues $x_1(A) \geq \cdots \geq x_n(A)$ and $x_1(B) \geq \cdots \geq x_n(B)$ respectively, we have

$$\sum_{i=1}^{n}(x_i(A) - x_i(B))^2 \leq \sum_{i,j=1}^{n} |A_{ij} - B_{ij}|^2.$$

In other words the map which associates to an $n \times n$ Hermitian matrix A its vector of eigenvalues $(x_1(A), \ldots, x_n(A)) \in \mathbb{R}^n$ is 1-Lipschitz for the Euclidean structure on $n \times n$ Hermitian matrices, given by $\langle A, B \rangle = \text{Trace}(AB)$. On the other hand, as we saw in Sect. 10.2.1, the Gaussian unitary ensemble is the image by this map of the Gaussian measure on $\mathbb{H}_n$ whose density is proportional to $e^{-(n/2)\text{Trace}(H^2)}$. The Poincaré constant of this Gaussian measure is $1/n$ so by the contraction principle the Poincaré constant of the GUE is $1/n$ at most. The argument works similarly for log-Sobolev and for the GOE.

10.3.4 Proof of Theorem 10.1.5

Proof of Theorem 10.1.5 The exponential decay of relative entropy (10.1.13) is a well-known consequence of the logarithmic Sobolev inequality, see for instance [7, Theorem 5.2.1]. The decay in Wasserstein distance follows from the Bakry–Émery machinery, see [7, Theorem 9.7.2]. Alternatively it can be seen using parallel coupling. We explain this argument briefly.

Let X and Y be two solutions of the stochastic differential equation (10.1.8) driven by the same Brownian motion:

$$dX_t = \sqrt{2}\, dB_t - \nabla U(X_t)\, dt + d\Phi_t$$

$$dY_t = \sqrt{2}\, dB_t - \nabla U(Y_t)\, dt + d\Psi_t,$$

where Φ and Ψ are the reflections at the boundary of the Weyl chamber of X and Y respectively, see Sect. 10.1.2 for a precise definition. Assume additionally that $X_0 \sim \nu_0$, $Y_0 \sim \nu_1$ and that

$$\mathbb{E}(|X_0 - Y_0|^p) = W_p(\nu_0, \nu_1)^p.$$

Observe that

$$d|X_t - Y_t|^2 = -2\langle X_t - Y_t, \nabla U(X_t) - \nabla U(Y_t)\rangle\, dt + 2\langle X_t - Y_t, d\Phi_t\rangle + 2\langle Y_t - X_t, d\Psi_t\rangle.$$

Since U is ρ-convex $\langle X_t - Y_t, \nabla U(X_t) - \nabla U(Y_t)\rangle \geq \rho|X_t - Y_t|^2$. Besides $d\Phi_t = -n_t dL_t$ where L is the local time of X at the boundary of the Weyl chamber D and n_t is an outer unit normal at X_t. Since $Y_t \in D$ and since D is convex we get in particular $\langle X_t - Y_t, d\Phi_t\rangle \leq 0$ for all t, and similarly $\langle Y_t - X_t, d\Psi_t\rangle \leq 0$. Thus $d|X_t - Y_t|^2 \leq -2\rho|X_t - Y_t|^2\, dt$, hence

$$|X_t - Y_t| \leq e^{-\rho t}|X_0 - Y_0|.$$

Taking the p-th power and expectation we get, in $[0, +\infty]$,

$$\mathbb{E}[|X_t - Y_t|^p]^{1/p} \leq e^{-\rho t}\, \mathbb{E}(|X_0 - Y_0|^p)^{1/p} = e^{-\rho t}W_p(\nu_0, \nu_1).$$

Moreover since $X_t \sim \nu_0 P_t$ and $Y_t \sim \nu_1 P_t$ we have by definition of W_p

$$W_p(\nu_0 P_t, \nu_1 P_t) \leq \mathbb{E}[|X_t - Y_t|^p]^{1/p}.$$

Hence the result.   □

Acknowledgements This work is linked with the French research project ANR-17-CE40-0030 - EFI - Entropy, flows, inequalities. A significant part was carried out during a stay at the Institute for Computational and Experimental Research in Mathematics (ICERM), during the 2018 Semester Program on "Point Configurations in Geometry, Physics and Computer Science", thanks to the kind invitation by Edward Saff and Sylvia Serfaty. We thank also Sergio Andraus, François Bolley, Nizar Demni, Peter Forrester, Nathaël Gozlan, Michel Ledoux, Mylène Maïda, and Elizabeth Meckes for useful discussions and some help to locate references. We would also like to thank the anonymous referee for his careful reading of the manuscript and his suggestion to use [27, 29, 30] to prove Lemma 10.3.1 and Theorem 10.1.3. This note takes its roots in the blog post [23].

References

1. G.W. Anderson, A. Guionnet, O. Zeitouni, An introduction to random matrices, in *Cambridge Studies in Advanced Mathematics*, vol 118 (Cambridge University Press, Cambridge, 2010)
2. C. Ané, S. Blachère, D. Chafaï, P. Fougères, I. Gentil, F. Malrieu, C. Roberto, G. Scheffer, Sur les inégalités de Sobolev logarithmiques, in *Panoramas et Synthèses*, vol 10 (Society Mathematics France, Paris, 2000)
3. A. Arnold, P. Markowich, G. Toscani, A. Unterreiter, On convex Sobolev inequalities and the rate of convergence to equilibrium for Fokker-Planck type equations. Comm. Partial Differ. Equ. **26**(1–2), 43–100 (2001)
4. G. Aubrun, A sharp small deviation inequality for the largest eigenvalue of a random matrix, in *Séminaire de Probabilités XXXVIII*. Lecture Notes in Mathematics, vol 1857, pp. 320–337 (Springer, Berlin, 2005)
5. T.H. Baker, P.J. Forrester, The Calogero-Sutherland model and generalized classical polynomials. Comm. Math. Phys. **188**(1), 175–216 (1997)
6. D. Bakry, M. Émery, Diffusions hypercontractives, in *Séminaire de probabilités, XIX, 1983/84*. Lecture Notes in Mathematics, vol 1123, pp. 177–206 (Springer, Berlin, 1985)
7. D. Bakry, I. Gentil, M. Ledoux, in *Analysis and Geometry of Markov Diffusion Operators*. Grundlehren der Mathematischen Wissenschaften, vol 348 (Springer, Cham, 2014)
8. K. Ball, An elementary introduction to modern convex geometry. in *Flavors of Geometry*. Mathematical Sciences Research Institute Publications, vol 31, pp. 1–58 (Cambridge University Press, Cambridge, 1997)
9. W. Beckner, A generalized Poincaré inequality for Gaussian measures. Proc. Am. Math. Soc. **105**(2), 397–400 (1989)
10. G. Ben Arous, A. Guionnet, Large deviations for Wigner's law and Voiculescu's non-commutative entropy. Probab. Theory Relat. Fields **108**(4), 517–542 (1997)
11. R. Bhatia, in *Matrix Analysis*. Graduate Texts in Mathematics, vol 169 (Springer, New York, 1997)
12. S.G. Bobkov, M. Ledoux, From Brunn-Minkowski to Brascamp-Lieb and to logarithmic Sobolev inequalities. Geom. Funct. Anal. **10**(5), 1028–1052 (2000)
13. H.J. Brascamp, E.H. Lieb, On extensions of the Brunn-Minkowski and Prékopa-Leindler theorems, including inequalities for log concave functions, and with an application to the diffusion equation. J. Funct. Anal. **22**(4), 366–389 (1976)
14. M.-F. Bru, Wishart processes. J. Theor. Probab. **4**(4), 725–751 (1991)

15. L.A. Caffarelli, Monotonicity properties of optimal transportation and the FKG and related inequalities. Comm. Math. Phys. **214**(3), 547–563 (2000)
16. L.A. Caffarelli, Erratum: "Monotonicity properties of optimal transportation and the FKG and related inequalities" [Comm. Math. Phys. **214** (2000), no. 3, 547–563; MR1800860 (2002c:60029)]. Comm. Math. Phys. **225**(2), 449–450 (2002)
17. F. Calogero, Solution of the one-dimensional N-body problems with quadratic and/or inversely quadratic pair potentials. J. Math. Phys. **12**, 419–436 (1971)
18. E.A. Carlen, Superadditivity of Fisher's information and logarithmic Sobolev inequalities. J. Funct. Anal. **101**(1), 194–211 (1991)
19. E. Cépa, Équations différentielles stochastiques multivoques, in *Séminaire de Probabilités, XXIX*. Lecture Notes in Mathematics, vol 1613, pp. 86–107 (Springer, Berlin, 1995)
20. E. Cépa, D. Lépingle, Diffusing particles with electrostatic repulsion. Probab. Theory Relat. Fields **107**(4), 429–449 (1997)
21. D. Chafaï, Glauber versus Kawasaki for spectral gap and logarithmic Sobolev inequalities of some unbounded conservative spin systems. Markov Process. Related Fields **9**(3), 341–362 (2003)
22. D. Chafaï, Entropies, convexity, and functional inequalities: on Φ-entropies and Φ-Sobolev inequalities. J. Math. Kyoto Univ. **44**(2), 325–363 (2004)
23. D. Chafaï, http://djalil.chafai.net/blog/2016/12/27/mind-the-gap/. blogpost (2016)
24. D. Chafaï, N. Gozlan, P.-A. Zitt, First-order global asymptotics for confined particles with singular pair repulsion. Ann. Appl. Probab. **24**(6), 2371–2413 (2014)
25. D. Chafaï, F. Bolley, J. Fontbona, *Dynamics of a Planar Coulomb Gas* (2017). Preprint arXiv:1706.08776 to appear in Ann. Appl. Probab.
26. D. Chafaï, G. Ferré, G. Stoltz, *Coulomb Gases Under Constraint: Some Theoretical and Numerical Results* (2019). Preprint
27. X. Cheng, D. Zhou, Eigenvalues of the drifted Laplacian on complete metric measure spaces. Commun. Contemp. Math. **19**(1), 1650001, 17 (2017)
28. L. Chris, G. Rogers, Z, Shi, Interacting Brownian particles and the Wigner law. Probab. Theory Relat. Fields **95**(4), 555–570 (1993)
29. T.A. Courtade, M. Fathi, *Stability of the Bakry-Émery theorem on* $\mathbb{R}^n$ (2018). Preprint arXiv:1807.09845
30. G. De Philippis, A. Figalli, Rigidity and stability of Caffarelli's log-concave perturbation theorem. Nonlinear Anal. **154**, 59–70 (2017)
31. N. Demni, *Radial Dunkl Processes: Existence and uniqueness, Hitting time, Beta Processes and Random Matrices* (2007). Preprint arxiv:0707.0367v1
32. N. Demni, First hitting time of the boundary of the Weyl chamber by radial Dunkl processes. SIGMA Symmetry Integrability Geom. Methods Appl. **4**, Paper 074, 14 (2008)
33. Y. Doumerc, Matrices aléatoires, Processus Stochastiques et Groupes de réflexions. Ph.D. thesis, Université de Toulouse, Toulouse, 2005
34. I. Dumitriu, A. Edelman, Matrix models for beta ensembles. J. Math. Phys. **43**(11), 5830–5847 (2002)
35. I. Dumitriu, A. Edelman, G. Shuman, MOPS: multivariate orthogonal polynomials (symbolically). J. Symbolic Comput. **42**(6), 587–620 (2007)
36. F.J. Dyson, A Brownian-motion model for the eigenvalues of a random matrix. J. Math. Phys. **3**, 1191–1198 (1962)
37. F.J. Dyson, The threefold way. Algebraic structure of symmetry groups and ensembles in quantum mechanics. J. Math. Phys. **3**, 1199–1215 (1962)
38. L. Erdös, H.-T. Yau, *A Dynamical Approach to Random Matrix Theory*. Courant Lecture Notes in Mathematics, vol 28 (Courant Institute of Mathematical Sciences/American Mathematical Society, New York/Providence, 2017)
39. L.C. Evans, in *Partial Differential Equations*, Graduate Studies in Mathematics, vol 19, 2nd edn. (American Mathematical Society, Providence, 2010)
40. P.J. Forrester, Log-gases and random matrices, in *London Mathematical Society Monographs Series*, vol 34 (Princeton University Press, Princeton, 2010)
41. N. Gozlan, private communication (2012)

42. D.J. Grabiner, Brownian motion in a Weyl chamber, non-colliding particles, and random matrices. Ann. Inst. H. Poincaré Probab. Statist. **35**(2), 177–204 (1999)
43. P. Graczyk, J. Mał ecki, E. Mayerhofer, A characterization of Wishart processes and Wishart distributions. Stoch. Process. Appl. **128**(4), 1386–1404 (2018)
44. A. Guionnet, in *Large Random Matrices: Lectures on Macroscopic Asymptotics*. Lecture Notes in Mathematics, vol 1957 (Springer, Berlin, 2009). Lectures from the 36th Probability Summer School held in Saint-Flour, 2006
45. A. Guionnet, O. Zeitouni, Concentration of the spectral measure for large matrices. Electron. Commun. Probab. **5**, 119–136 (2000)
46. J. Gustavsson, Gaussian fluctuations of eigenvalues in the GUE. Ann. Inst. H. Poincaré Probab. Stat. **41**(2), 151–178 (2005)
47. A.J. Hoffman, H.W. Wielandt, The variation of the spectrum of a normal matrix. Duke Math. J. **20**, 37–39 (1953)
48. D. Holcomb, E. *Paquette, Tridiagonal Models for Dyson Brownian Motion* (2017). preprint arXiv:1707.02700
49. R.A. Horn, C.R. Johnson, *Matrix Analysis*, 2nd edition (Cambridge University Press, Cambridge, 2013)
50. W. König, N. O'Connell, Eigenvalues of the Laguerre process as non-colliding squared Bessel processes. Electron. Comm. Probab. **6**, 107–114 (2001)
51. M. Lassalle, Polynômes de Hermite généralisés. C. R. Acad. Sci. Paris Sér. I Math. **313**(9), 579–582 (1991)
52. M. Lassalle, Polynômes de Jacobi généralisés. C. R. Acad. Sci. Paris Sér. I Math. **312**(6), 425–428 (1991)
53. M. Lassalle, Polynômes de Laguerre généralisés. C. R. Acad. Sci. Paris Sér. I Math. **312**(10), 725–728 (1991)
54. M. Ledoux, On an integral criterion for hypercontractivity of diffusion semigroups and extremal functions. J. Funct. Anal. **105**(2), 444–465 (1992)
55. M. Ledoux, in *The Concentration of Measure Phenomenon*. Mathematical Surveys and Monographs, vol 89 (American Mathematical Society, Providence, 2001)
56. M. Ledoux, A remark on hypercontractivity and tail inequalities for the largest eigenvalues of random matrices, in *Séminaire de Probabilités XXXVII*. Lecture Notes in Mathematics, vol 1832, pp. 360–369 (Springer, Berlin, 2003)
57. M. Ledoux, B. Rider, Small deviations for beta ensembles. Electron. J. Probab. **15**(41), 1319–1343 (2010)
58. F. Malrieu, Logarithmic Sobolev inequalities for some nonlinear PDE's. Stoch. Process. Appl. 95(1), 109–132 (2001)
59. O. Mazet, Classification des semi-groupes de diffusion sur **R** associés à une famille de polynômes orthogonaux, in *Séminaire de Probabilités, XXXI*. Lecture Notes in Mathematics, vol 1655, pp. 40–53 (Springer, Berlin, 1997)
60. E.S. Meckes, M.W. Meckes, Concentration and convergence rates for spectral measures of random matrices. Probab. Theory Relat. Fields **156**(1-2), 145–164 (2013)
61. E.S. Meckes, M.W. Meckes, Spectral measures of powers of random matrices. Electron. Commun. Probab. **18**(78), 13 (2013)
62. M.L. Mehta, in *Random Matrices*. Pure and Applied Mathematics (Amsterdam), vol 142 (Elsevier/Academic, Amsterdam, 2004)
63. J.A. Ramírez, B. Rider, B. Virág, Beta ensembles, stochastic Airy spectrum, and a diffusion. J. Am. Math. Soc. **24**(4), 919–944, 2011
64. M. Rösler, M. Voit, Markov processes related with Dunkl operators. Adv. Appl. Math. **21**(4), 575–643 (1998)
65. H. Tanaka, Stochastic differential equations with reflecting boundary condition in convex regions. Hiroshima Math. J. **9**(1), 163–177 (1979)
66. J.F. van Diejen, L. Vinet eds. *Calogero-Moser-Sutherland models*. CRM Series in Mathematical Physics. (Springer, New York, 2000)
67. M. Voit, J.H.C. Woerner, *Functional Central Limit Theorems for Multivariate Bessel Processes in the Freezing Regime* (2019). preprint arXiv:1901.08390

Chapter 11
Several Results Regarding
the (B)-Conjecture

Dario Cordero-Erausquin and Liran Rotem

Abstract In the first half of this note we construct Gaussian measures on $\mathbb{R}^n$ which do not satisfy a strong version of the (B)-property. In the second half we discuss equivalent functional formulations of the (B)-conjecture.

11.1 Introduction

By a convex body in $\mathbb{R}^n$ we mean a set $K \subseteq \mathbb{R}^n$ which is convex, compact and has non-empty interior. Our convex bodies will always be symmetric, in the sense that $K = -K$. We will denote the standard Gaussian measure on $\mathbb{R}^n$ by γ_n, or simply by γ if there is no possibility of confusion. The density of γ_n is

$$\frac{\mathrm{d}\gamma_n}{\mathrm{d}x} = \frac{1}{(2\pi)^{\frac{n}{2}}} e^{-|x|^2/2},$$

where $|\cdot|$ denotes the standard Euclidean norm.

Banaszczyk asked the following question, which was popularized by Latała [13]: Let $K \subseteq \mathbb{R}^n$ be a symmetric convex body and fix $a, b > 0$ and $0 \leq \lambda \leq 1$. Is it true that

$$\gamma \left(a^{1-\lambda} b^\lambda K \right) \geq \gamma(aK)^{1-\lambda} \gamma(bK)^\lambda? \tag{11.1}$$

Recall that a nonnegative function f on $\mathbb{R}^n$ is called log-concave if $f((1-\lambda)x + \lambda y) \geq f(x)^{1-\lambda} f(y)^\lambda$ for all $x, y \in \mathbb{R}^n$ and all $0 \leq \lambda \leq 1$, and that a Borel measure

---

D. Cordero-Erausquin
Institut de Mathématiques de Jussieu, Sorbonne Université, Paris, France
e-mail: dario.cordero@imj-prg.fr

L. Rotem (✉)
University of Minnesota, Minneapolis, MN, USA
e-mail: lrotem@umn.edu

© Springer Nature Switzerland AG 2020
B. Klartag, E. Milman (eds.), *Geometric Aspects of Functional Analysis*,
Lecture Notes in Mathematics 2256,
https://doi.org/10.1007/978-3-030-36020-7_11

μ on $\mathbb{R}^n$ is said to be log-concave if

$$\mu((1 - \lambda)A + \lambda B) \geq \mu(A)^{1-\lambda}\mu(B)^\lambda$$

for all Borel sets $A, B \subseteq \mathbb{R}^n$ and all $0 \leq \lambda \leq 1$. The addition of sets in the above definition is the Minkowski addition, defined by $A + B = \{a + b : a \in A, b \in B\}$. Borell proved the following relation between log-concave functions and measures [3, 4]: Let μ be a Borel measure on $\mathbb{R}^n$ which is not supported on any affine hyperplane. Then μ is log-concave if and only if μ has a log-concave density $f = \frac{d\mu}{dx}$.

In particular, the Gaussian measure γ is log-concave. Choosing $A = aK$ and $B = bK$ we see that

$$\gamma(((1 - \lambda)a + \lambda b)K) \geq \gamma(aK)^{1-\lambda}\gamma(bK)^\lambda. \qquad (11.2)$$

However, this inequality is strictly weaker than (11.1). Moreover, inequality (11.2) holds for any convex body K, symmetric or not, while the symmetry of K has to be used in order to prove (11.1)—Nayar and Tkocz [14] constructed a counter-example if one replaces the assumption that K is symmetric with the weaker assumption that $0 \in K$.

In [9], Cordero-Erausquin, Fradelizi and Maurey answered Banaszczyk question positively. In fact, they proved the following stronger result. For $t_1, t_2 \ldots, t_n \in \mathbb{R}$ let $\Delta(t_1, t_2, \ldots, t_n)$ denote the $n \times n$ diagonal matrix with $t_1, t_2, \ldots t_n$ on its diagonal.

Theorem 11.1 ([9]) *Let $K \subseteq \mathbb{R}^n$ be a symmetric convex body. Then the map*

$$(t_1, t_2, \ldots, t_n) \mapsto \gamma\left(e^{\Delta(t_1, t_2, \ldots, t_n)}K\right)$$

is log-concave.

Banaszczyk's original question is answered by restricting the above function to the line $t_1 = t_2 = \cdots = t_n$.

The main goal of this paper is to discuss extensions of Theorem 11.1 to other log-concave measures. Let us make the following definitions:

Definition 11.2 Let μ be a Borel measure on $\mathbb{R}^n$.

1. We say that μ satisfies the *(B)-property* if the function $t \mapsto \mu(e^t K)$ is log-concave for every symmetric convex body K.
2. We say that μ satisfies the *strong (B)-property* if the function $(t_1, t_2, \ldots, t_n) \mapsto \mu(e^{\Delta(t_1, t_2, \ldots, t_n)}K)$ is log-concave for every symmetric convex body K.

Theorem 11.1 states that the standard Gaussian measure has the strong (B)-property. Not many other examples are known in dimension $n \geq 3$. In [10], Eskenazis, Nayar and Tkocz proved that certain Gaussian mixtures satisfy the strong (B)-property. In particular, their result covers the case where μ has density $\frac{d\mu}{dx} = e^{-|x|^p}$ for $0 < p \leq 1$ (note that unless $p = 1$ these measures are not log-concave).

In dimension $n = 2$ much more is known. Livne Bar-on showed [2] that if $T \subseteq \mathbb{R}^2$ is a symmetric convex body and μ is the uniform measure on T then μ satisfies the (B)-property. Saroglou later showed [18] that every even log-concave measure μ on $\mathbb{R}^2$ satisfies the (B)-property.

Saroglou's proof uses the relation between the (B)-property and the log-Brunn-Minkowksi conjecture. In fact, at lot of the interest in the (B)-property comes from this relation. Recall that the support function h_K of a convex body K is defined by

$$h_K(y) = \sup_{x \in K} \langle x, y \rangle .$$

The λ-logarithmic mean of two convex bodies K and T is then defined by

$$L_\lambda(K, T) = \left\{ x \in \mathbb{R}^n : \langle x, y \rangle \leq h_K(y)^{1-\lambda} h_T(y)^\lambda \text{ for all } y \in \mathbb{R}^n \right\} .$$

In other words, $L = L_\lambda(K, T)$ is the largest convex body such that $h_L \leq h_K^{1-\lambda} h_T^\lambda$. We can also write

$$L_\lambda(K, T) = \bigcap_{s > 0} \left((1 - \lambda) s^{1/(1-\lambda)} K + \lambda s^{-1/\lambda} T \right).$$

since the support function of an intersection of convex bodies is the infimum of the support functions and $a^{1-\lambda} b^\lambda = \inf_{s>0} \{ (1 - \lambda) s^{1/(1-\lambda)} a + \lambda s^{-1/\lambda} b \}$ for $a, b \geq 0$. In [5], Böröczky, Lutwak, Yang and Zhang made the following conjecture:

Conjecture 11.3 (The Log-Brunn-Minkowski Conjecture) Let $K, T \subseteq \mathbb{R}^n$ be symmetric convex bodies. Then for every $0 \leq \lambda \leq 1$ we have

$$|L_\lambda(K, T)| \geq |K|^{1-\lambda} |T|^\lambda$$

where $|\cdot|$ denotes the Lebesgue volume.

Since $L_\lambda(K, T) \subseteq (1 - \lambda) K + \lambda T$, the log-Brunn-Minkowski conjecture is a strengthening of the Brunn-Minkowski inequality. Again, this strengthening can only hold under the extra assumption that K and T are convex and symmetric.

A considerable amount of work was done on the log-Brunn-Minkowski conjecture. In the original paper [5] it was proved by Böröczky, Lutwak, Yang and Zhang in dimension $n = 2$. Saroglou [17] proved the conjecture when K, T are unconditional, i.e. symmetric with respect to reflections by the coordinate hyperplanes. In [16] it was observed that a more general theorem from [8] implies the conjecture when K and T are unit balls of a complex normed space. Colesanti et al. [7] proved the conjecture when K and T are small C^2-perturbations of the Euclidean ball. Kolesnikov and Milman [12] proved a local form of the closely related L^p–Brunn–Minkowski inequalities for p close enough to 1. Based on their result Chen et al. [6] then proved the full L^p-Brunn–Minkowski inequality for the same values of p.

Saroglou also proved several connections between the log-Brunn-Minkowski inequality and the (B)-property. In one direction, he proved the following:

Theorem 11.4 ([18]) *Assume the log-Brunn-Minkowski conjecture holds in dimension n. Then*

$$\mu\left(L_\lambda(K, T)\right) \geq \mu(K)^{1-\lambda}\mu(T)^\lambda$$

for every symmetric $K, T \subseteq \mathbb{R}^n$, every $0 \leq \lambda \leq 1$, and every even log-concave measure μ on $\mathbb{R}^n$. In particular, by choosing K and T to be dilates of each other one could conclude that every *even log-concave measure μ on $\mathbb{R}^n$ has the (B)-property.*

This explains the previous claim that all even log-concave measures in dimension $n = 2$ have the (B)-property.

In the opposite direction, Saroglou also proved that the *strong* (B)-property can imply the log-Brunn-Minkowski conjecture. If Σ is an $n \times n$ positive definite matrix, let us denote by γ_Σ the Gaussian measure with covariance matrix Σ. More explicitly, γ_Σ has density

$$\frac{d\gamma_\Sigma}{dx} = \frac{1}{(2\pi)^{\frac{n}{2}}(\det\Sigma)^{\frac{1}{2}}} e^{-\frac{1}{2}\langle\Sigma^{-1}x, x\rangle}.$$

Let us also denote by $C_n = [-1, 1]^n$ the n-dimensional hypercube. Saroglou's result then implies:

Theorem 11.5 ([18]) *The following are equivalent:*

1. *The log-Brunn-Minkowski inequality holds in every dimension n.*
2. *For every dimension n, every $n \times n$ covariance matrix Σ and every diagonal matrix A the function*

$$t \mapsto \gamma_\Sigma\left(e^{tA} \cdot C_n\right)$$

is log-concave.

In fact, there is nothing special here about the Gaussian measure, and the same result holds if γ is replaced with any even log-concave measure together with all of its linear images. However, it is easy to see from Theorem 11.1 that the measures γ_Σ satisfy the (weak) (B)-property, so the Gaussian seems like the most natural choice.

In particular, Theorem 11.5 implies that if all measures γ_Σ satisfy the strong (B)-property then the log-Brunn-Minkowski conjecture is proved. One may conjecture that maybe *all* even log-concave measures have the strong (B)-property. However, an example of Nayar and Tkocz [15] shows that this is not the case. But the example from [15] is not Gaussian, so it does not contradict the idea above.

This paper has two mostly independent sections. In Sect. 11.2 we will show that not all measures γ_Σ satisfy the strong (B)-property. In fact, we will prove that in every dimension n there exist Gaussian measures γ_Σ arbitrarily close to the standard Gaussian γ_n, which nonetheless don't satisfy the strong (B)-property.

Theorem 11.6 *Fix $n \geq 2$. For every $\eta > 0$ there exists a positive-definite $n \times n$ matrix Σ and a symmetric convex body $K \subseteq \mathbb{R}^n$ such that:*

1. $\|\Sigma - I\| < \eta$, *where I denotes the $n \times n$ identity matrix.*
2. *The function $(t_1, t_2, \ldots, t_n) \mapsto \gamma_\Sigma \left(e^{\Delta(t_1, t_2, \ldots, t_n)} K \right)$ is not log-concave.*

Since all norms on a finite dimensional space are equivalent, the choice of the norm in property 1 is immaterial.

In Sect. 11.3 we turn our attention to the weak (B)-property. It is still a plausible conjecture that every even log-concave measure satisfies the (B)-property. We refer to this conjecture simply as the (B)-conjecture. The main goal of Sect. 11.3 is to prove several equivalent formulations of this conjecture. For example, it turns out that the (B)-conjecture is intimately related to correlation inequalities, and we prove the following result:

Theorem 11.7 *The following are equivalent:*

1. *Every even log-concave measure in any dimension satisfies the (B)-property.*
2. *For any dimension n, and any functions $\varphi, \psi : \mathbb{R}^n \to \mathbb{R}$ which are convex, even, C^2-smooth and homogeneous, one has*

$$\int \varphi \psi \, d\mu \geq \int \varphi \, d\mu \cdot \int \psi \, d\mu,$$

where μ is the probability measure with density $\frac{d\mu}{dx} = \frac{e^{-\varphi - \psi}}{\int e^{-\varphi - \psi}}$.

"Homogeneous" in the above theorem means homogeneous of an arbitrary degree. In other words, φ is homogeneous if there exists $p \geq 1$ such that $\varphi(\lambda x) = \lambda^p \varphi(x)$ for all $x \in \mathbb{R}^n$ and all $\lambda > 0$.

11.2   A Gaussian Counter-Example

In this section we prove Theorem 11.6. Recall that we denote by γ_Σ the Gaussian probability measure on $\mathbb{R}^n$ with covariance matrix Σ and simply write γ when $\Sigma = I$. For a convex body $K \subseteq \mathbb{R}^n$, γ_K denotes the standard Gaussian measure restricted to K:

$$\gamma_K(A) = \frac{\gamma(A \cap K)}{\gamma(K)}.$$

For $1 \leq i \leq n$, $\gamma_{(i)}$ denotes the one-dimensional standard Gaussian measure supported on the i-th axis in $\mathbb{R}^n$, that is, for a test function $f : \mathbb{R}^n \to \mathbb{R}$ we have $\int_{\mathbb{R}^n} f(x) d\gamma_{(i)}(x) = \int_{\mathbb{R}} f(te_i) d\gamma_1(t)$, where e_i is the i-th standard unit vector. We will also need the following straightforward fact, which allows us to approximate $\gamma_{(1)}$ by measures of the form γ_K:

Fact 11.8 Define $K(\varepsilon) = \left[-\frac{1}{\varepsilon}, \frac{1}{\varepsilon}\right] \times [-\varepsilon, \varepsilon]^{n-1} \subseteq \mathbb{R}^n$. Let $f : \mathbb{R}^n \to \mathbb{R}$ be a C^1 function, and assume that there exists a constant $C > 0$ such that $|f(x)| \le Ce^{C|x|}$ and $|\nabla f(x)| \le Ce^{C|x|}$ for all $x \in \mathbb{R}^n$. Then

$$\lim_{\varepsilon \to 0^+} \int f \, d\gamma_{K(\varepsilon)} = \int f \, d\gamma_{(1)}.$$

The assumptions on f in the lemma are not the minimal possible assumptions, but will more than suffice for our needs.

Proof Let us write a general point $x \in \mathbb{R}^n$ as $x = (t, y)$ where $t \in \mathbb{R}$ and $y \in \mathbb{R}^{n-1}$. Define $F, G : (0, \infty) \to \mathbb{R}$ by

$$F(\varepsilon) = \int f \, d\gamma_{K(\varepsilon)} = \frac{\int_{K(\varepsilon)} f(t, y) e^{-\frac{1}{2}(t^2 + |y|^2)} dt dy}{\int_{K(\varepsilon)} e^{-\frac{1}{2}(t^2 + |y|^2)} dt dy}$$

$$G(\varepsilon) = \frac{\int_{K(\varepsilon)} f(t, 0) e^{-\frac{1}{2}(t^2 + |y|^2)} dt dy}{\int_{K(\varepsilon)} e^{-\frac{1}{2}(t^2 + |y|^2)} dt dy} = \frac{\int_{-1/\varepsilon}^{1/\varepsilon} f(t, 0) e^{-\frac{1}{2}t^2} dt}{\int_{-1/\varepsilon}^{1/\varepsilon} e^{-\frac{1}{2}t^2} dt}.$$

Since $|f(x)| \le Ce^{C|x|}$ we may apply the dominated convergence theorem and conclude that $\lim_{\varepsilon \to 0^+} G(\varepsilon) = \int f \, d\gamma_{(1)}$.

For every point $(t, y) \in K(\varepsilon)$ we have $|y| \le \sqrt{n}\varepsilon$, and so $|f(t, y) - f(t, 0)| \le Ce^{C|(t,y)|} \cdot \left(\sqrt{n}\varepsilon\right)$. It follows that

$$|F(\varepsilon) - G(\varepsilon)| \le \frac{\int_{K(\varepsilon)} |f(t, y) - f(t, 0)| e^{-\frac{1}{2}|(t,y)|^2} dt dy}{\int_{K(\varepsilon)} e^{-\frac{1}{2}|(t,y)|^2} dt dy}$$

$$\le C\sqrt{n}\varepsilon \cdot \frac{\int_{K(\varepsilon)} e^{C|(t,y)| - \frac{1}{2}|(t,y)|^2} dt dy}{\int_{K(\varepsilon)} e^{-\frac{1}{2}|(t,y)|^2} dt dy}. \tag{11.3}$$

If we now assume further that $0 < \varepsilon < 1$ then $|y| \le \sqrt{n}$ so

$$\int_{K(\varepsilon)} e^{C|(t,y)| - \frac{1}{2}|(t,y)|^2} dt dy \le \int_{K(\varepsilon)} e^{C\sqrt{t^2 + n} - \frac{1}{2}t^2} dt dy$$

$$\le (2\varepsilon)^{n-1} \int_{-\infty}^{\infty} e^{C\sqrt{t^2 + n} - \frac{1}{2}t^2} dt = A_n \cdot \varepsilon^{n-1},$$

where A_n is a constant which depends on C and n, but not on ε. Similarly we have

$$\int_{K(\varepsilon)} e^{-\frac{1}{2}|(t,y)|^2}\,dt\,dy \geq \int_{K(\varepsilon)} e^{-\frac{1}{2}(t^2+n)}\,dt\,dy \geq (2\varepsilon)^{n-1}\int_{-1}^{1} e^{-\frac{1}{2}(t^2+n)}\,dt = B_n\varepsilon^{n-1}.$$

Plugging these estimates into (11.3) we see that $|F(\varepsilon) - G(\varepsilon)| \leq C\sqrt{n}\varepsilon \cdot \frac{A_n}{B_n} \xrightarrow{\varepsilon\to 0^+} 0$, so $\lim_{\varepsilon\to 0^+} F(\varepsilon) = \lim_{\varepsilon\to 0^+} G(\varepsilon) = \int f\,d\gamma_{(1)}$. $\qquad\square$

Using Fact 11.8 we can prove Theorem 11.6.

Proof of Theorem 11.6 Assume by contradiction that the function

$$(t_1, t_2, \ldots, t_n) \mapsto \gamma_\Sigma\left(e^{\Delta(t_1,t_2,\ldots,t_n)}K\right) \tag{11.4}$$

is log-concave for all symmetric convex bodies $K \subseteq \mathbb{R}^n$ and all positive-definite matrices Σ such that $\|\Sigma - I\|$ is small enough.

For a fixed $\eta > 0$, consider the $n \times n$ block matrix $P = \left(\begin{array}{cc|c} 1 & 2\eta & 0 \\ \eta & 1 & \\ \hline 0 & & I_{n-2} \end{array}\right)$, where I_{n-2} denotes the $(n-2) \times (n-2)$ identity matrix. Consider also the diagonal matrix $D = \Delta(2, 1, 1, \ldots, 1)$. A direct computation shows that

$$A = P^{-1}DP = \left(\begin{array}{cc|c} \frac{2-2\eta^2}{1-2\eta^2} & \frac{2\eta}{1-2\eta^2} & 0 \\ -\frac{\eta}{1-2\eta^2} & \frac{1-4\eta^2}{1-2\eta^2} & \\ \hline 0 & & I_{n-2} \end{array}\right).$$

We claim the function $F(t) := \gamma\left(e^{tA}K(\varepsilon)\right)$ is log-concave for all $\varepsilon > 0$ and small enough $\eta > 0$. Indeed, we have

$$F(t) = \gamma\left(P^{-1}e^{tD}P \cdot K(\varepsilon)\right) = \gamma_\Sigma\left(e^{tD}P \cdot K(\varepsilon)\right),$$

where $\Sigma = PP^T$. In other words, if we take $K = P \cdot K(\varepsilon)$, then the function F is a restriction of the function in (11.4) to a line. Since $\Sigma \to I$ as $\eta \to 0$, the claim follows.

Writing F more explicitly we have

$$F(t) = \frac{1}{(2\pi)^{\frac{n}{2}}}\int_{e^{tA}K(\varepsilon)} e^{-\frac{1}{2}|y|^2}\,dy = \frac{1}{(2\pi)^{\frac{n}{2}}}\int_{K(\varepsilon)} e^{-\frac{1}{2}|e^{tA}x|^2} \cdot e^{t\cdot\mathrm{Tr}A}\,dx.$$

As the term $e^{t \cdot \mathrm{Tr} A}$ is log-linear, it follows that the function $G(t) = \int_{K(\varepsilon)} e^{-\frac{1}{2}|e^{tA}x|^2} dx$ is log-concave, so in particular

$$(\log G)''(0) = \frac{G''(0)}{G(0)} - \left(\frac{G'(0)}{G(0)}\right)^2 \le 0. \tag{11.5}$$

An explicit calculation of the derivatives gives

$$G'(t) = -\int_{K(\varepsilon)} \left\langle Ae^{tA}x, e^{tA}x \right\rangle \cdot e^{-\frac{1}{2}|e^{tA}x|^2} dx$$

$$G''(t) = \int_{K(\varepsilon)} \left(\left\langle Ae^{tA}x, e^{tA}x \right\rangle^2 - \left\langle A^2 e^{tA}x, e^{tA}x \right\rangle - \left\langle Ae^{tA}x, Ae^{tA}x \right\rangle \right)$$

$$\cdot e^{-\frac{1}{2}|e^{tA}x|^2} dx,$$

So (11.5) reads

$$\int \left(\langle Ax, x \rangle^2 - \left\langle A^2 x, x \right\rangle - \langle Ax, Ax \rangle \right) d\gamma_{K(\varepsilon)}(x) - \left(\int \langle Ax, x \rangle \, d\gamma_{K(\varepsilon)}(x) \right)^2 \le 0.$$

Since this inequality is true for any $\varepsilon > 0$, we may let $\varepsilon \to 0$. Using Fact 11.8 we deduce that

$$\int \left(\langle Ax, x \rangle^2 - \left\langle A^2 x, x \right\rangle - \langle Ax, Ax \rangle \right) d\gamma_{(1)}(x) - \left(\int \langle Ax, x \rangle \, d\gamma_{(1)}(x) \right)^2 \le 0, \tag{11.6}$$

However, an explicit computation gives

$$\int \langle Ax, x \rangle^2 \, d\gamma_{(1)}(x) = \int \langle Ate_1, te_1 \rangle^2 \, d\gamma_1(t) = \langle Ae_1, e_1 \rangle^2 \cdot \int t^4 d\gamma_1(t)$$

$$= 3 \cdot \left(\frac{2 - 2\eta^2}{1 - 2\eta^2} \right)^2.$$

Computing the other three integrals in the same way, (11.6) reduces to

$$3 \cdot \left(\frac{2 - 2\eta^2}{1 - 2\eta^2} \right)^2 - \frac{4 - 2\eta^2}{1 - 2\eta^2} - \frac{4\eta^4 - 7\eta^2 + 4}{(1 - 2\eta^2)^2} - \left(\frac{2 - 2\eta^2}{1 - 2\eta^2} \right)^2 \le 0,$$

or equivalently $\frac{\eta^2}{(1-2\eta^2)^2} \le 0$. Since this is impossible for every $\eta > 0$, we arrived at a contradiction and the theorem is proved. $\square$

11.3   A Functional (B)-Conjecture

In this section we discuss several equivalent formulations of the (B)-conjecture. The usual statement of the (B)-conjecture, mentioned in the introduction, involves one log-concave measure μ, and one convex body K. However, it is possible and rather standard to state equivalent conjectures dealing only with bodies, or only with functions:

Theorem 11.9 *The following are equivalent:*

1. *For every dimension n and every symmetric convex bodies $K, T \subseteq \mathbb{R}^n$ the function $t \mapsto \left| e^t K \cap T \right|$ is log-concave.*
2. *For every dimension n, every symmetric convex body $K \subseteq \mathbb{R}^n$ and every even convex function $\varphi : \mathbb{R}^n \to (-\infty, \infty]$, the function*

$$t \mapsto \int_{e^t K} e^{-\varphi(x)} dx$$

is log-concave.
3. *For every dimension n and every even convex functions $\varphi, \psi : \mathbb{R}^n \to (-\infty, \infty]$, the function*

$$t \mapsto \int_{\mathbb{R}^n} e^{-\varphi(x) - \psi(e^t x)} dx$$

is log-concave.

Note that formulation 2. is exactly the standard (B)-conjecture.

Proof $(2 \Rightarrow 1)$ is obvious by taking

$$\varphi(x) = \mathbf{1}_T^\infty(x) = \begin{cases} 0 & \text{if } x \in T \\ \infty & \text{if } x \notin T \end{cases}.$$

Similarly, to see that $(3 \Rightarrow 2)$ one chooses $\psi = \mathbf{1}_K^\infty$ and applies the change of variables $t \mapsto -t$, which preserves log-concavity.

Next, we prove that $(1 \Rightarrow 2)$. Our first observation is that it is not important in 1. for K to be compact, as long as T is compact. Indeed, let us define $F_{K,T}(t) = \left| e^t K \cap T \right|$. If we denote the unit Euclidean ball in $\mathbb{R}^n$ by B_n then $F_{K \cap r B_n, T} \to F_{K,T}$ pointwise as $r \to \infty$. Since the pointwise limit of log-concave functions is log-concave, the claim follows.

To show that 2. holds, we will use a standard approximation argument similar to that of [1]. Fix a convex body $K \subseteq \mathbb{R}^n$ and a convex function $\varphi : \mathbb{R}^n \to (-\infty, \infty]$.

For every integer $m \geq 1$ we define

$$K_m = K \times \mathbb{R}^m \subseteq \mathbb{R}^{n+m}$$

$$T_m = \left\{ (x, y) \in \mathbb{R}^n \times \mathbb{R}^m : \varphi(x) \leq m \text{ and } |y| \leq c_m \cdot \left(1 - \frac{\varphi(x)}{m} \right) \right\} \subseteq \mathbb{R}^{n+m}.$$

Here c_m is a normalization constant chosen to have $|c_m B_m| = 1$. Obviously K_m is a symmetric convex set, and it easy to check that T_m is a symmetric convex body. Therefore by 1. the function F_{K_m, T_m} is log-concave. By Fubini's theorem we have

$$F_{K_m, T_m}(t) = \left| e^t K_m \cap T_m \right| = \int_{e^t K \cap [\varphi \leq m]} \left| c_m \left(1 - \frac{\varphi(x)}{m} \right) B_m \right| dx$$

$$= \int_{e^t K \cap [\varphi \leq m]} \left(1 - \frac{\varphi(x)}{m} \right)^m dx = \int_{e^t K} \left(1 - \frac{\varphi(x)}{m} \right)_+^m dx,$$

where $[\varphi \leq m] := \{x \in \mathbb{R}^n : \varphi(x) \leq m\}$ and $a_+ = \max\{a, 0\}$. The functions $g_m(x) = \left(1 - \frac{\varphi(x)}{m} \right)_+^m$ satisfy $g_m(x) \leq e^{-\varphi(x)}$ for all $x \in \mathbb{R}^n$ and $m \geq 1$, and $g_m \to e^{-\varphi}$ pointwise as $m \to \infty$. Hence by the dominated convergence theorem we have

$$\lim_{m \to \infty} F_{K_m, T_m}(t) = \int_{e^t K} e^{-\varphi(x)} dx$$

for every $t \in \mathbb{R}$. Since pointwise limits preserve log-concavity, 2. follows.

The proof that $(2 \Rightarrow 3)$ is similar. We are given that for every convex body $T \subseteq \mathbb{R}^n$ and convex function $\psi : \mathbb{R}^n \to (-\infty, \infty]$ the function

$$t \mapsto \int_{e^t T} e^{-\psi(x)} dx = \int_T e^{-\psi(e^t y)} \cdot e^{nt} dy$$

is log-concave. As e^{nt} is log-linear, we deduce that $G_{T, \psi}(t) = \int_T e^{-\psi(e^t x)} dx$ is log-concave.

Given $\varphi, \psi : \mathbb{R}^n \to (-\infty, \infty]$, we define functions $\psi_m : \mathbb{R}^{n+m} \to (-\infty, \infty]$ by $\psi_m(x, y) = \psi(x)$. The functions ψ_m do not satisfy $\int e^{-\psi_m} < \infty$, but this does not matter for the same reason as before. Using the bodies T_m from before and Fubini's theorem we have

$$G_{T_m, \psi_m}(t) = \int_{T_m} e^{-\psi_m(e^t x, e^t y)} dx dy = \int_{[\varphi \leq m]} \int_{|y| \leq c_m \left(1 - \frac{\varphi(x)}{m} \right)} e^{-\psi(e^t x)} dy dx$$

$$= \int e^{-\psi(e^t x)} \cdot \left(1 - \frac{\varphi(x)}{m} \right)_+^m dx \xrightarrow{m \to \infty} \int e^{-\varphi(x) - \psi(e^t x)} dx.$$

This completes the proof. □

Since the rest of this section deals with property 3. of Theorem 11.9, let us give this property a name:

Definition 11.10 We say that the functional (B)-conjecture holds in dimension n if for every even convex functions $\varphi, \psi : \mathbb{R}^n \to (-\infty, \infty]$ the function

$$t \mapsto \int_{\mathbb{R}^n} e^{-\varphi(x)-\psi(e^t x)} \mathrm{d}x$$

is log-concave.

By Theorem 11.9 the functional (B)-conjecture is equivalent to the standard (B)-conjecture, but only if one considers all dimensions simultaneously. For example, we saw in the introduction that the standard (B)-conjecture holds in dimension 2, but the same is unknown for the functional conjecture.

The functional (B)-conjecture is more general than the standard one, but it has one advantage: by a standard approximation argument one may assume that φ and ψ are as smooth as needed. This allows the use of analytic tools such as integration by parts. For example, one can use such tools to show that the (B)-conjecture is equivalent to a certain correlation inequality:

Definition 11.11 For a C^1-smooth function $f : \mathbb{R}^n \to \mathbb{R}$ we define its radial derivative $\mathrm{R}f : \mathbb{R}^n \to \mathbb{R}$ by

$$(\mathrm{R}f)(x) = \langle \nabla f(x), x \rangle.$$

Proposition 11.12 *The functional (B)-conjecture in dimension n is equivalent to the following: For every C^2-smooth even convex functions $\varphi, \psi : \mathbb{R}^n \to \mathbb{R}$ one has*

$$\int \mathrm{R}\varphi \cdot \mathrm{R}\psi \, \mathrm{d}\mu \geq \int \mathrm{R}\varphi \, \mathrm{d}\mu \cdot \int \mathrm{R}\psi \, \mathrm{d}\mu, \tag{11.7}$$

where $\frac{\mathrm{d}\mu}{\mathrm{d}x} = \frac{e^{-\varphi-\psi}}{\int e^{-\varphi-\psi}}$.

Proof By a standard approximation argument one may assume that φ and ψ are C^2-smooth (or C^∞-smooth if desired). The functional (B)-conjecture states that the function

$$F_{\varphi,\psi}(t) = \int e^{-\varphi(x)-\psi(e^t x)} \mathrm{d}x$$

is log-concave, which is equivalent to

$$\left(\log F_{\varphi,\psi}\right)''(t) = \frac{F''_{\varphi,\psi}(t)}{F_{\varphi,\psi}(t)} - \left(\frac{F'_{\varphi,\psi}(t)}{F_{\varphi,\psi}(t)}\right)^2 \leq 0.$$

If we define $\psi_s(x) = \psi(e^s x)$, then ψ_s is also an even convex function and $F_{\varphi,\psi}(t+s) = F_{\varphi,\psi_s}(t)$. This implies that $\left(\log F_{\varphi,\psi}\right)''(t) = \left(\log F_{\varphi,\psi_t}\right)''(0)$. Therefore the functional (B)-conjecture is equivalent to the inequality

$$\frac{F''_{\varphi,\psi}(0)}{F_{\varphi,\psi}(0)} - \left(\frac{F'_{\varphi,\psi}(0)}{F_{\varphi,\psi}(0)}\right)^2 \leq 0 \tag{11.8}$$

holding for all smooth even convex functions $\varphi, \psi : \mathbb{R}^n \to \mathbb{R}$.

Note that for every smooth $f : \mathbb{R}^n \to \mathbb{R}$ we have $\frac{d}{dt} f(e^t x) = R f(e^t x)$. Hence we have

$$F'_{\varphi,\psi}(t) = -\int R\psi(e^t x) \cdot e^{-\varphi(x)-\psi(e^t x)} dx$$

$$F''_{\varphi,\psi}(t) = \int \left((R\psi)^2 (e^t x) - R^2 \psi(e^t x) \right) e^{-\varphi(x)-\psi(e^t x)} dx,$$

and the inequality (11.8) becomes

$$\int \left((R\psi)^2 - R^2\psi \right) d\mu - \left(\int R\psi d\mu \right)^2 \leq 0. \tag{11.9}$$

To continue, we need to integrate by parts. For any smooth function f that doesn't grow too quickly we have

$$\int R f d\mu = \frac{\int \langle \nabla f, e^{-\varphi-\psi} \cdot x \rangle}{\int e^{-\varphi-\psi}} = -\frac{\int f \cdot \mathrm{div} \left(e^{-\varphi-\psi} \cdot x \right)}{\int e^{-\varphi-\psi}}$$

$$= -\frac{\int f \cdot \left(\langle \nabla e^{-\varphi-\psi}, x \rangle + e^{-\varphi-\psi} \mathrm{div} x \right)}{\int e^{-\varphi-\psi}}$$

$$= -\frac{\int f \cdot \left(\langle -\nabla\varphi - \nabla\psi, x \rangle + n \right) e^{-\varphi-\psi}}{\int e^{-\varphi-\psi}}$$

$$= \int f \cdot (R\varphi + R\psi - n) d\mu.$$

In particular, by taking $f = R\psi$ we see that

$$\int R^2 \psi d\mu = \int (R\psi)^2 d\mu + \int R\varphi \cdot R\psi d\mu - n \int R\psi d\mu,$$

so inequality (11.9) is equivalent to

$$-\left(\int R\varphi \cdot R\psi d\mu - n \int R\psi d\mu \right) - \left(\int R\psi d\mu \right)^2 \leq 0,$$

or

$$\int R\varphi \cdot R\psi d\mu \geq \int R\psi d\mu \cdot \left(n - \int R\psi d\mu\right). \qquad (11.10)$$

A second integration by parts shows that

$$\int R\psi d\mu = -\frac{\int \langle \nabla\left(e^{-\psi}\right), e^{-\varphi} \cdot x\rangle}{\int e^{-\varphi-\psi}} = \frac{\int e^{-\psi} \cdot \mathrm{div}\left(e^{-\varphi}x\right)}{\int e^{-\varphi-\psi}}$$

$$= \frac{\int e^{-\psi}\left(\langle\nabla e^{-\varphi}, x\rangle + e^{-\varphi}\mathrm{div}x\right)}{\int e^{-\varphi-\psi}} = \int (n - R\varphi)\, d\mu.$$

so (11.10) is equivalent to the correlation inequality

$$\int R\varphi \cdot R\psi d\mu \geq \int R\varphi d\mu \cdot \int R\psi d\mu,$$

and the proof is complete.   □

As a corollary we obtain:

Corollary 11.13 *The functional (B)-conjecture holds in dimension $n = 1$.*

Proof We should prove that for every smooth, even and convex functions φ, ψ : $\mathbb{R} \to \mathbb{R}$ we have

$$\int R\varphi \cdot R\psi d\mu \geq \int R\varphi d\mu \cdot \int R\psi d\mu,$$

where $\frac{d\mu}{dx} = \frac{e^{-\varphi-\psi}}{\int e^{-\varphi-\psi}}$. Since φ and ψ are even so are $R\varphi$ and $R\psi$, so one may replace the integrals over $\mathbb{R}$ by integrals over $[0, \infty)$:

$$\int_0^\infty R\varphi \cdot R\psi d\widetilde{\mu} \geq \int_0^\infty R\varphi d\widetilde{\mu} \cdot \int_0^\infty R\psi d\widetilde{\mu},$$

where $\frac{d\widetilde{\mu}}{dx} = \frac{e^{-\varphi-\psi}}{\int_0^\infty e^{-\varphi-\psi}}\mathbf{1}_{[0,\infty)}$.

Since φ and ψ are convex and increasing on $[0, \infty)$, $R\varphi$ and $R\psi$ are also increasing on $[0, \infty)$. The assertion follows by Chebyshev's correlation inequality (see, e.g. [11]. In fact the inequality is true for *any* probability measure $\widetilde{\mu}$).

□

In dimension $n \geq 2$ it is no longer true that $R\varphi$ and $R\psi$ are correlated with respect to an arbitrary probability measure μ, even if we further assume that μ is log-concave with respect to $e^{-\varphi-\psi}$. It is not clear how to use the special choice of μ in the inequality.

We conclude this section by proving Theorem 11.7, which can be seen as a strengthening of Theorem 11.9 which allows one to check (11.7) only for homogeneous functions φ and ψ. This may be useful since if φ is homogeneous of degree d then $\mathrm{R}\varphi = d\varphi$. Therefore for homogeneous functions the inequality (11.7) no longer involves any derivatives.

Proof of Theorem 11.7 In one direction, assume the (B)-conjecture holds in any dimension. By Theorem 11.9 the functional (B)-conjecture also holds in any dimension. By Proposition 11.12 we deduce that

$$\int \mathrm{R}\varphi \cdot \mathrm{R}\psi \, \mathrm{d}\mu \geq \int \mathrm{R}\varphi \mathrm{d}\mu \cdot \int \mathrm{R}\psi \mathrm{d}\mu.$$

However, if φ is homogeneous of some degree d_1 then $\mathrm{R}\varphi = d_1\varphi$. Similarly if ψ is homogeneous of degree d_2 then $\mathrm{R}\psi = d_2\psi$. Hence we have

$$d_1 d_2 \int \varphi\psi \mathrm{d}\mu \geq \left(d_1 \int \varphi \mathrm{d}\mu \right) \cdot \left(d_2 \int \psi \mathrm{d}\mu \right),$$

which is what we wanted.

In the other direction, assume property 2. in the Theorem holds. We will prove Theorem 11.9's formulation 1. of the (B)-conjecture. Let K and T be even convex bodies. Recall the definition of the Minkowski functional

$$\|x\|_K = \inf \{\lambda > 0 : \ x \in \lambda K\}.$$

By approximating K and T, we may assume without loss of generality that $\|x\|_K$ and $\|x\|_T$ are C^2-smooth on $\mathbb{R}^n \setminus \{0\}$ (see, e.g., Section 2.5 of [19]). It follows that the functions $\varphi_m(x) = \|x\|_K^m$ and $\psi_m(x) = \|x\|_T^m$ are even, convex, C^2-smooth and homogeneous for all $m \geq 2$. The same is obviously true for the functions $\psi_{m,t}(x) = \psi_m(e^t x)$. By our assumption we have

$$\int \mathrm{R}\varphi_m \cdot \mathrm{R}\psi_{m,t} \mathrm{d}\mu = m^2 \cdot \int \varphi_m \psi_{m,t} \mathrm{d}\mu \geq m^2 \int \varphi_m \mathrm{d}\mu \cdot \int \psi_{m,t} \mathrm{d}\mu$$

$$= \int \mathrm{R}\varphi_m \mathrm{d}\mu \cdot \int \mathrm{R}\psi_{m,t} \mathrm{d}\mu.$$

As we saw in the proof of Proposition 11.12, this inequality is equivalent to

$$\left(\log F_{\varphi_m, \psi_m} \right)'' (t) = \left(\log F_{\varphi_m, \psi_{m,t}} \right)'' (0) \leq 0,$$

so F_{φ_m, ψ_m} is log-concave.

For any fixed $t \in \mathbb{R}$ we have

$$\lim_{m \to \infty} e^{-\varphi_m(x) - \psi_m(e^t x)} = \mathbf{1}_{K \cap e^{-t} T}(x)$$

for almost every $x \in \mathbb{R}^n$. More precisely, the convergence holds for every $x \notin \partial K \cup \partial \left(e^{-t}T\right)$. Moreover,

$$e^{-\varphi_m(x)-\psi_m(e^t x)} \leq e^{-\varphi_m(x)} \leq \max\left\{\mathbf{1}_K, e^{-\|x\|_K}\right\},$$

which is an integrable function. Hence by dominated convergence we have

$$\lim_{m\to\infty} F_{\varphi_m,\psi_m}(t) = \left|K \cap e^{-t}T\right|.$$

It follows that the function $t \mapsto \left|K \cap e^{-t}T\right|$ is log-concave, which is what we wanted to prove up to an immaterial change of sign. □

References

1. S. Artstein-Avidan, B. Klartag, V. Milman, The Santaló point of a function, and a functional form of the Santaló inequality. Mathematika **51**(1–2), 33–48 (2010)
2. A.L. Bar-on, The (B) conjecture for uniform measures in the plane, in *Bo'az Klartag and Emanuel Milman* (eds.) Geometric Aspects of Functional Analysis, Israel Seminar 2011–2013. Lecture Notes in Mathematics, vol 2116, pp. 341–353 (Springer, Cham, 2014)
3. C. Borell, Convex measures on locally convex spaces. Arkiv för matematik **12**(1-2), 239–252 (1974)
4. C. Borell, Convex set functions in d-space. Period. Math. Hung. **6**(2), 111–136 (1975)
5. K.J. Böröczky, E. Lutwak, D. Yang, G. Zhang, The log-Brunn-Minkowski inequality. Adv. Math. **231**(3–4), 1974–1997 (2012)
6. S. Chen, Y. Huang, Q.-R. Li, J. Liu, L_p-*Brunn-Minkowski inequality for* $p \in (1 - \frac{c}{n^{\frac{3}{2}}}, 1)$ (2018). arxiv:1811.10181
7. A. Colesanti, G. Livshyts, A. Marsiglietti, On the stability of Brunn-Minkowski type inequalities. J. Funct. Anal. **273**(3), 1120–1139 (2017)
8. D. Cordero-Erausquin, Santaló's inequality on $\mathbb{C}^n$ by complex interpolation. C.R. Math. **334**(9), 767–772 (2002)
9. D. Cordero-Erausquin, M. Fradelizi, B. Maurey, The (B) conjecture for the Gaussian measure of dilates of symmetric convex sets and related problems. J. Funct. Anal. **214**(2), 410–427 (2004)
10. A. Eskenazis, P. Nayar, T. Tkocz, Gaussian mixtures: Entropy and geometric inequalities. Ann. Probab. **46**(5), 2908–2945 (2018)
11. F. Franklin, Proof of a theorem of Tchebycheff's on definite integrals. Am. J. Math. **7**(4), 377 (1885)
12. A. Kolesnikov, E. Milman, *Local* L^p-*Brunn-Minkowski Inequalities for* $p < 1$ (2017). arxiv:1711.01089
13. R. Latała, On some inequalities for Gaussian measures, in *Proceedings of the International Congress of Mathematicians, Beijing*, vol II, pp. 813–822 (Higher Ed. Press, Beijing, 2002)
14. P. Nayar, T. Tkocz, A note on a Brunn-Minkowski inequality for the Gaussian measure. Proc. Am. Math. Soc. **141**(11), 4027–4030 (2013)
15. P. Nayar, T. Tkocz, *On a Convexity Property of Sections of the Cross-Polytope* (2018). arxiv:1810.02038
16. L. Rotem, *A Letter: The Log-Brunn-Minkowski Inequality for Complex Bodies* (2014). arXiv:1412.5321

17. C. Saroglou, Remarks on the conjectured log-Brunn-Minkowski inequality. Geom. Dedicata. **177**(1), 353–365 (2014)
18. C. Saroglou, More on logarithmic sums of convex bodies. Mathematika, **62**(03), 818–841 (2016)
19. R. Schneider, in *Convex Bodies: The Brunn-Minkowski Theory*. Encyclopedia of Mathematics and its Applications, 2nd edition (Cambridge University Press, Cambridge, 2014)

Chapter 12
A Dimension-Free Reverse Logarithmic Sobolev Inequality for Low-Complexity Functions in Gaussian Space

Ronen Eldan and Michel Ledoux

Abstract We discuss new proofs, and new forms, of a reverse logarithmic Sobolev inequality, with respect to the standard Gaussian measure, for low-complexity functions, measured in terms of Gaussian-width. In particular, we provide a dimension-free improvement for a related result given in Eldan (Geom Funct Anal 28:1548–1596, 2018).

12.1 A Reverse Logarithmic Sobolev Inequality

The recent work [5] has put forward a reverse logarithmic Sobolev inequality, with respect to the standard Gaussian measure, for low-complexity functions measured in terms of Gaussian-width. To briefly recall this inequality, we take again the notation from [5].

Let γ denote the standard Gaussian measure on $\mathbb{R}^n$, and let ν be the probability measure $d\nu = e^f d\gamma$ where $f : \mathbb{R}^n \to \mathbb{R}$ is twice-differentiable. Let

$$
D_{KL}(\nu \,||\, \gamma) = \int_{\mathbb{R}^n} f \, d\nu \quad \text{and} \quad \mathcal{I}(\nu) = \int_{\mathbb{R}^n} |\nabla f|^2 d\nu
$$

Incumbent of the Elaine Blond career development chair. Supported by a European Research Commission Starting Grant and by an Israel Science Foundation grant no. 715/16.

R. Eldan (✉)
Department of Mathematics, Weizmann Institute of Science, Rehovot, Israel

M. Ledoux
Institut de Mathématiques de Toulouse, Université de Toulouse – Paul-Sabatier, Toulouse, France

Institut Universitaire de France, Paris, France
e-mail: ledoux@math.univ-toulouse.fr

B. Klartag, E. Milman (eds.), *Geometric Aspects of Functional Analysis*,
Lecture Notes in Mathematics 2256,
https://doi.org/10.1007/978-3-030-36020-7_12

be respectively the Kullback-Leibler divergence (relative entropy) and the Fisher information of ν with respect to γ, assumed to be finite in the following. In particular ν has a second moment. The standard logarithmic Sobolev inequality of L. Gross (cf. e.g. [3, 11]) ensures that

$$D_{KL}(\nu \| \gamma) \leq \frac{1}{2}\mathcal{I}(\nu). \qquad (12.1)$$

Let

$$\mathcal{D}(\nu) = \mathbf{GW}(K) = \int_{\mathbb{R}^n} \sup_{t \in K} \langle y, t \rangle \, d\gamma(y)$$

be the Gaussian-width of the set $K = \{\nabla f(x) ; x \in \mathbb{R}^n\}$. Remark that the above expression is integrable if and only if f is a Lipschitz function, which is the only case in which we obtain meaningful results; Otherwise, set $\mathcal{D}(\nu) = \infty$. The quantity $\mathcal{D}(\nu)$ is a measure of the complexity of ν (rather the gradient-complexity of f). The following reverse logarithmic Sobolev inequality for measures of low-complexity has been established in [5, Theorem 4]. Set $M(\nu) = -\inf_{x \in \mathbb{R}^n} \Delta f(x)$, assumed to be finite. Then one has

$$\frac{1}{2}\mathcal{I}(\nu) \leq D_{KL}(\nu \| \gamma) + \frac{1}{2}M(\nu)_+ + \mathcal{D}(\nu)^{2/3}\mathcal{I}(\nu)^{1/3}, \qquad (12.2)$$

where $M(\nu)_+ = \max(M(\nu), 0)$.

Our first theorem gives the following related bound.

Theorem 12.1 *In the preceding notation,*

$$\frac{1}{2}\mathcal{I}(\nu) \leq D_{KL}(\nu \| \gamma) + M(\nu) + \mathcal{D}(\nu).$$

As discussed in [5], the inequality is sharp on the extremal functions $f(x) = \langle \alpha, x \rangle$, $x \in \mathbb{R}^n$, $\alpha \in \mathbb{R}^n$, of the logarithmic Sobolev inequality which have complexity $M(\nu) = \mathcal{D}(\nu) = 0$.

To compare between Theorem 12.1 and the bound (12.2), observe that the latter trivially holds true in the case that $\mathcal{I}(\nu) \leq \mathcal{D}(\nu)$, thus we may generally assume that $\mathcal{D}(\nu) \leq \mathcal{D}(\nu)^{2/3}\mathcal{I}(\nu)^{1/3}$. Unlike inequality (12.2), the bound of the theorem has the feature that both sides of the inequality are additive with respect to taking products and in this sense it is dimension-free. In Sect. 12.2 below, we give a slightly different form which improves on (12.2), and improves on Theorem 12.1 when, for example, the potential f is convex.

Such a reverse logarithmic Sobolev inequality is of theoretical interest in the study of approximations of partition functions and of low-complexity Gibbs measures on product spaces (cf. [1, 5]). An analogous definition of low-complexity for Boolean functions was considered in [5], where it is shown that a low-complexity condition implies that the measure can be decomposed as a mixture of approximate product measures.

In fact, it was very recently shown [7] that if a measure ν satisfies a reverse logarithmic Sobolev inequality, then it is close, in transportation distance, to a mixture of translated Gaussian measures (see also [2]). The combination of such a result with Theorem 12.1 gives a structure theorem for measures of low-complexity, analogous to the one given in [5], but for the Gaussian setting. We formulate this as a corollary.

Recall the quadratic Kantorovich metric $W_2(\nu, \gamma)$ between ν and γ defined by

$$W_2^2(\nu, \gamma) = \inf \int_{\mathbb{R}^n \times \mathbb{R}^n} |x - y|^2 d\pi(x, y)$$

where the infimum is taken over all couplings π on $\mathbb{R}^n \times \mathbb{R}^n$ with respective marginals ν and γ. A combination of Theorem 12.1 with [7, Theorem 5] yields the following statement.

Corollary 12.2 *In the preceding notation, there exists a probability measure μ such that*

$$W_2^2(\nu, \gamma \star \mu) \leq 16n^{1/3}\big(M(\nu) + \mathcal{D}(\nu)\big)^{2/3}.$$

In particular, the above corollary gives a meaningful result whenever $M(\nu) + \mathcal{D}(\nu) = o(n)$. It is also conjectured that a dimension-free analogue of [7, Theorem 5] should hold true, which, combined with our bound would imply the existence of a probability measure μ such that

$$W_2^2(\nu, \gamma \star \mu) \leq C\big(M(\nu) + \mathcal{D}(\nu)\big),$$

where $C > 0$ is a universal constant.

The proof of (12.2) strongly relies on a construction coming from stochastic control theory, of an entropy-optimal coupling of the measure ν to a Brownian motion. We will come back to it in Sect. 12.2. In contrast, our proof of Theorem 12.1 follows a simple and direct approach.

Proof of Theorem 12.1 Recall that we may assume that f is Lipschitz, since otherwise $\mathcal{D}(\nu) = \infty$. Thus we have $\nabla f(x) \frac{d\gamma}{dx} \to 0$ uniformly as $|x| \to \infty$ and therefore, by integration by parts with respect to the Gaussian measure γ,

$$\mathcal{I}(\nu) = \int_{\mathbb{R}^n} |\nabla f|^2 d\nu = \int_{\mathbb{R}^n} |\nabla f|^2 e^f d\gamma$$

$$= \int_{\mathbb{R}^n} \langle \nabla(e^f), \nabla f \rangle \, d\gamma$$

$$= -\int_{\mathbb{R}^n} e^f \, \mathrm{L}f \, d\gamma = -\int_{\mathbb{R}^n} \mathrm{L}f \, d\nu$$

where $L = \Delta f - \langle x, \nabla f \rangle$ is the Ornstein-Uhlenbeck operator. Therefore

$$\mathcal{I}(v) = -\int_{\mathbb{R}^n} \Delta f \, dv + \int_{\mathbb{R}^n} \langle x, \nabla f \rangle \, dv \leq M(v) + \int_{\mathbb{R}^n} \langle x, \nabla f \rangle \, dv. \qquad (12.3)$$

Let π be a coupling on $\mathbb{R}^n \times \mathbb{R}^n$ with respective marginals v and γ. Then,

$$\int_{\mathbb{R}^n} \langle x, \nabla f(x) \rangle \, dv(x) = \int_{\mathbb{R}^n \times \mathbb{R}^n} \langle x, \nabla f(x) \rangle \, d\pi(x, y)$$

$$= \int_{\mathbb{R}^n \times \mathbb{R}^n} \langle y, \nabla f(x) \rangle \, d\pi(x, y)$$

$$+ \int_{\mathbb{R}^n \times \mathbb{R}^n} \langle x - y, \nabla f(x) \rangle \, d\pi(x, y).$$

Now, on the one hand,

$$\int_{\mathbb{R}^n \times \mathbb{R}^n} \langle y, \nabla f(x) \rangle \, d\pi(x, y) \leq \int_{\mathbb{R}^n \times \mathbb{R}^n} \sup_{t \in K} \langle y, t \rangle \, d\pi(x, y) = \int_{\mathbb{R}^n} \sup_{t \in K} \langle y, t \rangle \, d\gamma(y).$$

On the other hand, by the standard quadratic inequality,

$$\int_{\mathbb{R}^n \times \mathbb{R}^n} \langle x - y, \nabla f(x) \rangle \, d\pi(x, y)$$

$$\leq \frac{1}{2} \int_{\mathbb{R}^n \times \mathbb{R}^n} |\nabla f(x)|^2 d\pi(x, y) + \frac{1}{2} \int_{\mathbb{R}^n \times \mathbb{R}^n} |x - y|^2 d\pi(x, y)$$

$$= \frac{1}{2} \int_{\mathbb{R}^n} |\nabla f(x)|^2 dv(x) + \frac{1}{2} \int_{\mathbb{R}^n \times \mathbb{R}^n} |x - y|^2 d\pi(x, y).$$

Taking the infimum over all couplings π with respective marginals v and γ, it holds true that

$$\int_{\mathbb{R}^n} \langle x, \nabla f \rangle \, dv \leq \int_{\mathbb{R}^n} \sup_{t \in K} \langle y, t \rangle \, d\gamma(y) + \frac{1}{2} \int_{\mathbb{R}^n} |\nabla f(x)|^2 dv(x) + \frac{1}{2} W_2^2(v, \gamma).$$

Therefore, together with (12.3),

$$\frac{1}{2} \mathcal{I}(v) \leq M(v) + \mathcal{D}(v) + \frac{1}{2} W_2^2(v, \gamma). \qquad (12.4)$$

It remains to recall the quadratic transportation cost inequality by M. Talagrand (cf. [3, 11])

$$\frac{1}{2} W_2^2(v, \gamma) \leq D_{KL}(v \, \| \, \gamma) \qquad (12.5)$$

and the proof is complete. $\square$

Together with the logarithmic Sobolev inequality (12.1) $D_{KL}(\nu||\gamma) \leq \frac{1}{2}\mathcal{I}(\nu)$, the step (12.4) of the preceding proof actually also yields a reverse transportation cost inequality

$$D_{KL}(\nu \,||\, \gamma) \leq \frac{1}{2} W_2^2(\nu, \gamma) + M(\nu) + \mathcal{D}(\nu). \tag{12.6}$$

Theorem 12.1 may also be deduced from a classical integrability result for the supremum of a Gaussian process. Given a bounded set $K \subset \mathbb{R}^n$, observe that $x \mapsto \sup_{t \in K} \langle x, t \rangle$ is integrable with respect to γ. Setting $Z = Z(x) = \sup_{t \in K} [\langle x, t \rangle - \frac{1}{2}|t|^2]$, $x \in \mathbb{R}^n$, it holds true that

$$\int_{\mathbb{R}^n} e^Z d\gamma \leq e^{\int_{\mathbb{R}^n} \sup_{t \in K} \langle x, t \rangle d\gamma}. \tag{12.7}$$

This inequality was originally put forward in [10, 12] in the context of concentration properties of suprema of Gaussian processes. Now, the classical entropic inequality (Gibbs variational principle) expresses that

$$\int_{\mathbb{R}^n} Z \, d\nu \leq D_{KL}(\nu \,||\, \gamma) + \log \int_{\mathbb{R}^n} e^Z d\gamma.$$

With $K = \{\nabla f(x) ; x \in \mathbb{R}^n\}$, it therefore follows that

$$\int_{\mathbb{R}^n} \left[\langle x, \nabla f \rangle - \frac{1}{2}|\nabla f|^2 \right] d\nu \leq \int_{\mathbb{R}^n} Z \, d\nu \leq D_{KL}(\nu \,||\, \gamma) + \int_{\mathbb{R}^n} \sup_{t \in K} \langle x, t \rangle \, d\gamma.$$

Again, together with (12.3), this yields the conclusion of the theorem.

At the same time, the integrability inequality (12.7) may be seen as a consequence of the transportation cost inequality (12.5) and the Kantorovich duality. The argument actually works for any probability μ on the Borel sets of $\mathbb{R}^n$ satisfying the transportation cost inequality

$$\frac{1}{2C} W_2^2(\mu', \mu) \leq D_{KL}(\mu' \,||\, \mu) \tag{12.8}$$

for some constant $C > 0$ and every $\mu' \ll \mu$ ($C = 1$ for $\mu = \gamma$). Namely, the Kantorovich duality (cf. [11]) expresses that

$$\frac{1}{2} W_2^2(\mu', \mu) = \sup \left[\int_{\mathbb{R}^n} \varphi \, d\mu' + \int_{\mathbb{R}^n} \psi \, d\mu \right]$$

where the supremum runs over the set of measurable functions $(\varphi, \psi) \in L^1(\mu') \times L^1(\mu)$ satisfying

$$\varphi(x) + \psi(y) \leq \frac{1}{2} |x - y|^2 \tag{12.9}$$

for $d\mu'$-almost all $x \in \mathbb{R}^n$ and $d\mu$-almost all $y \in \mathbb{R}^n$. Given then a couple of functions (φ, ψ) satisfying (12.9), the choice in (12.8) of $\frac{d\mu'}{d\mu} = \frac{e^g}{\int_{\mathbb{R}^n} e^g d\mu}$ where $g = \frac{1}{C}[\varphi + \int_{\mathbb{R}^n} \psi\, d\mu]$ yields that $\log \int_{\mathbb{R}^n} e^g d\mu \leq 0$, that is

$$\int_{\mathbb{R}^n} e^{\frac{1}{C}\varphi} d\mu \leq e^{-\frac{1}{C} \int_{\mathbb{R}^n} \psi d\mu}.$$

For every $x, y \in \mathbb{R}^n$, and $t \in K$,

$$\langle x, t \rangle - \frac{1}{2}|t|^2 = \langle y, t \rangle + \langle x - y, t \rangle - \frac{1}{2}|t|^2 \leq \langle y, t \rangle + \frac{1}{2}|x - y|^2.$$

Therefore, if $\varphi(x) = \sup_{t \in K} \left[\langle x, t \rangle - \frac{1}{2}|t|^2\right]$, $x \in \mathbb{R}^n$, then $\psi(y) = -\sup_{t \in K} \langle y, t \rangle$ is a valid candidate for (12.9). Hence

$$\int_{\mathbb{R}^n} e^{\frac{1}{C}Z} d\mu \leq e^{\frac{1}{C} \int_{\mathbb{R}^n} \sup_{t \in K} \langle x, t \rangle d\mu}$$

which amounts to (12.7) when $\mu = \gamma$.

12.2 Stochastic Calculus and the Föllmer Process

As mentioned above, the proof of (12.2) developed in [5] uses tools from stochastic control theory, and in particular the so-called Föllmer process [8] to achieve an entropy-optimal coupling of the measure ν to a Brownian motion. This argument has already been proved useful in the study of various functional inequalities [4, 6, 9].

To summarize a few facts from [5, 9], let $(B_t)_{t \geq 0}$ be standard Brownian motion in $\mathbb{R}^n$ (starting from the origin) adapted to a filtration $(\mathcal{F}_t)_{t \geq 0}$. Set $v(t, x) = \nabla \log Z(t, x)$, $t \in [0, 1]$, $x \in \mathbb{R}^n$, where

$$Z(t, x) = \mathbb{E}\big([e^f](x + B_{1-t})\big).$$

The Föllmer process $(X_t)_{t \in [0,1]}$ solves the stochastic differential equation

$$X_0 = 0, \quad dX_t = dB_t + v_t dt$$

where $v_t = v(t, X_t)$. Amongst its relevant properties, the random variable X_1 has distribution ν, $(v_t)_{t \in [0,1]}$ is a martingale, and

$$\mathbb{E}\left(\int_0^1 |v_t|^2 dt\right) = 2\, \mathrm{D_{KL}}(\nu \,\|\, \gamma).$$

The arguments developed in [5] thus make use of these properties towards a proof of the inequality (12.2). Now, actually, a small variation in the same spirit allows for the following inequality.

Theorem 12.3 *In the notation of Sect. 12.1, assume that v has a finite second moment. Then*

$$\int_{\mathbb{R}^n} |x|^2 dv - \int_{\mathbb{R}^n} |x|^2 d\gamma \leq 2\, \mathrm{D_{KL}}\big(v \,||\, \gamma\big) + \mathcal{D}(v).$$

For the sake of intuition, let us consider an equivalent form of the bound provided by this theorem. Denote

$$\mathrm{H}(v) = -\int_{\mathbb{R}^n} \log\left(\frac{dv}{dx}\right) dv,$$

the differential entropy of v. It is then straightforward to check that the inequality of Theorem 12.3 is equivalent to

$$\mathrm{H}(v) - \mathrm{H}(\gamma) \leq \frac{1}{2}\mathcal{D}(v). \tag{12.10}$$

Note that, in the special case that v has the form $\log\frac{dv}{dx} = \sup_{t \in K}\left[\langle x, t\rangle - \frac{1}{2}|t|^2\right] + $ const, this bound becomes somewhat similar to the bound (12.7).

Proof of Theorem 12.3 Observe first that by integration by parts

$$\int_{\mathbb{R}^n} |x|^2 dv - \int_{\mathbb{R}^n} |x|^2 d\gamma = \int_{\mathbb{R}^n} |x|^2 dv - n = \int_{\mathbb{R}^n} \langle x, \nabla f\rangle \, dv$$

where the boundary term vanishes due to the legitimate assumption that f is Lipschitz (otherwise $\mathcal{D}(v) = \infty$ and there is nothing to prove). The inequality of the theorem amounts to

$$\int_{\mathbb{R}^n} \langle x, \nabla f\rangle \, dv \leq 2\, \mathrm{D_{KL}}\big(v \,||\, \gamma\big) + \mathcal{D}(v). \tag{12.11}$$

Recall $K = \{\nabla f(x)\,;\, x \in \mathbb{R}^n\}$. Arguing as for the proof of Theorem 12.1, for any coupling π with respective marginals v and γ,

$$\int_{\mathbb{R}^n} \langle x, \nabla f\rangle \, dv = \int_{\mathbb{R}^n \times \mathbb{R}^n} \langle x, \nabla f(x)\rangle \, d\pi(x, y)$$

$$= \int_{\mathbb{R}^n \times \mathbb{R}^n} \langle x - y, \nabla f(x)\rangle \, d\pi(x, y) + \int_{\mathbb{R}^n \times \mathbb{R}^n} \langle y, \nabla f(x)\rangle \, d\pi(x, y)$$

$$\leq \int_{\mathbb{R}^n \times \mathbb{R}^n} \langle x - y, \nabla f(x) \rangle \, d\pi(x, y) + \int_{\mathbb{R}^n \times \mathbb{R}^n} \sup_{t \in K} \langle y, t \rangle \, d\pi(x, y)$$

$$= \int_{\mathbb{R}^n \times \mathbb{R}^n} \langle x - y, \nabla f(x) \rangle \, d\pi(x, y) + \mathcal{D}(\nu).$$

The inequality (12.11) would then follow if for some coupling π,

$$\int_{\mathbb{R}^n \times \mathbb{R}^n} \langle x - y, \nabla f(x) \rangle \, d\pi(x, y) \leq 2 \, \mathrm{D}_{\mathrm{KL}}(\nu \,\|\, \gamma).$$

But the point is that the Föllmer process actually produces an exact coupling for this identity to hold. Namely, by the definition and properties of $(X_t)_{t \in [0,1]}$, X_1 has law ν, B_1 has law γ and

$$d\langle X_t - B_t, v_t \rangle = |v_t|^2 dt + \langle X_t - B_t, dv_t \rangle.$$

Since $(v_t)_{t \in [0,1]}$ is a martingale,

$$\mathbb{E}\big(\langle X_1 - B_1, v_1 \rangle\big) = \mathbb{E}\left(\int_0^1 |v_t|^2 dt \right) = 2 \, \mathrm{D}_{\mathrm{KL}}(\nu \,\|\, \gamma)$$

from which the claim follows since $v_1 = \nabla f(X_1)$.   □

It remains to connect Theorem 12.3, or rather inequality (12.11), to Theorem 12.1. By the definition of the Fisher information (cf. (12.3))

$$\mathcal{I}(\nu) = - \int_{\mathbb{R}^n} \Delta f \, d\nu + \int_{\mathbb{R}^n} \langle x, \nabla f \rangle \, d\nu,$$

so that (12.11) expresses that

$$\frac{1}{2} \mathcal{I}(\nu) \leq \mathrm{D}_{\mathrm{KL}}(\nu \,\|\, \gamma) - \frac{1}{2} \int_{\mathbb{R}^n} \Delta f \, d\nu + \frac{1}{2} \mathcal{D}(\nu). \qquad (12.12)$$

As a first minor observation, note that, according to the discussion following Theorem 12.1, the above inequality improves upon (12.2) with the correct factor $\frac{1}{2}$ in front of $M(\nu)_+$ (unlike Theorem 12.1).

While presented and established with the quantity $M(\nu) = -\inf_{x \in \mathbb{R}^n} \Delta f(x)$, the proof of Theorem 12.1 shows in the same way that

$$\frac{1}{2} \mathcal{I}(\nu) \leq \mathrm{D}_{\mathrm{KL}}(\nu \,\|\, \gamma) - \int_{\mathbb{R}^n} \Delta f \, d\nu + \mathcal{D}(\nu). \qquad (12.13)$$

Hence, if $\mathcal{D}(\nu) \geq \int_{\mathbb{R}^n} \Delta f \, d\nu$ (which holds in particular for convex potentials), the inequality (12.12) improves upon (12.13). On the other hand, it does not seem possible to reach (12.12) as simply as (12.13), and in any case, the inequality of Theorem 12.3, even up to a constant, may not be deduced from Theorem 12.1.

References

1. T. Austin, The Structure of Low-complexity Gibbs Measures on Product Spaces. Ann. Probab. **47**(6), 4002–4023 (2019)
2. T. Austin, *Multi-Variate Correlation and Mixtures of Product Measures* (2018). Arxiv:1809.10272
3. D. Bakry, I. Gentil, M. Ledoux, Analysis and geometry of Markov diffusion operators, in *Grundlehren der mathematischen Wissenschaften*, vol 348 (Springer, Berlin, 2014)
4. C. Borell, Isoperimetry, log-concavity, and elasticity of option prices, in *New Directions in Mathematical Finance*, pp. 73–91 (Wiley, New York, 2002)
5. R. Eldan, Gaussian-width gradient complexity, reverse log-Sobolev inequalities and nonlinear large deviations. Geom. Funct. Anal. **28**, 1548–1596 (2018)
6. R. Eldan, J.R. Lee, Regularization under diffusion and anti-concentration of the information content. Duke Math. J. **167**, 969–993 (2018)
7. R. Eldan, J. Lehec, Y. Shenfeld, *The Logarithmic Sobolev Inequality via the Föllmer Process* (2018). Arxiv:1903.04522
8. H. Föllmer, An entropy approach to the time reversal of diffusion processes, in *Stochastic Differential Systems (Marseille-Luminy, 1984)*. Lecture Notes in Control and Information Science, vol 69, pp. 156–163 (Springer, Berlin, 1985)
9. J. Lehec, Representation formula for the entropy and functional inequalities. Ann. Inst. Henri Poincaré Probab. Stat. **49**, 885–899 (2013)
10. B.S. Tsirel'son, A geometric approach to maximum likelihood estimation for an infinite-dimensional Gaussian location. I. Teor. Veroyatnost. i Primenen. **27**, 388–395 (1982)
11. C. Villani, Topics in optimal transportation, in *Studies in Mathematics*, vol 58 (American Mathematical Society, Providence, 2003)
12. R. Vitale, The Wills functional and Gaussian processes. Ann. Probab. **24**, 2172–2178 (1996)

Chapter 13
Information and Dimensionality of Anisotropic Random Geometric Graphs

Ronen Eldan and Dan Mikulincer

Abstract This paper deals with the problem of detecting non-isotropic high-dimensional geometric structure in random graphs. Namely, we study a model of a random geometric graph in which vertices correspond to points generated randomly and independently from a non-isotropic d-dimensional Gaussian distribution, and two vertices are connected if the distance between them is smaller than some pre-specified threshold. We derive new notions of dimensionality which depend upon the eigenvalues of the covariance of the Gaussian distribution. If α denotes the vector of eigenvalues, and n is the number of vertices, then the quantities $\left(\frac{\|\alpha\|_2}{\|\alpha\|_3}\right)^6 / n^3$ and $\left(\frac{\|\alpha\|_2}{\|\alpha\|_4}\right)^4 / n^3$ determine upper and lower bounds for the possibility of detection. This generalizes a recent result by Bubeck, Ding, Rácz and the first named author from Bubeck et al. (Random Struct Algoritm 49(3):503–532, 2016) which shows that the quantity d/n^3 determines the boundary of detection for isotropic geometry. Our methods involve Fourier analysis and the theory of characteristic functions to investigate the underlying probabilities of the model. The proof of the lower bound uses information theoretic tools, based on the method presented in Bubeck and Ganguly (Int Math Res Not 2018(2):588–606, 2016).

13.1 Introduction

This study continues a line of work initiated by Bubeck, Ding, Rácz and the first named author [3], in which the problem of detecting geometric structure in large graphs was studied. In other words, given a large graph one is interested in determining whether or not it was generated using a latent geometric structure. The main contribution of this study is a generalization of the results to the anisotropic case.

R. Eldan (✉) · D. Mikulincer
Weizmann Institute of Science, Rehovot, Israel

© Springer Nature Switzerland AG 2020   273
B. Klartag, E. Milman (eds.), *Geometric Aspects of Functional Analysis*,
Lecture Notes in Mathematics 2256,
https://doi.org/10.1007/978-3-030-36020-7_13

Extracting information from large graphs is an extensively studied statistical task. In many cases, a given network, or graph, reflects some underlying structure; for example, a biological neuronal network is likely to reflect certain characteristics of its functionality such as physical location and cell structure. The objective of this paper is thus the detection of such an underlying geometric structure.

As a motivating example, consider the graph representing a large social network. It may be assumed that each node (or user) is described by a set of numerical parameters representing its properties (such as geographical location, age, political association, interests, etc.). It is plausible to assume that two nodes are more likely to be connected when their two respective points in parameter space are more correlated. Adopting this assumption, the nodes of such a graph may be thought of as points in a Euclidean space, with links appearing between two nodes when their distance is small enough. A natural question in this context would be: What can be said about the geometric structure by inspection of the graph itself? Specifically, can one distinguish between such a graph and a graph with no underlying geometric structure?

In statistical terms, given a graph G on n vertices, our null hypothesis is that G is an instance of the standard Erdős-Rényi random graph $G(n, p)$ [8], where the presence of each edge is determined independently, with probability p:

$$H_0 : G \sim G(n, p).$$

On the other hand, for the alternative, we consider the so-called random geometric graph. In this model each vertex is a point in some metric space and an edge is present between two points if the distance between them is smaller than some predefined threshold. Perhaps the most well-studied setting of this model is the isotropic Euclidean model, where the vertices are generated uniformly on the d-dimensional sphere or simply from the standard normal d-dimensional distribution. However, it seems that this model is too simplistic to reflect real world social networks. One particular problem, which we intend to tackle in this study, is the isotropicity assumption, which amounts to the fact that all of the properties associated with a node have the same significance in determining the network structure. It is clear that some parameters, such as geographic location, can be more significant than others. We therefore propose to extend this model to a non-isotropic setting. Roughly speaking, we replace the sphere with an ellipsoid; Instead of generating vertices from $\mathcal{N}(0, I_n)$, they will be generated from $\mathcal{N}(0, D_\alpha)$ for some diagonal matrix D_α with non-negative entries. We denote the model by $G(n, p, \alpha)$ where p is the probability of an edge appearing, and the diagonal of D_α is given by a vector $\alpha \in \mathbb{R}^d$. Formally, let $X_1, \ldots, X_n$ be i.i.d. points generated from $\mathcal{N}(0, D_\alpha)$. In $G(n, p, \alpha)$ vertices correspond to $X_1, \ldots, X_n$ and two distinct vertices are joined by an edge if and only if $\langle X_i, X_j \rangle \geq t_{p,\alpha}$, where $t_{p,\alpha}$ is the unique number satisfying $\mathbb{P}(\langle X_1, X_2 \rangle \geq t_{p,\alpha}) = p$. Our alternative hypothesis is thus

$$H_1 : G \sim G(n, p, \alpha).$$

In this paper, we will focus on the high-dimensional regime of the problem. Namely, we assume that the dimension and covariance matrix can depend on n. This point of view becomes highly relevant when considering recent developments in data sciences, where big data and high-dimensional feature spaces are becoming more prevalent. We will focus on the dense regime, where p is a constant independent of n and α.

13.1.1 Previous Work

This paper can be seen a direct follow-up of [3], which, as noted above, deals with the isotropic model of $G(n, p, d)$ in which $D_\alpha = I_d$. In the dense regime, it was shown that the total variation between the models depends asymptotically on the ratio $\frac{d}{n^3}$. The dependence is such that if $d >> n^3$, then $G(n, p, d)$ converges in total variation to $G(n, p)$. Conversely, on the other hand, if $d << n^3$ the total variation converges to 1.

Our starting point is thus the result of [3] stated as follows:

Theorem 13.1

(a) Let $p \in (0, 1)$ be fixed and assume that $d/n^3 \to 0$. Then,

$$\mathrm{TV}(G(n, p), G(n, p, d)) \to 1.$$

(b) Furthermore, if $d/n^3 \to \infty$ then

$$\mathrm{TV}(G(n, p), G(n, p, d)) \to 0.$$

One of the fundamental differences between $G(n, p)$ and $G(n, p, d)$ is a consequence of the triangle inequality. That is, if two points u and v are both close to a point w, then u and v cannot be too far apart. This roughly means that if both u and v are connected to w, then there is an increased probability of u being connected to v, unlike the case of the Erdős-Rényi graph where there is no dependence between the edges. Thus, counting the number of triangles in a graph seems to be a natural test to uncover geometric structure.

The idea of using triangles was extended in [3] and a variant was proposed: the *signed triangle*. This statistic was successfully used to completely characterize the asymptotics of $\mathrm{TV}(G(n, p), G(n, p, d))$ in the isotropic case. To understand the idea behind signed triangles, we first note that if A is the adjacency matrix of G then the number of triangles in G is given by $\mathrm{Tr}(A^3)$. The "number" of signed triangles is represented by $\mathrm{Tr}((A - p\mathbf{1})^3)$ where $\mathbf{1}$ is the matrix whose entries are all equal to 1. It turns out that the variance of signed triangles is significantly smaller than the corresponding quantity for regular triangles.

The methods used in [3] relied heavily on the symmetries of the sphere. As mentioned, our goal is to generalize this to the non-isotropic case, which requires us to apply different methods. The dimension d of the isotropic space arises as a natural parameter when discussing the underlying probabilities of Theorem 13.1. Clearly, however, when different coordinates of the space have different scales, the dimension by itself has little meaning. For example, consider a d-dimensional ellipsoid with one axis being large and the rest being much smaller. This ellipsoid behaves more like a one-dimensional sphere rather than a d-dimensional one, in the sense mentioned above. It would stand to reason the more anisotropic the ellipsoid is, the smaller its effective dimension would be.

13.1.2 Main Results and Ideas

In accordance to the above, our first task is to find a suitable notion of dimensionality for our model. For $q \geq 1$, let $\|\cdot\|_q$ stand for the q-norm. We derive the quantities $\left(\frac{\|\alpha\|_2}{\|\alpha\|_3}\right)^6$ and $\left(\frac{\|\alpha\|_2}{\|\alpha\|_4}\right)^4$ as the new notions of dimension, where α parametrizes the eigenvalues of D_α, the covariance matrix of the normal distribution, and is considered as a d-dimensional vector. We note that, in the isotropic case, those quantities reduce to d which also maximizes the expressions.

This notion of dimension allows us to tackle the main objective of this paper; studying the total variation distance between $G(n, p)$ and $G(n, p, \alpha)$. Considering what we know about the isotropic case our question becomes: What conditions are required from α, so that the total variation remains bounded away from 0? The following theorem provides a sufficient condition on α as well as a necessary one:

Theorem 13.2

(a) Let $p \in (0, 1)$ be fixed and assume that $\left(\frac{\|\alpha\|_2}{\|\alpha\|_3}\right)^6 / n^3 \to 0$. Then,

$$\mathrm{TV}(G(n, p), G(n, p, \alpha)) \to 1.$$

(b) Furthermore, if $\left(\frac{\|\alpha\|_2}{\|\alpha\|_4}\right)^4 / n^3 \to \infty$, then

$$\mathrm{TV}(G(n, p), G(n, p, \alpha)) \to 0.$$

Note that there is a gap between the bounds 2(a) and 2(b) (for example, if $\alpha_i \sim \frac{1}{\sqrt[3]{i}}$, then $\left(\frac{\|\alpha\|_2}{\|\alpha\|_3}\right)^6$ is order of $\frac{d}{\ln^2(d)}$, while $\left(\frac{\|\alpha\|_2}{\|\alpha\|_4}\right)^4$ is about $d^{\frac{2}{3}}$). We conjecture that the bound 2(a) is tight:

Conjecture 13.1 Let $p \in (0, 1)$ be fixed and assume that $\left(\frac{\|\alpha\|_2}{\|\alpha\|_3}\right)^6 / n^3 \to \infty$. Then

$$\mathrm{TV}(G(n, p), G(n, p, \alpha)) \to 0$$

In the following we describe some of the ideas used to prove Theorem 13.2.

As discussed, the main idea underlying this work has to do with counting triangles. Given a graph G we denote by $T(G)$ the number of triangles in the graph. It is easy to verify that $\mathbb{E} \, T(G(n, p)) = \binom{n}{3}p^3$ and $\mathrm{Var}(T(G(n, p)))$ is of order n^4. In the isotropic case, standard calculations show that the expected number of triangles in $G(n, p, d)$ is boosted by a factor proportional to $1 + \frac{1}{\sqrt{d}}$. The first difficulty that arises is to find a precise estimate for the probability increment in the non-isotropic case. In this case, we show that there is a constant δ_p depending only on p such that $\mathbb{E} \, T(G(n, p, \alpha)) \geq \binom{n}{3}p^3 \left(1 + \delta_p \left(\frac{\|\alpha\|_3}{\|\alpha\|_2}\right)^3\right)$. This would imply a non-negligible total variation distance as long as $\binom{n}{3} \left(\frac{\|\alpha\|_3}{\|\alpha\|_2}\right)^3$ is bigger than the standard deviation of $T(G(n, p))$. We incorporate the idea of using *signed triangles* which attain a similar difference between expected values but have a smaller variance. The number of *signed triangles* is defined as:

$$\tau(G) = \sum_{\{i,j,k\} \in \binom{[n]}{3}} (A_{i,j} - p)(A_{i,k} - p)(A_{j,k} - p),$$

where A is the adjacency matrix of G, which is proportional to $\mathrm{Tr}((A - p\mathbf{1})^3)$. It is known that $\mathrm{Var}(\tau(G(n, p)))$ is only of order n^3. Resolving the value of $\mathrm{Var}(\tau(G(n, p, \alpha)))$ leads to the following result (which implies Theorem 13.2(a)):

Theorem 13.3 *Let $p \in (0, 1)$ be fixed and assume that $\left(\frac{\|\alpha\|_2}{\|\alpha\|_3}\right)^6 / n^3 \to 0$. Then*

$$\mathrm{TV}(\tau(G(n, p)), \tau(G(n, p, \alpha))) \to 1$$

To prove Theorem 13.2(b) we may view the random graph $G(n, p, \alpha)$ as a measurable function of a random $n \times n$ matrix $W(n, \alpha)$ with entries proportional to $\langle \gamma_i, \gamma_j \rangle$ where γ_i are drawn *i.i.d.* from $\mathcal{N}(0, D_\alpha)$ and $D_\alpha = \mathrm{diag}(\alpha)$. Similarly, $G(n, p)$ can be viewed as a function of an $n \times n$ GOE random matrix denoted by $M(n)$. In [3] Theorem 13.1(b) was proven using direct calculations on the densities of the involved distributions. However, in our case, no simple formula exists, which makes their method inapplicable. The premise is instead proven using information theoretic tools, adopting ideas from [2]. The main idea is to use Pinsker's inequality to bound the total variation distance by the respective relative entropy. Thus we are interested in

$$\mathrm{Ent}\,[W(n, \alpha)\|M(n)].$$

Theorem 13.2(b) will then follow from the next result:

Theorem 13.4 *Let $p \in (0, 1)$ be fixed and assume that $\left(\frac{\|\alpha\|_2}{\|\alpha\|_4}\right)^4 / n^3 \to \infty$. Then*

$$\mathrm{Ent}\,[W(n, \alpha)\|M(n)] \to 0.$$

We suspect, as stated in Conjecture 13.1, that Theorem 13.2(b) does not give a tight characterization of the lower bound. Indeed, in the dense regime of the isotropic case, *signed triangles* act as an optimal statistic. It would seem to reason that deforming the sphere shouldn't affect the utility of such a local tool.

13.2 Preliminaries

We work in $\mathbb{R}^n$, equipped with the standard Euclidean structure $\langle \cdot, \cdot \rangle$. For $q \geq 1$, we denote by $\|\cdot\|_q$ the corresponding q-norm. That is, for $(v_1, \ldots, v_n) = v \in \mathbb{R}^n$, $\|v\|_q = \left(\sum_{i=1}^{n} |v_i|^q\right)^{\frac{1}{q}}$. If $\alpha = \{\alpha_i\}_{i=1}^{d}$ is a multi-set with elements from $\mathbb{R}$, we adopt the same notation for $\|\alpha\|_q$. We abbreviate $\|\cdot\| := \|\cdot\|_2$, the usual Euclidean norm and denote by $\mathbb{S}^{n-1}$ the unit sphere under this norm. In our proofs, we will allow ourselves to use the letters c, C, c', C', c_1, C_1, etc. to denote absolute positive constants whose values may change between appearances. The letters x, y, z will usually denote spatial variables while a, b, c will denote the corresponding frequencies in the Fourier domain. The letters X, Y, Z will usually be used as random variables and vectors. The imaginary unit will be denoted as $\mathbf{i}$.

Let X be a real valued random variable. The characteristic function of X is a function $\varphi_X : \mathbb{R} \to \mathbb{R}$, given by

$$\varphi_X(t) = \mathbb{E}[e^{\mathbf{i}tX}].$$

More generally, if X is an n-dimensional random vector, then the characteristic function of X is a function $\varphi_X : \mathbb{R}^n \to \mathbb{R}$ given by

$$\varphi_X(t) = \mathbb{E}[e^{\mathbf{i}\langle t, X \rangle}].$$

By elementary Fourier analysis, one can use the characteristic function to recover the distribution, whenever the random vector is integrable. We will be interested in the specific case where the dimension of X is 3. Assume $X = (X^1, X^2, X^3)$ has a density, denoted by f, a characteristic function, denoted by φ and (reflected) cumulative distribution function

$$F(t_1, t_2, t_3) = \mathbb{P}(X^1 > t_1, X^2 > t_2, X^3 > t_3),$$

with marginals onto the first 1 or 2 coordinates denoted as $F(t_1, t_2)$ and $F(t_1)$ respectively (remark that this is slightly a different definition than the usual notion of the cumulative distribution function, which results in a change of sign for the next identity). Then e.g., [13, Theorem 5] states that

$$\frac{\mathbf{i}}{\pi^3} \fint_{\mathbb{R}^3} \frac{\varphi(a, b, c)e^{-\mathbf{i}(at_1 + bt_2 + ct_3)}}{abc} da\,db\,dc \tag{13.1}$$

$$= 8F(t_1, t_2, t_3) - 4(F(t_1, t_2) + F(t_2, t_3) + F(t_1, t_3)) + 2(F(t_1) + F(t_2) + F(t_3)) - 1,$$

where the integral is taken as a Cauchy principal value; In $\mathbb{R}^3$, the Cauchy principal value of a function g, which we henceforth denote by $\fint_{\mathbb{R}^3} g$, is defined as

$$\int_0^\infty \int_0^\infty \int_0^\infty \Delta_c \Delta_b \Delta_a g(a, b, c)da\,db\,dc,$$

where $\Delta_a g(a, b, c) := g(a, b, c) + g(-a, b, c)$ and likewise for b, c. In the following, for multivariate functions, we interpret the definition of an *odd* (resp. *even*) function in the following sense: g is odd (resp. even) if it is antisymmetric (resp. symmetric) under change of sign of any coordinate, while keeping the values of the rest of the coordinates intact. We note that the principal value of an odd function vanishes, and if g is integrable then $\fint_{\mathbb{R}^3} g = \int_{\mathbb{R}^3} g$. Furthermore, by denoting

$$\mathrm{sgn}_{(t_1, t_2, t_3)}(x, y, z) = \mathrm{sgn}(x - t_1)\mathrm{sgn}(y - t_2)\mathrm{sgn}(z - t_3),$$

a simple calculation shows the following equality:

$$\int_{\mathbb{R}^3} f(x, y, z) \cdot \mathrm{sgn}_{(t_1, t_2, t_3)}(x, y, z)dx\,dy\,dz$$

$$= 8F(t_1, t_2, t_3) - 4(F(t_1, t_2) + F(t_2, t_3) + F(t_1, t_3)) + 2(F(t_1) + F(t_2) + F(t_3)) - 1.$$

Since the Fourier transform is an isometry we have that

$$\int_{\mathbb{R}^3} f \cdot \mathrm{sgn}_{(t_1, t_2, t_3)} = \frac{1}{\pi^3} \fint_{\mathbb{R}^3} \varphi \cdot \widehat{\mathrm{sgn}}_{(t_1, t_2, t_3)}, \tag{13.2}$$

where $\widehat{\mathrm{sgn}}_{(t_1, t_2, t_3)}$ is the Fourier transform of $\mathrm{sgn}_{(t_1, t_2, t_3)}$, when considered as a tempered distribution (for more information on the topic, see [9]). Putting all of the above together yields

$$\widehat{\mathrm{sgn}}_{(t_1,t_2,t_3)}(a,b,c) = \frac{\mathbf{i}e^{-\mathbf{i}(at_1+bt_2+ct_3)}}{abc}. \tag{13.3}$$

For a positive semi-definite $n \times n$ matrix Σ, we denote by $\mathcal{N}(0, \Sigma)$ the law of the centered Gaussian distribution with covariance Σ. If $X \sim \mathcal{N}(0, \Sigma)$ then $X^T X$ has the law $\mathcal{W}_n(\Sigma, 1)$ of the Wishart distribution with 1 degree of freedom. The characteristic function of $X^T X$ is known (see [7]) and given by

$$\Theta \rightarrow \det(\mathrm{I} - 2\mathbf{i}\Theta\Sigma)^{-\frac{1}{2}}. \tag{13.4}$$

If Z is distributed as a standard Gaussian random variable, then Z^2 has the χ^2 distribution with 1 degree of freedom. For such a distribution, we have $\mathbb{E}[\chi^2] = 1$ and $\mathrm{Var}(\chi^2) = 2$. The χ^2 distribution has a sub-exponential tail which may be bounded using a Bernstein's type inequality [15], in the following way. If $\{\chi_i^2\}_{i=1}^n$, are independent χ^2 random variables, then for every $(v_1, \ldots, v_n) = v \in \mathbb{R}^n$ and every $t > 0$

$$\mathbb{P}\left(\left|\sum v_i \chi_i^2 - \sum v_i\right| \geq t\right) \leq 2\exp\left(-\min\left(\frac{t}{2\|v\|_\infty}, \frac{t^2}{4\|v\|_2^2}\right)\right). \tag{13.5}$$

Let $X_1, \ldots, X_n$ be independent random variables with 0 mean and variance $\mathbb{E}[X_i^2] = \sigma_i^2$. Define $s_n = \sqrt{\sum_{i=1}^n \sigma_i^2}$ and $S_n = \sum_{i=1}^n \frac{X_i}{s_n}$. Under appropriate regularity conditions the central limit theorem states that S_n converges in distribution to $\mathcal{N}(0, 1)$.

Berry-Esseen's inequality [12] quantifies this convergence. Suppose that the absolute third moments of X_i exist and $\mathbb{E}[|X_i|^3] = \rho_i$. If we denote by Z a standard Gaussian and define S_n as above then, for every $x \in \mathbb{R}$,

$$|\mathbb{P}(S_n < x) - \mathbb{P}(Z < x)| \leq \frac{\sum\limits_{i=1}^n \rho_i}{s_n^3}. \tag{13.6}$$

This can be generalized to higher dimensions, as found in [1, Theorem 1.1]. In that case assume $X_1, \ldots, X_n$ are independent random vectors in $\mathbb{R}^d$ and $S_n = \sum_{i=1}^n X_i$ has covariance Σ^2. Assume that Σ is invertible and denote $\mathbb{E}[|\Sigma^{-1}X_i|^3] = \rho_i$. If Z_d is a d-dimensional standard Gaussian vector, then there exists a universal constant $C_{\mathrm{be}} > 0$, such that for any convex set A:

$$|\mathbb{P}(\Sigma^{-1}S_n \in A) - \mathbb{P}(Z_d \in A)| \leq C_{\mathrm{be}}d^{\frac{1}{4}}\sum_i \rho_i. \tag{13.7}$$

For a random vector X on $\mathbb{R}^n$ with density f, the differential entropy of X is defined

$$\text{Ent}[X] = -\int_{\mathbb{R}^n} f(x) \ln(f(x)) dx.$$

If Y is another random vector with density g, the relative entropy of X with respect to Y is

$$\text{Ent}[X||Y] = \int_{\mathbb{R}^n} f(x) \ln\left(\frac{f(x)}{g(x)}\right) dx.$$

Pinsker's inequality connects between the relative entropy and the total variation distance,

$$\text{TV}(X, Y) \leq \sqrt{\frac{1}{2}\text{Ent}[X||Y]}. \tag{13.8}$$

The chain rule for relative entropy states that for any random vectors X_1, X_2, Y_1, Y_2,

$$\text{Ent}[(X_1, X_2)||(Y_1, Y_2)] = \text{Ent}[X_1||Y_1] + \mathbb{E}_{x \sim \lambda_1} \text{Ent}[X_2|X_1 = x||Y_2|Y_1 = x], \tag{13.9}$$

where λ_1 is the marginal of X_1, and $X_2|X_1 = x$ is the distribution of X_2 conditioned on the event $X_1 = x$ (similarly for $Y_2|Y_1 = x$).

13.3 Estimates for a Triangle in a Random Geometric Graph

In this section we derive a lower bound for the probability that an induced subgraph, of size 3, of a random geometric graph forms a triangle. This calculation is instrumental for the derivation of Theorem 13.2(a). Using the notation of the introduction, let $X_1, X_2, X_3 \sim \mathcal{N}(0, D_\alpha)$ be independent normal random vectors with coordinates X_1^i, X_2^i, X_3^i for $1 \leq i \leq d$. We denote by f the joint density of $(\langle X_1, X_2\rangle, \langle X_1, X_3\rangle, \langle X_2, X_3\rangle)$. Consider the event

$$E_p = \{\langle X_1, X_2\rangle \geq t_{p,\alpha}, \langle X_1, X_3\rangle \geq t_{p,\alpha}, \langle X_2, X_3\rangle \geq t_{p,\alpha}\},$$

that the corresponding vertices form a triangle in $G(n, p, \alpha)$. The main result of this section is the following theorem.

Theorem 13.5 *Let $p \in (0, 1)$ and assume $\|\alpha\|_\infty = 1$. One has*

$$p^3 + \Delta \left(\frac{\|\alpha\|_3}{\|\alpha\|_2} \right)^3 \geq \mathbb{P}(E_p) \geq p^3 + \delta_p \left(\frac{\|\alpha\|_3}{\|\alpha\|_2} \right)^3$$

whenever $\|\alpha\|_2 > c_p$, for constants $\Delta, \delta_p, c_p > 0$ which may depend only on p.

13.3.1 Lower Bound: The Case $p = \frac{1}{2}$

It will be instructive to begin the discussion with the (easier) case $p = \frac{1}{2}$, in which $t_{p,\alpha} = 0$. We are thus interested in the probability that $\langle X_1, X_2 \rangle, \langle X_1, X_3 \rangle, \langle X_2, X_3 \rangle > 0$. Note that the triplet $(\langle X_1, X_2 \rangle, \langle X_1, X_3 \rangle, \langle X_2, X_3 \rangle)$ can be realized as a linear combination of upper off-diagonal elements taken from d independent three-dimensional Wishart random matrices (see below for an elaborated explanation). Unfortunately, there is no known closed expression for the density of such a distribution. The following lemma utilizes the characteristic function of the joint distribution to derive a closed expression for the desired probability.

Lemma 13.1

$$\mathbb{P}\left(E_{\frac{1}{2}} \right) = \frac{1}{8} + \oint_{\mathbb{R}^3} \frac{\mathbf{i}}{8abc\pi^3} \left(\prod_i (1 + \alpha_i^2(a^2 + b^2 + c^2) + 2\alpha_i^3 abc\mathbf{i})^{-\frac{1}{2}} \right) dadbdc.$$

$$(13.10)$$

Proof Consider the event $\{\langle X_1, X_2 \rangle > 0, \langle X_1, X_3 \rangle < 0, \langle X_2, X_3 \rangle < 0\}$. The map $(x, y, z) \mapsto (x, y, -z)$ preserves the law of (X_1, X_2, X_3). Thus,

$\mathbb{P}(\{\langle X_1, X_2 \rangle > 0, \langle X_1, X_3 \rangle < 0, \langle X_2, X_3 \rangle < 0\})$

$\quad = \mathbb{P}(\{\langle X_1, X_2 \rangle > 0, \langle X_1, X_3 \rangle > 0, \langle X_2, X_3 \rangle > 0\}).$

By the same argument,

$\mathbb{P}(\{\langle X_1, X_2 \rangle > 0, \langle X_1, X_3 \rangle > 0, \langle X_2, X_3 \rangle < 0\})$

$\quad = \mathbb{P}(\{\langle X_1, X_2 \rangle < 0, \langle X_1, X_3 \rangle < 0, \langle X_2, X_3 \rangle < 0\}).$

We denote the event on the right side by $\mathbb{P}\left(I_{\frac{1}{2}} \right)$, the probability of an induced independent set on 3 vertices.

From the above observation, it is clear that $4\left(\mathbb{P}(E_{\frac{1}{2}}) + \mathbb{P}(I_{\frac{1}{2}})\right) = 1$. Also, we may note that $\int_{\mathbb{R}^3} \mathrm{sgn}(xyz) \cdot f(x,y,z)\,dxdydz = 4\left(\mathbb{P}(E_{\frac{1}{2}}) - \mathbb{P}(I_{\frac{1}{2}})\right)$. Combining the two equalities yields $\mathbb{P}(E_{\frac{1}{2}}) = \frac{1}{8} + \frac{1}{8}\int_{\mathbb{R}^3} \mathrm{sgn}(xyz) \cdot f(x,y,z)\,dxdydz$. As noted, no closed expression for f is known, so the calculation of the above integral cannot be carried out in a straightforward manner. Instead, (13.2) allows us to rewrite the integral as

$$\int_{\mathbb{R}^3} f(x,y,z) \cdot \mathrm{sgn}(xyz)\,dxdydz = \frac{1}{\pi^3} \oint_{\mathbb{R}^3} \varphi(a,b,c) \cdot \widehat{\mathrm{sgn}}(abc)\,dadbdc,$$

where φ is the characteristic function of f, and $\widehat{\mathrm{sgn}}$ is the Fourier transform of $\mathrm{sgn}_{(0,0,0)}$ as in (13.3).

Thus, we are required to calculate $\varphi(a,b,c)$. Consider three independent normal random variables, X, Y, Z, with mean 0 and variance σ^2, the characteristic function of (XY, XZ, YZ) is defined by $(a,b,c) \rightarrow E[\exp(i(a \cdot XY + b \cdot XZ + c \cdot YZ))]$. We have that

$$a \cdot XY + b \cdot XZ + c \cdot YZ = \mathrm{Tr}\left(\begin{bmatrix} 0 & \frac{a}{2} & \frac{b}{2} \\ \frac{a}{2} & 0 & \frac{c}{2} \\ \frac{b}{2} & \frac{c}{2} & 0 \end{bmatrix} \cdot \begin{bmatrix} X^2 & XY & XZ \\ XY & Y^2 & YZ \\ XZ & YZ & Z^2 \end{bmatrix}\right).$$

If we consider the Wishart distribution $\mathcal{W}_3(\Sigma_\sigma, 1)$, where Σ_σ is a σ^2 scalar matrix, we note that the above function equals the characteristic function of $\mathcal{W}_3(\Sigma_\sigma, 1)$ on the matrix $\begin{bmatrix} 0 & \frac{a}{2} & \frac{b}{2} \\ \frac{a}{2} & 0 & \frac{c}{2} \\ \frac{b}{2} & \frac{c}{2} & 0 \end{bmatrix}$. Using the formula (13.4), this equals

$\det\left(\begin{bmatrix} 1 & -i\sigma^2 a & -i\sigma^2 b \\ -i\sigma^2 a & 1 & -i\sigma^2 c \\ -i\sigma^2 b & -i\sigma^2 c & 1 \end{bmatrix}\right)^{-\frac{1}{2}}$, which may be written otherwise as $(1 + (\sigma^2)^2(a^2 + b^2 + c^2) + 2(\sigma^2)^3 abc\mathbf{i})^{-\frac{1}{2}}$.

By the convolution-multiplication theorem [6, Theorem 3.3.2], the characteristic function of a sum of independent variables is the multiplication of their characteristic functions, it then follows that:

$$\varphi(a,b,c) = \prod_{i=1}^{d}(1 + \alpha_i^2(a^2 + b^2 + c^2) + 2\alpha_i^3 abc\mathbf{i})^{-\frac{1}{2}}, \tag{13.11}$$

which results in:

$$\oint_{\mathbb{R}^3} \varphi(a, b, c) \cdot \widehat{\mathrm{sgn}}(abc) \, da \, db \, dc$$

$$= \oint_{\mathbb{R}^3} \frac{\mathbf{i}}{abc} \prod_i (1 + \alpha_i^2(a^2 + b^2 + c^2) + 2\alpha_i^3 abc\mathbf{i})^{-\frac{1}{2}} \, da \, db \, dc.$$

This concludes the proof. □

In view of the above, it suffices to estimate the integral in (13.10). We will show that the integral of the expression

$$\mathrm{Re}\left(\frac{\mathbf{i}}{abc} \prod_i (1 + \alpha_i^2(a^2 + b^2 + c^2) + 2\alpha_i^3 abc\mathbf{i})^{-\frac{1}{2}}\right),$$

is concentrated in a ball of radius $\frac{1}{\|\alpha\|_2}$, and that inside this ball, the above expression is very close in value to $\|\alpha\|_3^3$. From this, it will follow that

$$\mathbb{P}\left(E_{\frac{1}{2}}\right) \simeq \frac{1}{8} + \left(\frac{\|\alpha\|_3}{\|\alpha\|_2}\right)^3.$$

The next result will be used to control the integral outside of the aforementioned ball.

Lemma 13.2 *Let $n \geq 3$ and $\gamma = \{\gamma_i\}_{i=1}^d$, suppose that $\gamma_i \in [0, 1]$ for $1 \leq i \leq d$. Define*

$$I(T) = \int_T^\infty \frac{r^2 \, dr}{\sqrt{\prod_i (1 + \gamma_i^2 r^2)}}, \quad \forall T \geq 1,$$

and denote $\|\gamma\|_2^2 = \sum_i \gamma_i^2$, then there exist constants $c_n, C_n > 0$, depending only on n, such that whenever $\|\gamma\|_2^2 > c_n$ we have that $I(T) \leq C_n \left(\frac{1}{\|\gamma\|_2^2}\right)^{\frac{n}{2}} \frac{1}{T^{n-3}}$.

Proof Indeed, assume $\|\gamma\|_2^2 > n$. Note that necessarily $d \geq n$ in this case. Thus we can give a non trivial lower bound of $\prod_i (1 + \gamma_i^2 r^2)$ by considering the sum of all products of n different elements of γ. That is

$$\prod_i \left(1 + \gamma_i^2 r^2\right) \geq \left(\sum_{\substack{S \subset \gamma \\ |S|=n}} \prod_{\gamma_j \in S} \gamma_j^2\right) r^{2n}.$$

We claim now that:

$$\sum_{\substack{S \subset \gamma \\ |S|=n}} \prod_{\gamma_j \in S} \gamma_j^2 \geq \frac{1}{n!} \prod_{k=0}^{n-1} \left(\|\gamma\|_2^2 - k\right). \tag{13.12}$$

To see that, we may rewrite

$$\sum_{\substack{S \subset \gamma \\ |S|=n}} \prod_{\gamma_i \in S} \gamma_i^2 = \frac{1}{n} \sum_i \gamma_i^2 \sum_{\substack{S \subset \gamma \setminus \{\gamma_i\} \\ |S|=n-1}} \prod_{\gamma_j \in S} \gamma_j^2,$$

where we have counted each $S \subset \gamma$, n times. But, $\gamma_i \leq 1$ for every $1 \leq i \leq d$, and so $\|\gamma \setminus \{\gamma_i\}\|_2^2 > \|\gamma\|_2^2 - 1$. Equation (13.12) now follows by induction, since

$$\frac{1}{n} \sum_i \gamma_i^2 \sum_{\substack{S \subset \gamma \setminus \{\gamma_i\} \\ |S|=n-1}} \prod_{\gamma_j \in S} \gamma_j^2 \geq \frac{1}{n} \sum_i \gamma_i^2 \frac{1}{(n-1)!} \prod_{k=0}^{n-2} (\|\gamma\|_2^2 - 1 - k)$$

$$= \frac{1}{n!} \prod_{k=0}^{n-1} \left(\|\gamma\|_2^2 - k\right)$$

If we further assume that $\|\gamma\|_2^2 \geq 2n$, then $\|\gamma\|_2^2 - k > \frac{1}{2} \|\gamma\|_2^2$, for every $0 \leq k \leq n - 1$. Plugging this into (13.12) produces

$$\prod_i \left(1 + \gamma_i^2 r^2\right) \geq \left(\frac{\|\gamma\|_2^2}{n!2}\right)^n r^{2n},$$

which implies

$$I(T) \leq \left(\frac{n!2}{\|\gamma\|_2^2}\right)^{\frac{n}{2}} \int_T^\infty \frac{dr}{r^{n-2}} = \frac{(n!2)^n}{n-3} \left(\frac{1}{\|\gamma\|_2^2}\right)^{\frac{n}{2}} \frac{1}{T^{n-3}},$$

as desired. $\square$

Remark The constants obtained in the above proof are far from optimal, but will suffice for our needs.

We will use the above result in order to bound from below the integral in formula (13.10). For this, we will assume W.L.O.G. that α is normalized in the following way:

$$\alpha_1 = 1 \text{ and } \alpha_i \in [0, 1] \text{ for } 1 \leq i \leq d. \tag{13.13}$$

We note that this normalization yields the following properties for $n, m \in \mathbb{N}$, which we shall use freely:

- For every $k > 0$, $\|\alpha\|_k^k \geq 1$ and thus $\left(\|\alpha\|_k^k\right)^n \leq \left(\|\alpha\|_k^k\right)^m$ when $n \leq m$.
- $\alpha_i^n \geq \alpha_i^m$ and $\|\alpha\|_n^n \geq \|\alpha\|_m^m$ when $n \leq m$.
- For any $n > 2$ and $\varepsilon > 0$ there exists $c > 0$ such that whenever $\|\alpha\|_2^2 > c$ we have $\left(\frac{\|\alpha\|_n}{\|\alpha\|_2}\right)^n < \varepsilon$.

Lemma 13.3 *There exists a constant $c_{1/2} > 0$ such that whenever $\|\alpha\|_2^2 > c_{1/2}$ then*

$$\oint_{\mathbb{R}^3} \frac{\mathbf{i}}{abc} \prod_i \left(1 + \alpha_i^2(a^2 + b^2 + c^2) + 2\alpha_i^3 abc\mathbf{i}\right)^{-\frac{1}{2}} dadbdc \geq \frac{1}{100} \left(\frac{\|\alpha\|_3}{\|\alpha\|_2}\right)^3.$$

Proof First, we have the privilege of knowing the integral evaluates to some probability. Therefore, the principal value of it's imaginary part must vanish. This becomes evident by noting that the imaginary part is an odd function. Thus, we are interested in:

$$\text{Re}\left(\oint_{\mathbb{R}^3} \frac{\mathbf{i}}{abc} \prod_i (1 + \alpha_i^2(a^2 + b^2 + c^2) + 2\alpha_i^3 abc\mathbf{i})^{-\frac{1}{2}} dadbdc\right)$$

$$= \oint_{\mathbb{R}^3} \frac{-1}{abc} \text{Im}\left(\prod_i (1 + \alpha_i^2(a^2 + b^2 + c^2) + 2\alpha_i^3 abc\mathbf{i})^{-\frac{1}{2}}\right) dadbdc$$

$$= \oint_{\mathbb{R}^3} \frac{-\sin\left(\arg\left(\prod_i \left(1 + \alpha_i^2(a^2 + b^2 + c^2) + 2\alpha_i^3 abc\mathbf{i}\right)^{-\frac{1}{2}}\right)\right)}{abc \left|\prod_i \left(1 + \alpha_i^2(a^2 + b^2 + c^2) + 2\alpha_i^3 abc\mathbf{i}\right)^{\frac{1}{2}}\right|} dadbdc$$

$$
= \oint_{\mathbb{R}^3} \frac{\sin\left(\frac{1}{2}\sum_i \arctan\left(\frac{2\alpha_i^3 abc}{1+\alpha_i^2(a^2+b^2+c^2)}\right)\right)}{abc \prod_i \left(\left(1+\alpha_i^2(a^2+b^2+c^2)\right)^2 + 4\alpha_i^6 a^2 b^2 c^2\right)^{\frac{1}{4}}} dadbdc
$$

$$
= \oint_{\mathbb{R}^3} \frac{-\mathrm{Im}(\varphi(a,b,c))}{abc} dadbdc = \int_{\mathbb{R}^3} \frac{-\mathrm{Im}(\varphi(a,b,c))}{abc} dadbdc, \qquad (13.14)
$$

where φ is as in (13.11). It is straightforward to verify that $\mathrm{Im}(\varphi(a,b,c)) = O(abc)$, which implies that the above integrand is actually integrable, and thus justifies the last equality. We will estimate the above integral in several steps.

Step 1—The Integral is Bounded from Below on $B_1 = \{x \in \mathbb{R}^3 : \|x\|^2 \le \frac{1}{\|\alpha\|_2^2}\}$, the Ball of Radius $\frac{1}{\|\alpha\|_2}$

First, we will prove that the following holds:

$$
\sin\left(\frac{1}{2}\sum_i \arctan\left(\frac{2\alpha_i^3 abc}{1+\alpha_i^2(a^2+b^2+c^2)}\right)\right) \ge \sum_i \frac{\alpha_i^3 abc}{1+\alpha_i^2(a^2+b^2+c^2)}
$$
$$
- 3\|\alpha\|_3^6 (abc)^2. \qquad (13.15)
$$

Indeed, since $\sin(x) \ge x - x^2$ we have that

$$
\sin\left(\frac{1}{2}\sum_i \arctan\left(\frac{2\alpha_i^3 abc}{1+\alpha_i^2(a^2+b^2+c^2)}\right)\right)
$$

$$
\ge \frac{1}{2}\sum_i \arctan\left(\frac{2\alpha_i^3 abc}{1+\alpha_i^2(a^2+b^2+c^2)}\right)
$$

$$
-\frac{1}{4}\left(\sum_i \arctan\left(\frac{2\alpha_i^3 abc}{1+\alpha_i^2(a^2+b^2+c^2)}\right)\right)^2
$$

$$
\ge \frac{1}{2}\sum_i \arctan\left(\frac{2\alpha_i^3 abc}{1+\alpha_i^2(a^2+b^2+c^2)}\right) - \left(\sum_i \alpha_i^3\right)^2 (abc)^2.
$$

With the last inequality following from the fact that $\arctan^2(x) \leq x^2$. Now, using the inequality $\arctan(x) \geq x - x^2$ yields

$$\frac{1}{2} \sum_i \arctan\left(\frac{2\alpha_i^3 abc}{1 + \alpha_i^2(a^2 + b^2 + c^2)}\right) - \left(\sum_i \alpha_i^3\right)^2 (abc)^2$$

$$\geq \sum_i \frac{\alpha_i^3 abc}{1 + \alpha_i^2(a^2 + b^2 + c^2)} - 2\left(\sum_i \alpha_i^6\right)(abc)^2 - \left(\sum_i \alpha_i^3\right)^2 (abc)^2$$

$$\geq \sum_i \frac{\alpha_i^3 abc}{1 + \alpha_i^2(a^2 + b^2 + c^2)} - 3\|\alpha\|_3^6 (abc)^2.$$

When $(a, b, c) \in B_1$, then $\alpha_i^2(a^2 + b^2 + c^2) \leq \frac{\alpha_i^2}{\|\alpha\|_2^2} \leq 1$ and we have

$$\sum_i \frac{\alpha_i^3 abc}{1 + \alpha_i^2(a^2 + b^2 + c^2)} - 3\|\alpha\|_3^6 (abc)^2 \geq \frac{1}{2}\|\alpha\|_3^3 abc - 3\|\alpha\|_3^6 (abc)^2.$$

$$(13.16)$$

Next, we note that for $(a, b, c) \in B_1$:

$$1 \geq \frac{1}{\prod_i \left[(1 + \alpha_i^2(a^2 + b^2 + c^2))^2 + 4\alpha_i^6 a^2 b^2 c^2\right]^{\frac{1}{4}}} \geq \frac{1}{\prod_i \left[\left(1 + \frac{\alpha_i^2}{\|\alpha\|_2^2}\right)^2 + \frac{4\alpha_i^6}{\|\alpha\|_2^6}\right]^{\frac{1}{4}}}.$$

Since, in (13.13), we've assumed that $\alpha_i \leq 1$ for each i while $\sum_i \alpha_i^2 \geq 1$, we may now lower bound the above by $\dfrac{1}{\prod_i\left(1 + \frac{7\alpha_i^2}{\|\alpha\|_2^2}\right)^{-\frac{1}{4}}}$, and since $\ln\left(\prod_i \left(1 + \frac{7\alpha_i^2}{\|\alpha\|_2^2}\right)\right) \leq$

$\frac{7}{\|\alpha\|_2^2} \sum_i \alpha_i^2 = 7$, we have

$$\frac{1}{\prod_i \left(1 + \frac{7\alpha_i^2}{\|\alpha\|_2^2}\right)^{\frac{1}{4}}} \geq e^{-2}.$$

$$(13.17)$$

By combining (13.16) and (13.17) into (13.11) we may see for $(a, b, c) \in B_1$ the following holds:

$$\text{Im}\,(\varphi(a, b, c)) \geq \left(\frac{1}{2}\|\alpha\|_3^3 abc - 3\|\alpha\|_3^6 (abc)^2\right) e^{-2} \quad \text{when } abc > 0.$$

Also, it is not hard to see that $\text{Im}(\varphi)$ is an odd function, which makes $\frac{\text{Im}(\varphi(a,b,c))}{abc}$ even. Hence, if $H = \{(a, b, c) \in \mathbb{R}^3 | abc > 0\}$, then

$$\int_{B_1} \frac{\text{Im}(\varphi(a, b, c))}{abc} \, dadbdc = 2 \int_{B_1 \cap H} \frac{\text{Im}(\varphi(a, b, c))}{abc} \, dadbdc.$$

Finally, since the volume of B_1 is $\frac{4\pi}{3\|\alpha\|_2^3}$, and as long as $\|\alpha\|_2^2$ is large enough:

$$\int_{B_1 \cap H} \frac{-\text{Im}(\varphi(a, b, c))}{abc} \, dadbdc \geq \frac{1}{e^2} \int_{B_1 \cap H} \left(\frac{1}{2} \|\alpha\|_3^3 - 3 \|\alpha\|_3^6 \, abc \right) dadbdc$$

$$\geq \frac{\pi}{3e^2} \left(\frac{\|\alpha\|_3}{\|\alpha\|_2} \right)^3 - \frac{3 \|\alpha\|_3^6}{e^2} \int_{B_1} |abc| \, dadbdc \geq \frac{\pi}{3e^2} \left(\frac{\|\alpha\|_3}{\|\alpha\|_2} \right)^3 - \frac{3}{e^2} \left(\frac{\|\alpha\|_3}{\|\alpha\|_2} \right)^6,$$

where the last inequality uses the fact

$$\int_{B_1} |abc| \, dadbdc \leq \frac{1}{\|\alpha\|_2^6}.$$

Now, by using the properties of the normalization (13.13), $\|\alpha\|_3^3 \leq \|\alpha\|_2^2$. Thus,

$$\left(\frac{\|\alpha\|_3}{\|\alpha\|_2} \right)^6 \leq \frac{1}{\|\alpha\|_2} \left(\frac{\|\alpha\|_3}{\|\alpha\|_2} \right)^3,$$

and there exists a constant $c_1 > 0$ such that whenever $\|\alpha\|_2^2 > c_1$ then

$$\int_{B_1} \frac{-\text{Im}(\varphi(a, b, c))}{abc} \, dadbdc > \frac{\pi}{4e^2} \left(\frac{\|\alpha\|_3}{\|\alpha\|_2} \right)^3 > \frac{1}{10} \left(\frac{\|\alpha\|_3}{\|\alpha\|_2} \right)^3.$$

Step 2—The Integrand is Positive on $B_2 = \left\{ x \in \mathbb{R}^3 : \|x\|^2 \leq \frac{1}{\|\alpha\|_2^{22/12}} \right\}$, the Ball of Radius $\frac{1}{\|\alpha\|_2^{11/12}}$

We first note that whenever $\left| \sum_i \arctan \left(\frac{2\alpha_i^3 abc}{1 + \alpha_i^2 (a^2 + b^2 + c^2)} \right) \right| < \pi$, then

$$\sin \left(\arg \left(\prod_i \left[1 + \alpha_i^2 (a^2 + b^2 + c^2) + 2\alpha_i^3 abci \right] \right) \right)$$

has the same sign as that of abc, which in turn implies that $\frac{-\text{Im}(\varphi(a,b,c))}{abc} > 0$.
Thus, it will suffice to show that whenever $(a, b, c) \in B_2$ and $abc > 0$, we have

$$\sum_i \arctan\left(\frac{2\alpha_i^3 abc}{1+\alpha_i^2(a^2+b^2+c^2)}\right) < \pi.$$

Indeed, for $(a, b, c) \in B_2$, $abc < \left(\|\alpha\|_2^{-11/12}\right)^3 \leq \frac{1}{\|\alpha\|_2^3}$ which, under the assumption $abc > 0$, results in

$$\sum_i \arctan\left(\frac{2\alpha_i^3 abc}{1 + \alpha_i^2(a^2 + b^2 + c^2)}\right) \leq \sum_i \frac{2\alpha_i^3 abc}{1+\alpha_i^2(a^2 + b^2 + c^2)} \leq \frac{2\|\alpha\|_3^3}{\|\alpha\|_2^3} < 2 < \pi,$$

as desired.

Step 3—The Absolute Value of the Integrand Is Negligible on the Spherical Shell $B \setminus B_2$ Where B Is the Unit Ball in $\mathbb{R}^3$

Observe that,

$$\left|\frac{\sin\left(\frac{1}{2}\sum_i \arctan\left(\frac{2\alpha^3 abc}{1+\alpha_i^2(a^2+b^2+c^2)}\right)\right)}{abc}\right| \leq \frac{1}{2}\sum_i \frac{\frac{2\alpha_i^3|abc|}{1+\alpha_i^2(a^2+b^2+c^2)}}{|abc|} \leq \|\alpha\|_3^3.$$

$$(13.18)$$

On the other hand, for $(a, b, c) \notin B_2$ we have that:

$$\frac{1}{\prod_i\left[\left(1 + \alpha_i^2(a^2 + b^2 + c^2)\right)^2 + 4\alpha_i^6 a^2 b^2 c^2\right]^{\frac{1}{4}}} \leq \frac{1}{\prod_i(1 + \alpha_i^2(a^2 + b^2 + c^2))^{\frac{1}{2}}}$$

$$\leq \prod_i\left(1 + \frac{\alpha_i^2}{\|\alpha\|_2^{22/12}}\right)^{-\frac{1}{2}}.$$

Using the elementary inequality $\ln(1 + x) \geq x - \frac{x^2}{2}$ for $x > 0$ yields:

$$\ln\left(\prod_i\left(1 + \frac{\alpha_i^2}{\|\alpha\|_2^{22/12}}\right)\right) = \sum_i \ln\left(1 + \frac{\alpha_i^2}{\|\alpha\|_2^{22/12}}\right) \geq \|\alpha\|_2^{2/12} - \frac{\|\alpha\|_4^4}{2\|\alpha\|_2^{44/12}}$$

$$\geq \|\alpha\|_2^{2/12} - 1$$

where the last inequality follows from the fact that $\|\alpha\|_4^4 \leq \|\alpha\|_2^2$. In turn, this implies

$$\prod_i \left(1 + \frac{\alpha_i^2}{\|\alpha\|_2^{22/12}}\right)^{-\frac{1}{2}} \leq e^{-\frac{\|\alpha\|_2^{2/12}-1}{2}}.$$

Finally, since the volume of the unit ball is $\frac{4\pi}{3}$, this gives

$$\int_{B \setminus B_2} \left|\frac{\mathrm{Im}(\varphi(a,b,c))}{abc}\right| da\, db\, dc < \frac{4\pi}{3} \|\alpha\|_3^3 \, e^{-\frac{\|\alpha\|_2^{2/12}-1}{2}}. \tag{13.19}$$

Consequently, there is a constant c_2 such that whenever $\|\alpha\|_2^2 > c_2$ then

$$\int_{B \setminus B_2} \left|\frac{\mathrm{Im}(\varphi(a,b,c))}{abc}\right| da\, db\, dc \leq \frac{1}{100}\left(\frac{\|\alpha\|_3}{\|\alpha\|_2}\right)^3.$$

Step 4—The Integral Is Negligible Outside of B

For $(a,b,c) \notin B$ we use (13.18) to achieve

$$\frac{\sin\left(\frac{1}{2}\sum_i \arctan\left(\frac{2\alpha_i^3 abc}{1+\alpha_i^2(a^2+b^2+c^2)}\right)\right)}{abc \prod_i \left(\left(1+\alpha_i^2(a^2+b^2+c^2)\right)^2 + 4\alpha_i^6 a^2 b^2 c^2\right)^{\frac{1}{4}}} < \frac{\|\alpha\|_3^3}{\prod_i \left(1+\alpha_i^2(a^2+b^2+c^2)\right)^{\frac{1}{2}}}.$$

By passing to spherical coordinates we obtain:

$$\int_{\mathbb{R}^3 \setminus B} \frac{1}{\prod_i \left(1+\alpha_i^2(a^2+b^2+c^2)\right)^{\frac{1}{2}}} da\, db\, dc = 4\pi \int_1^\infty \frac{r^2\, dr}{\prod_i (1+\alpha_i^2 r^2)^{\frac{1}{2}}}.$$

Applying Lemma 13.2 with $n = 4$ and $T = 1$, shows the existence of constants $C, c_3' > 0$ such that whenever $\|\alpha\|_2^2 > c_3'$,

$$\int_1^\infty \frac{r^2\, dr}{\prod_i (1+\alpha_i^2 r^2)^{\frac{1}{2}}} \leq C\left(\frac{1}{\|\alpha\|_2^2}\right)^2 = C\frac{1}{\|\alpha\|_2^4}.$$

Thus, there exists a constant $c_3 = \max(c_3', (16C)^2)$ such that whenever $\|\alpha\|_2^2 > c_3$ then

$$\int_{\mathbb{R}^3 \setminus B} \left| \frac{\mathrm{Im}(\varphi(a,b,c))}{abc} \right| dadbdc \leq \frac{1}{100} \left(\frac{\|\alpha\|_3}{\|\alpha\|_2} \right)^3.$$

Final Step—$\int_{\mathbb{R}^3} \frac{-\mathbf{Im}(\varphi(a,b,c))}{abc} dadbdc \geq \frac{1}{100} \left(\frac{\|\alpha\|_3}{\|\alpha\|_2} \right)^3$

We may now decompose the integral

$$\int_{\mathbb{R}^3} \frac{-\mathrm{Im}(\varphi(a,b,c))}{abc} dadbdc = \int_{B_2} \frac{-\mathrm{Im}(\varphi(a,b,c))}{abc} dadbdc$$

$$+ \int_{\mathbb{R}^3 \setminus B_2} \frac{-\mathrm{Im}(\varphi(a,b,c))}{abc} dadbdc$$

Letting $\|\alpha\|_2^2 > \max(c_1, c_2, c_3)$ steps 1 and 2 show that

$$\int_{B_2} \frac{-\mathrm{Im}(\varphi(a,b,c))}{abc} dadbdc \geq \frac{1}{100} \left(\frac{\|\alpha\|_3}{\|\alpha\|_2} \right)^3, \tag{13.20}$$

while steps 2 and 3 show

$$\int_{\mathbb{R}^3 \setminus B_2} \left| \frac{\mathrm{Im}(\varphi(a,b,c))}{abc} \right| dadbdc \leq \frac{2}{100} \left(\frac{\|\alpha\|_3}{\|\alpha\|_2} \right)^3.$$

The required bound then follows by combining the above two estimates.

$\square$

13.3.2 Arbitrary $0 < p < 1$

We now consider the case for arbitrary p. First, we would like to derive bounds on the behavior of $t_{p,\alpha}$, which constitute the following lemma.

Lemma 13.4 *Let $p \in (0,1)$ and denote by Φ the cumulative distribution function of the standard Gaussian. If $t_p = \Phi^{-1}(p)$ then $\|\alpha\|_2 t_p - k_p \leq t_{p,\alpha} \leq \|\alpha\|_2 t_p + k_p$,* *for a constant k_p depending only on p. Furthermore, if $p' := \Phi\left(\frac{t_{p,\alpha}}{\|\alpha\|_2}\right)$ then $|p - p'| \leq 3\left(\frac{\|\alpha\|_3}{\|\alpha\|_2}\right)^3$.*

Proof Let $W = \frac{\langle X_1, X_2 \rangle}{\|\alpha\|_2}$ where X_1, X_2 are defined as in the beginning of the section. We may consider $\langle X_1, X_2 \rangle$ as sum of independent random variables $X_1^i \cdot X_2^i$, where for each $1 \leq i \leq d$, X_1^i and X_2^i are independently distributed as $\mathcal{N}(0, \alpha_i)$. It then holds that $\mathbb{E}[X_1^i \cdot X_2^i] = 0$, $\mathbb{E}[(X_1^i \cdot X_2^i)^2] = \alpha_i^2$. The absolute third moments are given as a product of absolute third moments of Gaussians. That is, $\mathbb{E}[|X_1^i \cdot X_2^i|^3] = \frac{8\alpha_i^3}{\pi} < 3\alpha_i^3$.

Let t be such that $p = \mathbb{P}(W \geq t)$, in which case we also have $t_{p,\alpha} = t\|\alpha\|_2$. Note that

$$\frac{\sum_i \mathbb{E}[|X_1^i \cdot X_2^i|^3]}{\left(\sum_i \mathbb{E}[(X_1^i \cdot X_2^i)^2]\right)^{3/2}} \leq \frac{3\|\alpha\|_3^3}{\|\alpha\|_2^3}.$$

Thus, if Z is a standard normal random variable, Berry-Esseen's inequality, (13.6), yields for every $s \in \mathbb{R}$:

$$|\mathbb{P}(W > s) - \mathbb{P}(Z > s)| \leq \frac{3\|\alpha\|_3^3}{\|\alpha\|_2^3}.$$

If $t_p = \Phi^{-1}(p)$ then $\mathbb{P}(Z > t_p) = p$ and

$$|\Phi(t_p) - \Phi(t)| = |\mathbb{P}(Z > t_p) - \mathbb{P}(Z > t)| = |\mathbb{P}(W > t) - \mathbb{P}(Z > t)| \leq \frac{3\|\alpha\|_3^3}{\|\alpha\|_2^3}.$$

Since $|p - p'| = |\Phi(t_p) - \Phi(t)|$, this shows the second part of the statement. To finish the proof, denote $m = \inf_{s \in [t_p, t]}(\Phi'(s))$. By Lagrange's theorem

$$m|t_p - t| \leq |\Phi(t_p) - \Phi(t)| \leq \frac{3\|\alpha\|_3^3}{\|\alpha\|_2^3} \leq \frac{3}{\|\alpha\|_2},$$

which shows $t_{p,\alpha} \in \|\alpha\|_2 t_p \pm \frac{3}{m}$. $\qquad\square$

Before proceeding, we need some further definitions. Let X_1', X_2', X_3' be independent copies of X_1, X_2, X_3 and consider the joint distribution $(\langle X_1, X_2 \rangle, \langle X_1', X_3 \rangle, \langle X_2', X_3' \rangle)$. This distribution has independent coordinates. Denote its density by g and corresponding characteristic function by ψ. If N_1, N_2 are two independent standard Gaussians then the characteristic function of their product can be derived from (13.4) as $\mathbb{E}e^{it\,N_1N_2} = (1+t^2)^{-\frac{1}{2}}$. From this, it follows that the characteristic function of $\langle X_1, X_2 \rangle$ is $\mathbb{E}e^{it\langle X_1, X_2 \rangle} = \prod_i (1 + \alpha_i^2 t^2)^{-\frac{1}{2}}$, and we have, by independence

$$\psi(a,b,c) = \prod_i \left((1 + \alpha_i^2 a^2)(1 + \alpha_i^2 b^2)(1 + \alpha_i^2 c^2) \right)^{-\frac{1}{2}}. \tag{13.21}$$

We denote by $\psi_1(a', b', c') = \psi\left(\frac{a'}{\|\alpha\|_2}, \frac{b'}{\|\alpha\|_2}, \frac{c'}{\|\alpha\|_2} \right)$ and $\varphi_1(a', b', c') = \varphi\left(\frac{a'}{\|\alpha\|_2}, \frac{b'}{\|\alpha\|_2}, \frac{c'}{\|\alpha\|_2} \right)$ for the characteristic function φ, (13.11). The following result will help us relate the independent version of the distribution and the original one.

Lemma 13.5 *There exist absolute constants $c, C, \varepsilon > 0$ such that whenever $\|\alpha\|_2^2 > c$ then*

$$\int_{\mathbb{R}^3} |\text{Re}(\varphi_1) - \psi_1| da' db' dc' \leq C \left(\frac{\|\alpha\|_3}{\|\alpha\|_2} \right)^{3+\varepsilon}.$$

Proof Note that since ψ_1 and φ_1 are characteristic functions, then $|\psi_1|, |\text{Re}(\varphi_1)| \leq 1$. So, $|\psi_1 - \text{Re}(\varphi_1)| \leq |\ln(\psi_1) - \ln(\text{Re}(\varphi_1))|$. Now, let

$$B_{0.01} = \left\{ x \in \mathbb{R}^3 : \|x\|^2 \leq \left(\frac{\|\alpha\|_2}{\|\alpha\|_3} \right)^{0.01} \right\}.$$

Clearly, $|\text{Re}(\varphi_1)| \leq |\varphi_1| = \prod_i \left((1 + \frac{\alpha_i^2}{\|\alpha\|_2^2}(a'^2 + b'^2 + c'^2))^2 + 4\frac{\alpha_i^6}{\|\alpha\|_2^6} a'^2 b'^2 c'^2 \right)^{-\frac{1}{4}}$, and since

$$|a'b'c'| \leq \left(a'^2 + b'^2 + c'^2 \right)^{\frac{3}{2}} \leq \left(\frac{\|\alpha\|_2}{\|\alpha\|_3} \right)^{0.015} \quad \text{for } (a', b', c') \in B_{0.01},$$

we have

$$\left| \arg\left(1 + \frac{\alpha_i^2}{\|\alpha\|_2^2}(a'^2 + b'^2 + c'^2) + 2\frac{\alpha_i^3}{\|\alpha\|_2^3} a'b'c'\mathbf{i} \right) \right| \leq 2\frac{\alpha_i^3}{\|\alpha\|_2^3} |a'b'c'|$$

$$\leq 2\frac{\alpha_i^3}{\|\alpha\|_2^3} \left(\frac{\|\alpha\|_2}{\|\alpha\|_3} \right)^{0.015}.$$

By using the inequality $\cos(x) \geq 1 - x^2$, we achieve

$$\mathrm{Re}(\varphi_1) \geq \cos\left(2\frac{\|\alpha\|_3^3}{\|\alpha\|_2^3}\left(\frac{\|\alpha\|_2}{\|\alpha\|_3}\right)^{0.015}\right)|\varphi_1| \geq \left(1 - 4\frac{\|\alpha\|_3^6}{\|\alpha\|_2^6}\left(\frac{\|\alpha\|_2}{\|\alpha\|_3}\right)^{0.03}\right)|\varphi_1|.$$

Using the above, together with the triangle inequality gives

$$|\ln(\psi_1) - \ln(\mathrm{Re}(\varphi_1))| \leq |\ln(\psi_1) - \ln(|\varphi_1|)| + \left|\ln\left(1 - 4\frac{\|\alpha\|_3^6}{\|\alpha\|_2^6}\left(\frac{\|\alpha\|_2}{\|\alpha\|_3}\right)^{0.03}\right)\right|.$$

$$(13.22)$$

For $x \in (0, \frac{1}{2})$ we have the inequality $|\ln(1 - x)| \leq 2x$, thus, as long as $\|\alpha\|_2^2$ is large enough

$$\left|\ln\left(1 - 4\frac{\|\alpha\|_3^6}{\|\alpha\|_2^6}\left(\frac{\|\alpha\|_2}{\|\alpha\|_3}\right)^{0.03}\right)\right| \leq 8\frac{\|\alpha\|_3^6}{\|\alpha\|_2^6}\left(\frac{\|\alpha\|_2}{\|\alpha\|_3}\right)^{0.03},$$

and

$$8 \int_{B_{0.01}} \frac{\|\alpha\|_3^6}{\|\alpha\|_2^6}\left(\frac{\|\alpha\|_2}{\|\alpha\|_3}\right)^{0.03} da'db'dc' \leq 32\pi \frac{\|\alpha\|_3^6}{\|\alpha\|_2^6}\left(\frac{\|\alpha\|_2}{\|\alpha\|_3}\right)^{0.045}$$

$$= 32\pi \left(\frac{\|\alpha\|_3}{\|\alpha\|_2}\right)^{5.955}. \qquad (13.23)$$

By using the inequality $|\ln(1 + x) - x| \leq x^2$ for $x > 0$ we bound $\ln(\psi_1)$ with

$$\ln(\psi_1(a', b', c')) = -\frac{1}{2}\sum_i\left[\ln\left(1 + \frac{\alpha_i^2 a'^2}{\|\alpha\|_2^2}\right) + \ln\left(1 + \frac{\alpha_i^2 b'^2}{\|\alpha\|_2^2}\right) + \ln\left(1 + \frac{\alpha_i^2 c'^2}{\|\alpha\|_2^2}\right)\right]$$

$$= -\frac{1}{2}\left(a'^2 + b'^2 + c'^2\right) + O\left(\frac{\|\alpha\|_4^4}{\|\alpha\|_2^4}\right)\left(a'^4 + b'^4 + c'^4\right).$$

Similar considerations show

$$\ln(|\varphi_1|) = -\frac{1}{4}\sum \ln\left(1 + \frac{2\alpha_i^2}{\|\alpha\|_2^2}(a'^2 + b'^2 + c'^2)\right.$$

$$\left. + \frac{\alpha_i^4}{\|\alpha\|_2^4}(a'^2 + b'^2 + c'^2)^2 + 4\frac{\alpha_i^6}{\|\alpha\|_2^6}a'^2b'^2c'^2\right)$$

$$= -\frac{1}{2}\left(a'^2 + b'^2 + c'^2\right) - \frac{\|\alpha\|_4^4}{4\|\alpha\|_2^4}(a'^2 + b'^2 + c'^2)^2 - \frac{\|\alpha\|_6^6}{\|\alpha\|_2^6}a'^2b'^2c'^2$$

$$+ O\left(\frac{\|\alpha\|_4^4}{\|\alpha\|_2^4}\right)\left((a'^2 + b'^2 + c'^2)^2 + \left(a'^2 + b'^2 + c'^2\right)^4 + a'^4b'^4c'^4\right)$$

$$= -\frac{1}{2}\left(a'^2 + b'^2 + c'^2\right) + O\left(\frac{\|\alpha\|_4^4}{\|\alpha\|_2^4}\right)\left(1 + \left(a'^2 + b'^2 + c'^2\right)^6\right).$$

$$\tag{13.24}$$

The above shows the existence of a constant $C > 0$ such that

$$\int_{B_{0.01}} |\ln(\psi_1) - \ln(|\varphi_1|)| \le C\left(\frac{\|\alpha\|_4}{\|\alpha\|_2}\right)^4 \int_{B_{0.01}} (a'^2 + b'^2 + c'^2)^6 \, da'db'dc'$$

$$= 4\pi C \left(\frac{\|\alpha\|_4}{\|\alpha\|_2}\right)^4 \left(\frac{\|\alpha\|_2}{\|\alpha\|_3}\right)^{0.075} \le 4\pi C \left(\frac{\|\alpha\|_3}{\|\alpha\|_2}\right)^4 \left(\frac{\|\alpha\|_2}{\|\alpha\|_3}\right)^{0.075}$$

$$= 4\pi C \left(\frac{\|\alpha\|_3}{\|\alpha\|_2}\right)^{3.925}. \tag{13.25}$$

By combining (13.23), (13.25) and (13.22), we obtain

$$\int_{B_{0.01}} |\psi_1 - \mathrm{Re}(\varphi_1)|da'db'dc' \le \pi(4C + 32)\left(\frac{\|\alpha\|_3}{\|\alpha\|_2}\right)^{3.925}.$$

To bound the integral in $\mathbb{R}^3 \setminus B_{0.01}$ we proceed in similar fashion to step 3 in Lemma 13.3. First, note that

$$|\varphi_1|, |\psi_1| \le \frac{1}{\prod_i \left(1 + \frac{\alpha_i^2}{\|\alpha\|_2^2}(a'^2 + b'^2 + c'^2)\right)^{\frac{1}{2}}}.$$

Denoting $r = \sqrt{a'^2 + b'^2 + c'^2}$, $T = \left(\frac{\|\alpha\|_2}{\|\alpha\|_3}\right)^{0.005}$ and passing to spherical coordinates yields

$$\int_{\mathbb{R}^3 \setminus B_{0.01}} |\mathrm{Re}(\varphi_1) - \psi_1|da'db'dc' \le \int_{\mathbb{R}^3 \setminus B_{0.01}} |\mathrm{Re}(\varphi_1)| + |\psi_1|da'db'dc'$$

$$\le 8\pi \int_T^\infty \frac{r^2 \, dr}{\prod_i \left(1 + \frac{\alpha_i^2}{\|\alpha\|_2^2}r^2\right)^{\frac{1}{2}}}.$$

Invoking Lemma 13.2 with $n > 606$ shows the existence of constants $C, c > 0$ such that

$$\int_T^\infty \frac{r^2 \, dr}{\prod_i \left(1 + \frac{\alpha_i^2}{\|\alpha\|_2^2} r^2\right)^{\frac{1}{2}}} \leq CT^{-603} = C \left(\frac{\|\alpha\|_3}{\|\alpha\|_2}\right)^{3.015},$$

whenever $\|\alpha\|_2^2 > c$. This concludes the proof when we take $\varepsilon = 0.015$. □

We are now ready to bound from below the probability of an induced triangle occurring in the general setting. Set $p \in (0, 1)$ and $t := t_{p,\alpha}$. We are interested in the event

$$\left\{ \min\left(\langle X_1, X_2 \rangle, \langle X_1, X_3 \rangle, \langle X_2, X_3 \rangle\right) > t \right\}.$$

As before, let f be the joint density of $(\langle X_1, X_2 \rangle, \langle X_1, X_3 \rangle, \langle X_2, X_3 \rangle)$ and consider the integral:

$$I_p := \int_{\mathbb{R}^3} f(x, y, z) \text{sgn}(x - t) \text{sgn}(y - t) \text{sgn}(z - t) \, dx dy dz.$$

Note that, in the above formula, replacing f with g, the density of the coordinate-independent version, as defined above, would yield $I_p = p^3 + 3(1 - p)^2 p - 3(1 - p) p^2 - (1 - p)^3 = (2p - 1)^3$.

For the rest of this section, our goal will be to show that I_p is large, compared to $(2p - 1)^3$. That is, the dependency between the coordinates induces an increased probability for triangles and induced edges. As in (13.1), we may write the Fourier transform of $\text{sgn}(x - t)\text{sgn}(y - t)\text{sgn}(z - t)$ as $\widehat{\text{sgn}}(a, b, c)e^{-2\pi i t(a+b+c)}$. Thus, by (13.2), we have the equality

$$I_p = \frac{1}{\pi^3} \oint_{\mathbb{R}^3} \varphi(a, b, c)\widehat{\text{sgn}}(a, b, c)e^{-2\pi i t(a+b+c)} \, da db dc,$$

where φ, as in (13.11), is the characteristic function of f. Since I_p represents a real number, we only need to consider the real part of the integral:

$$I_p = \frac{1}{\pi^3} \oint \text{Re}\left(\varphi(a, b, c)\widehat{\text{sgn}}(a, b, c)\right) \cos(2\pi t(a + b + c)) da db dc$$

$$+ \frac{1}{\pi^3} \oint \text{Im}\left(\varphi(a, b, c)\widehat{\text{sgn}}(a, b, c)\right) \sin(2\pi t(a + b + c)) da db dc$$

$$= \frac{1}{\pi^3} \oint \frac{-\mathrm{Im}\,(\varphi(a,b,c))}{abc} \cos(2\pi t(a+b+c))dadbdc$$

$$+ \frac{1}{\pi^3} \oint \frac{\mathrm{Re}\,(\varphi(a,b,c))}{abc} \sin(2\pi t(a+b+c))dadbdc.$$

We denote

$$I'_p = \frac{1}{\pi^3} \oint \frac{\mathrm{Re}\,(\varphi(a,b,c))}{abc} \sin(2\pi t(a+b+c))dadbdc$$

and

$$I''_p = \frac{1}{\pi^3} \oint \frac{-\mathrm{Im}\,(\varphi(a,b,c))}{abc} \cos(2\pi t(a+b+c))dadbdc.$$

In the proof of Lemma 13.3 we have seen that

$$\oint \frac{-\mathrm{Im}\,(\varphi(a,b,c))}{abc}dadbdc,$$

is mostly concentrated near the origin. Thus, we should expect that

$$\oint \frac{-\mathrm{Im}\,(\varphi(a,b,c))}{abc} \cos(2\pi t(a+b+c))dadbdc \simeq \oint \frac{-\mathrm{Im}\,(\varphi(a,b,c))}{abc}dadbdc.$$

So that I''_p represents the increase in probability. From Lemma 13.5, we know that $\mathrm{Re}(\varphi)$ is close to ψ, the characteristic function of the coordinate-independent version, which means that I'_p should be close to $(2p-1)^3$. The next two claims will formalize this intuition. We begin by showing that I''_p is large.

Claim 13.1 Fix $p \in (0,1)$. There exist constants $\delta'_p, c_p > 0$ depending only on p such that whenever $\|\alpha\|_2^2 > c_p$ then $I''_p \geq 2\delta'_p \left(\frac{\|\alpha\|_3}{\|\alpha\|_2} \right)^3$.

Proof First, it is not hard to see that the integrand in I''_p is continuous, up to a removable discontinuity, and we may pass to standard integration. Let R be an arbitrary orthogonal transformation which takes $(1,0,0)$ to $\frac{1}{\sqrt{3}}(1,1,1)$. Consider the set

$$K = R\left(\left[-\frac{1}{\|\alpha\|_2^{11/12}}, \frac{1}{\|\alpha\|_2^{11/12}} \right] \times \left[-\frac{1}{\|\alpha\|_2^{11/12}}, \frac{1}{\|\alpha\|_2^{11/12}} \right] \right.$$

$$\left. \times \left[-\frac{1}{\|\alpha\|_2^{11/12}}, \frac{1}{\|\alpha\|_2^{11/12}} \right] \right).$$

Note that if $B_2 = \left\{ x \in \mathbb{R}^3 \mid \|x\|^2 \leq \frac{1}{\|\alpha\|_2^{22/12}} \right\}$ and $B_2' = \left\{ x \in \mathbb{R}^3 \mid \|x\|^2 \leq \frac{4}{\|\alpha\|_2^{22/12}} \right\}$ then,

$$B_2 \subset K \subset B_2'.$$

Now, recall from (13.14) that,

$$\frac{-\mathrm{Im}(\varphi(a,b,c))}{abc} = \frac{\sin\left(\frac{1}{2}\sum_i \arctan\left(\frac{2\alpha_i^3 abc}{1+\alpha_i^2(a^2+b^2+c^2)}\right)\right)}{abc \prod_i \left((1+\alpha_i^2(a^2+b^2+c^2))^2 + 4\alpha_i^6 a^2 b^2 c^2\right)^{\frac{1}{4}}}.$$

From (13.18) and (13.15), we have

$$\|\alpha\|_3^3 \geq \frac{\sin\left(\frac{1}{2}\sum_i \arctan\left(\frac{2\alpha_i^3 abc}{1+\alpha_i^2(a^2+b^2+c^2)}\right)\right)}{abc} \geq \sum_i \frac{\alpha_i^3}{1+\alpha_i^2(a^2+b^2+c^2)}$$

$$- 3\|\alpha\|_3^6 |abc|.$$

Along with the inequality $\frac{\alpha_i^3}{1+\alpha_i^2(a^2+b^2+c^2)} \geq \alpha_i^3 \left(1 - \alpha_i^2(a^2+b^2+c^2)\right)$, the above yields

$$\left| \frac{\sin\left(\frac{1}{2}\sum_i \arctan\left(\frac{2\alpha_i^3 abc}{1+\alpha_i^2(a^2+b^2+c^2)}\right)\right)}{abc} - \|\alpha\|_3^3 \right| \leq \|\alpha\|_5^5 (a^2+b^2+c^2) - 3\|\alpha\|_3^6 |abc|.$$

Therefore

$$\int_K \frac{-\mathrm{Im}(\varphi(a,b,c))}{abc} \cos(2\pi t(a+b+c))\,da\,db\,dc$$

$$\geq \|\alpha\|_3^3 \int_K \frac{\cos(2\pi t(a+b+c))\,da\,db\,dc}{\prod_i \left((1+\alpha_i^2(a^2+b^2+c^2))^2 + 4\alpha_i^6 a^2 b^2 c^2\right)^{\frac{1}{4}}}$$

$$- 3\|\alpha\|_3^6 \int_K |abc|\,da\,db\,dc - \|\alpha\|_5^5 \int_K (a^2+b^2+c^2)\,da\,db\,dc, \qquad (13.26)$$

with

$$3\,\|\alpha\|_3^6 \int_K |abc| da\, db\, dc \leq C_1 \frac{\|\alpha\|_3^6}{\|\alpha\|_2^{5.5}} = C_1 \left(\frac{\|\alpha\|_3}{\|\alpha\|_2}\right)^3 \frac{\|\alpha\|_3^3}{\|\alpha\|_2^2} \frac{1}{\|\alpha\|_2^{0.5}}$$

$$\leq C_1 \left(\frac{\|\alpha\|_3}{\|\alpha\|_2}\right)^3 \frac{1}{\|\alpha\|_2^{0.5}},$$

$$\|\alpha\|_5^5 \int_K (a^2 + b^2 + c^2) da\, db\, dc \leq C_1 \frac{\|\alpha\|_5^5}{\|\alpha\|_2^{55/12}} \leq C_1 \left(\frac{\|\alpha\|_3}{\|\alpha\|_2}\right)^3 \frac{1}{\|\alpha\|_2},$$

for an absolute constant $C_1 > 0$. Recalling that

$$|\varphi(a, b, c)| = \prod_i \left(\left(1 + \alpha_i^2 (a^2 + b^2 + c^2)\right)^2 + 4\alpha_i^6 a^2 b^2 c^2\right)^{-\frac{1}{4}},$$

we would like approximate $|\varphi(a, b, c)|$ by $e^{-\frac{\|\alpha\|_2^2}{2}(a^2+b^2+c^2)}$. For that, we note that

$$\left| |\varphi(a, b, c)| - e^{-\frac{\|\alpha\|_2^2}{2}(a^2+b^2+c^2)} \right| \leq \left| \ln\left(|\varphi(a, b, c)|\right) - \ln\left(e^{-\frac{\|\alpha\|_2^2}{2}(a^2+b^2+c^2)}\right) \right|.$$

Since $|\ln(x + 1) - x| \leq x^2$, similar considerations as in (13.24), show for $(a, b, c) \in K$:

$$\ln(|\varphi|) = -\frac{1}{4} \sum_i \ln\left(1 + 2\alpha_i^2 \left(a^2 + b^2 + c^2\right) + \alpha_i^4 \left(a^2 + b^2 + c^2\right)^2 + 4\alpha_i^6 a^2 b^2 c^2\right)$$

$$= -\frac{\|\alpha\|_2^2}{2} \left(a^2 + b^2 + c^2\right) - \frac{\|\alpha\|_4^4}{4} \left(a^2 + b^2 + c^2\right)^2 - \|\alpha\|_6^6 a^2 b^2 c^2$$

$$+ O\left(\|\alpha\|_4^4\right) \left(\left(a^2 + b^2 + c^2\right)^2 + \left(a^2 + b^2 + c^2\right)^4 + a^4 b^4 c^4\right)$$

$$= -\frac{\|\alpha\|_2^2}{2} \left(a^2 + b^2 + c^2\right) + O\left(\|\alpha\|_4^4\right) \left(a^2 + b^2 + c^2\right)^2.$$

This shows the existence of an absolute constant $C_2 > 0$ such that for $(a, b, c) \in K$

$$\left| |\varphi(a, b, c)| - e^{-\frac{\|\alpha\|_2^2}{2}(a^2+b^2+c^2)} \right| \leq C_2 \|\alpha\|_4^4 \left(a^2 + b^2 + c^2\right)^2.$$

Hence

$$\int_K |\varphi(a,b,c)| \cos(2\pi t(a+b+c)) dadbdc$$

$$\geq \int_K e^{-\frac{\|\alpha\|_2^2}{2}(a^2+b^2+c^2)} \cos(2\pi t(a+b+c)) dadbdc$$

$$- C_2 \|\alpha\|_4^4 \int_K \left(a^2+b^2+c^2\right)^2 dadbdc, \qquad (13.27)$$

and

$$C_2 \|\alpha\|_4^4 \int_K \left(a^2+b^2+c^2\right)^2 dadbdc \leq C_3 \frac{\|\alpha\|_4^4}{\|\alpha\|_2^{77/12}} \leq C_3 \left(\frac{\|\alpha\|_3}{\|\alpha\|_2}\right)^3 \frac{1}{\|\alpha\|_2^3},$$

for an absolute constant $C_3 > 0$. By rotational invariance of $e^{-\frac{\|\alpha\|_2^2}{2}(a^2+b^2+c^2)}$, we may apply R as a unitary coordinate change, which shows

$$\int_K e^{-\frac{\|\alpha\|_2^2}{2}(a^2+b^2+c^2)} \cos(2\pi t(a+b+c)) dadbdc$$

$$= \int_{R^{-1}K} e^{-\frac{\|\alpha\|_2^2}{2}(a^2+b^2+c^2)} \cos(2\sqrt{3}\pi ta) dadbdc$$

$$= \int_{-\frac{1}{\|\alpha\|_2^{11/12}}}^{\frac{1}{\|\alpha\|_2^{11/12}}} e^{-\frac{\|\alpha\|_2^2}{2}c^2} dc \int_{-\frac{1}{\|\alpha\|_2^{11/12}}}^{\frac{1}{\|\alpha\|_2^{11/12}}} e^{-\frac{\|\alpha\|_2^2}{2}b^2} db \int_{-\frac{1}{\|\alpha\|_2^{11/12}}}^{\frac{1}{\|\alpha\|_2^{11/12}}} e^{-\frac{\|\alpha\|_2^2}{2}a^2} \cos(\sqrt{12}\pi ta) da$$

$$= \frac{1}{\|\alpha\|_2^3} \int_{-\|\alpha\|_2^{1/12}}^{\|\alpha\|_2^{1/12}} e^{-\frac{c^2}{2}} dc \int_{-\|\alpha\|_2^{1/12}}^{\|\alpha\|_2^{1/12}} e^{-\frac{b^2}{2}} db \int_{-\|\alpha\|_2^{1/12}}^{\|\alpha\|_2^{1/12}} e^{-\frac{a^2}{2}} \cos\left(\sqrt{12}\pi \frac{t}{\|\alpha\|_2} a\right) da,$$

$$(13.28)$$

where the last equality is a result of a second coordinate change. By Lemma 13.4, we know that

$$|t_p| - \frac{k_p}{\|\alpha\|_2} \leq \left|\frac{t}{\|\alpha\|_2}\right| \leq |t_p| + \frac{k_p}{\|\alpha\|_2}$$

for constants k_p, t_p depending on p. Also, a calculation shows that

$$\int_{-\infty}^{\infty} e^{-\frac{a^2}{2}} \cos\left(\sqrt{12}\pi \frac{t}{\|\alpha\|_2} a\right) da = \sqrt{2\pi} e^{-6\pi^2 \frac{t^2}{\|\alpha\|_2^2}}.$$

Note that if $\|\alpha\|_2^{2/12} \geq 12\pi^2 \left(|t_p| + k_p\right)^2 + 2$, then

$$\int_{|a| > \|\alpha\|_2^{1/12}} e^{-\frac{a^2}{2}} da \leq 2e^{-\frac{\|\alpha\|_2^{2/12}}{2}} \leq \frac{2}{e} e^{-12\pi^2 \frac{(|t_p| + k_p)^2}{2}} \leq \frac{1}{2}\sqrt{2\pi} e^{-6\pi^2 \frac{t^2}{\|\alpha\|_2^2}}.$$

That is, if $\|\alpha\|_2^{2/12}$ is larger than some constant, which depends only on t, we have

$$\int_{-\|\alpha\|_2^{1/12}}^{\|\alpha\|_2^{1/12}} e^{-\frac{a^2}{2}} \cos\left(\sqrt{12}\pi \frac{t}{\|\alpha\|_2} a\right) da \geq \frac{1}{2}\sqrt{2\pi} e^{-6\pi^2 \frac{t^2}{\|\alpha\|_2^2}}.$$

Together with the observation $\int_{-1}^{1} e^{\frac{-x^2}{2}} dx > 1$, this shows that the expression

(13.28) is bounded from below by $\frac{1}{2}\sqrt{2\pi} e^{-6\pi^2 \frac{t^2}{\|\alpha\|_2^2}}$. Combining the above, along with (13.26) and (13.27) shows

$$\int_K \frac{-\mathrm{Im}\,(\varphi(a, b, c))}{abc} \cos(2\pi t(a + b + c)) da\,db\,dc$$

$$\geq \|\alpha\|_3^3 \int_K \cos(2\pi t(a + b + c)) \, |\varphi(a, b, c)| \, da\,db\,dc - 2C_1 \left(\frac{\|\alpha\|_3}{\|\alpha\|_2}\right)^3 \frac{1}{\|\alpha\|_2^{0.5}}$$

$$\geq \|\alpha\|_3^3 \int_K e^{-\frac{\|\alpha\|_2^2}{2}(a^2 + b^2 + c^2)} \cos(2\pi t(a + b + c)) da\,db\,dc - C_3 \left(\frac{\|\alpha\|_3}{\|\alpha\|_2}\right)^6$$

$$- 2C_1 \left(\frac{\|\alpha\|_3}{\|\alpha\|_2}\right)^3 \frac{1}{\|\alpha\|_2^{0.5}}$$

$$\geq \frac{1}{2}\sqrt{2\pi} e^{-6\pi^2 \frac{t^2}{\|\alpha\|_2^2}} \left(\frac{\|\alpha\|_3}{\|\alpha\|_2}\right)^3 - C_3 \left(\frac{\|\alpha\|_3}{\|\alpha\|_2}\right)^6 - 2C_1 \left(\frac{\|\alpha\|_3}{\|\alpha\|_2}\right)^3 \frac{1}{\|\alpha\|_2^{0.5}}$$

$$\geq 4\delta_p' \left(\frac{\|\alpha\|_3}{\|\alpha\|_2}\right)^3.$$

whenever $\|\alpha\|_2^2 > c_p''$, for c_p'', δ_p' constants, depending only on p. From (13.19), we can choose a constant $c_p' > c_p'' > 0$ such that

$$\int_{\mathbb{R}^3 \setminus B_2} |\mathrm{Re}\,(\varphi(a, b, c)\widehat{\mathrm{sgn}}(a, b, c))|\, dadbdc < 2\delta_p' \left(\frac{\|\alpha\|_3}{\|\alpha\|_2} \right)^3,$$

whenever $\|\alpha\|_2^2 > c_p'$. Thus

$$I_p'' > \int_K \frac{-\mathrm{Im}\,(\varphi(a, b, c))}{abc} dadbdc - \int_{\mathbb{R}^3 \setminus B_2} \left| \frac{\mathrm{Im}\,(\varphi(a, b, c))}{abc} \right| dadbdc$$

$$\geq 2\delta_p' \left(\frac{\|\alpha\|_3}{\|\alpha\|_2} \right)^3.$$

$\square$

It now remains to show that the difference between I_p' and $(2p - 1)^3$ is small, compared to I_p''.

Claim 13.2 Fix $p \in (0, 1)$, there exists a constant $c_p > 0$ depending only on p such that whenever $\|\alpha\|_2^2 > c_p$ then $|I_p' - (2p - 1)^3| \leq \delta_p' \left(\frac{\|\alpha\|_3}{\|\alpha\|_2} \right)^3$, where δ_p' is the same as in Claim 13.1.

Proof Let g be the density of the coordinate free version of f, as in Lemma 13.5, and let ψ be its characteristic function (13.21). Evidently, we have the equality:

$$\frac{1}{\pi^3} \oint_{\mathbb{R}^3} \psi(a, b, c)\widehat{\mathrm{sgn}}(a, b, c)e^{-2\pi i t(a+b+c)} dadbdc = (2p - 1)^3.$$

Thus, by rewriting I_p' as

$$\frac{1}{\pi^3} \oint_{\mathbb{R}^3} (\mathrm{Re}\,(\varphi(a, b, c)) + \psi(a, b, c) - \psi(a, b, c)) \frac{\sin(2\pi t(a + b + c))}{abc} dadbdc,$$

we obtain

$$I_p' = (2p - 1)^3 + \frac{1}{\pi^3} \oint_{\mathbb{R}^3} (\mathrm{Re}\,(\varphi(a, b, c)) - \psi(a, b, c)) \frac{\sin(2\pi t(a + b + c))}{abc} dadbdc.$$

Next, we rewrite $\sin(2\pi t(a + b + c))$ as:

$$\sin(2\pi ta) \sin(2\pi tb) \sin(2\pi tc) + \cos(2\pi ta) \cos(2\pi tb) \sin(2\pi tc)$$

$$+ \cos(2\pi ta) \sin(2\pi tb) \cos(2\pi tc) + \sin(2\pi ta) \cos(2\pi tb) \cos(2\pi tc).$$

Recall

$$\varphi(a, b, c) = \prod_{i=1}^{d} (1 + \alpha_i^2(a^2 + b^2 + c^2) + 2\alpha_i^3 abc\mathbf{i})^{-\frac{1}{2}},$$

$$\psi(a, b, c) = \prod_i \left((1 + \alpha_i^2 a^2)(1 + \alpha_i^2 b^2)(1 + \alpha_i^2 c^2) \right)^{-\frac{1}{2}}.$$

One may now verify that $\mathrm{Re}(\varphi(a, b, c) - \psi(a, b, c))\frac{1}{abc}$ is an odd function. For a function h, we've defined $\Delta_c h(a, b, c) = h(a, b, c) + h(a, b, -c)$. Thus,

$$\Delta_c \left(\frac{\mathrm{Re}(\varphi(a, b, c)) - \psi(a, b, c)}{abc} \sin(2\pi ta) \cos(2\pi tb) \cos(2\pi tc) \right) = 0.$$

Looking at the principal value, we see that:

$$\fint_{\mathbb{R}^3} \frac{\mathrm{Re}(\varphi(a, b, c)) - \psi(a, b, c)}{abc} \sin(2\pi ta) \cos(2\pi tb) \cos(2\pi tc) da db dc = 0,$$

and the same can be said for the other similar terms. We are then left to consider an integrable function:

$$I_p' - (2p - 1)^3 = \int_{\mathbb{R}^3} \frac{\sin(2\pi ta) \sin(2\pi tb) \sin(2\pi tc)}{abc}$$

$$\times (\mathrm{Re}(\varphi(a, b, c) - \psi(a, b, c)) da db dc.$$

By making the substitution $a' = \|\alpha\|_2 a, b' = \|\alpha\|_2 b, c' = \|\alpha\|_2 c$, and denoting $t' = \frac{t}{\|\alpha\|_2}$ the above equals

$$\int_{\mathbb{R}^3} \frac{\sin(2\pi t'a') \sin(2\pi t'b') \sin(2\pi t'c')}{a'b'c'}$$

$$\times \left(\mathrm{Re}(\varphi_1(a', b', c') - \psi_1(a', b', c')) \right) da' db' dc',$$

where φ_1 and ψ_1 are as in Lemma 13.5. By Lemma 13.4, we know that $|t'| < |t_p| + \frac{k_p}{\|\alpha\|_2}$. Thus

$$\sup_{(a',b',c')\in\mathbb{R}^3} \left| \left(\frac{\sin(2\pi t'a') \sin(2\pi t'b') \sin(2\pi t'c')}{a'b'c'} \right) \right| \leq \left(2\pi \left(|t_p| + \frac{k_p}{\|\alpha\|_2} \right) \right)^3.$$

And so

$$|I'_p - (2p-1)^3| \leq \left(2\pi \left(|t_p| + \frac{k_p}{\|\alpha\|_2}\right)\right)^3 \int_{\mathbb{R}^3} |Re(\varphi_1(a',b',c'))$$

$$-\psi_1(a',b',c')| \, da'db'dc'.$$

Lemma 13.5 asserts that $\int_{\mathbb{R}^3} |Re(\varphi_1) - \psi_1| \leq C \left(\frac{\|\alpha\|_3}{\|\alpha\|_2}\right)^{3+\varepsilon}$ for large enough $\|\alpha\|_2^2$.
Thus,

$$I'_p - (2p-1)^3 \leq \left(2\pi \left(|t_p| + \frac{k_p}{\|\alpha\|_2}\right)\right)^3 C \left(\frac{\|\alpha\|_3}{\|\alpha\|_2}\right)^{3+\varepsilon}.$$

Since we've assumed α to be normalized as in (13.13), $\frac{\|\alpha\|_3}{\|\alpha\|_2}$ can be made as small as needed. The proof concludes by choosing $c_p > c'_p$ to be such that

$$\left(2\pi \left(|t_p| + \frac{k_p}{\|\alpha\|_2}\right)\right)^3 C \left(\frac{\|\alpha\|_3}{\|\alpha\|_2}\right)^{3+\varepsilon} < \delta'_p \left(\frac{\|\alpha\|_3}{\|\alpha\|_2}\right)^3 \text{ whenever } \|\alpha\|_2^2 > c_p.$$

$\square$

Combining Claims 13.2 and 13.1, we have thus established

Lemma 13.6 *Fix $p \in (0, 1)$. There exist constants $\delta'_p, c_p > 0$ depending only on p such that whenever $\|\alpha\|_2^2 > c_p$ then $I_p \geq (2p-1)^3 + \delta'_p \left(\frac{\|\alpha\|_3}{\|\alpha\|_2}\right)^3$.*

Now, by definition $\mathbb{P}(\langle X_1, X_2 \rangle > t_{p,\alpha}) = p$ and $\mathbb{P}(\langle X_1, X_2 \rangle > t_{p,\alpha}, \langle X_1, X_3 \rangle > t_{p,\alpha}) = p^2$. We note that Lemma 13.6, along with (13.1) produces:

$$(2p-1)^3 + \delta'_p \left(\frac{\|\alpha\|_3}{\|\alpha\|_2}\right)^3 \leq 8\mathbb{P}(E_p) - 12p^2 + 6p - 1.$$

This proves the lower bound of Theorem 13.5:

$$p^3 + \frac{\delta'_p}{8} \left(\frac{\|\alpha\|_3}{\|\alpha\|_2}\right)^3 \leq \mathbb{P}(E_p).$$

13.3.3 Upper Bound

To finish the proof of Theorem 13.5 it remains to prove the upper bound. This is done in the following lemma.

Lemma 13.7 *Let* $p \in (0, 1)$, $\mathbb{P}(E_p) - p^3 \leq \Delta \left(\frac{\|\alpha\|_3}{\|\alpha\|_2} \right)^3$, *for a universal constant* $\Delta > 0$.

Proof The proof of this lemma will use the higher dimensional analogue of the Berry-Esseen's inequality.

Define the random vector $V = (\langle X_1, X_2 \rangle, \langle X_1, X_3 \rangle, \langle X_2, X_3 \rangle)$. It is straightforward to check that the covariance matrix of V is $\|\alpha\|_2^2 I_3$ where I_3 is the identity matrix. We decompose V into $V_i = \left(X_1^i X_2^i, X_1^i X_3^i, X_2^i X_3^i \right)$. Clearly $V = \sum_{i=1}^{d} V_i$ and, since X_1^i, X_2^i, X_3^i are i.i.d. Gaussians,

$$
\begin{aligned}
\mathbb{E} \|V_i\|^3 &\leq \sqrt{\mathbb{E}\left[\left((X_1^i X_2^i)^2 + (X_1^i X_3^i)^2 + (X_2^i X_3^i)^2 \right)^3 \right]} \\
&= \sqrt{3\mathbb{E}[(X_1^i X_2^i)^6] + 18\mathbb{E}[(X_1^i)^6 (X_2^i)^4 (X_3^i)^2] + 6\mathbb{E}[(X_1^i)^4 (X_2^i)^4 (X_3^i)^4]} \\
&\leq 50\sqrt{\alpha_i^6} = 50\alpha_i^3.
\end{aligned}
$$

Thus, if Z_3 a three-dimensional standard Gaussian random vector, by (13.7) there is a constant C_{be} such that for any convex set $K \subset \mathbb{R}^3$ we have that

$$
|\mathbb{P}(V / \|\alpha\|_2 \in K) - \mathbb{P}(Z_3 \in K)| \leq 100 C_{be} \left(\frac{\|\alpha\|_3}{\|\alpha\|_2} \right)^3.
$$

In particular, this holds for the convex set

$$
E_p = \left\{ (x, y, z) \in \mathbb{R}^3 \,\middle|\, x > \frac{t_{p,\alpha}}{\|\alpha\|_2}, \; y > \frac{t_{p,\alpha}}{\|\alpha\|_2}, \; z > \frac{t_{p,\alpha}}{\|\alpha\|_2} \right\}.
$$

If we denote $p' = \Phi^{-1}(\frac{t_{p,\alpha}}{\|\alpha\|_2})$, the above shows

$$
|\mathbb{P}(V / \|\alpha\|_2 \in E_p) - p'^3| \leq 100 C_{be} \left(\frac{\|\alpha\|_3}{\|\alpha\|_2} \right)^3.
$$

By Lemma 13.4, $|p - p'| \leq 3 \left(\frac{\|\alpha\|_3}{\|\alpha\|_2} \right)^3$. Also

$$
|p^3 - p'^3| = |p - p'|(p^2 + pp' + p'^2) \leq 9 \left(\frac{\|\alpha\|_3}{\|\alpha\|_2} \right)^3.
$$

We then have

$$|\mathbb{P}(E_p) - p^3| \leq |\mathbb{P}(E_p) - p'^3| + |p^3 - p'^3| \leq (9 + 100 C_{be}) \left(\frac{\|\alpha\|_3}{\|\alpha\|_2} \right)^3$$

as desired. $\qquad\qquad\qquad\qquad\qquad\qquad\qquad\qquad\qquad\qquad\qquad\square$

13.4 Proof of Theorem 13.3

Recall from the introduction that $\tau(G)$ denotes the number of signed triangles of a graph G. If A is the adjacency matrix of G with entries $A_{i,j}$ we denote the centered adjacency matrix of G as $\bar{A}$ with entries $\bar{A}_{i,j} := A_{i,j} - \mathbb{E}[A_{i,j}]$. Given three distinct vertices i, j and k, the signed triangle induced by those 3 vertices is $\tau_G(i, j, k) := \bar{A}_{i,j} \bar{A}_{i,k} \bar{A}_{j,k}$. It then holds that for a graph $G = (V, E)$ the number of signed triangles is given by:

$$\tau(G) := \sum_{\{i,j,k\} \in \binom{V}{3}} \tau_G(i, j, k).$$

Analysis of $\tau(G(n, p))$ was done in [3], where it was shown that $\mathbb{E}\tau(G(n, p)) = 0$ while $\mathrm{Var}(\tau(G(n, p))) \leq n^3$.

To prove Theorem 13.3 it will suffice to show that $\mathbb{E}\tau(G(n, p, \alpha))$ is asymptotically bigger than both the standard deviation of $\tau(G(n, p))$ and of $\tau(G(n, p, \alpha))$, provided that $\left(\frac{\|\alpha\|_2}{\|\alpha\|_3} \right)^6 << n^3$.

For this aim we first prove some technical lemmas:

Lemma 13.8 *Let* $p \in (0, 1)$, *then*

$$\mathbb{E}A_{1,2} A_{2,3} \leq p^2 + 8 \left(\frac{\|\alpha\|_4}{\|\alpha\|_2} \right)^4.$$

Proof Let X, Y, Z be i.i.d. random variables generated from $\mathcal{N}(0, D_\alpha)$, then, conditioning on Y yields the expression

$$\mathbb{E}A_{1,2} A_{2,3} = \mathbb{E}\left[\mathbb{P}\left(\langle X, Y \rangle \geq t_{p,\alpha} \right) \mathbb{P}\left(\langle Z, Y \rangle \geq t_{p,\alpha} \right) | Y \right] = \mathbb{E}\left[\Phi \left(\frac{t_{p,\alpha}}{\sqrt{\sum \alpha_i Y_i^2}} \right)^2 \right],$$

where Φ is the standard Gaussian cumulative distribution function. By the same argument, we also have

$$\mathbb{E}\left[\Phi\left(\frac{t_{p,\alpha}}{\sqrt{\sum \alpha_i Y_i^2}}\right)\right]^2 = \mathbb{E}A_{1,2}\mathbb{E}A_{2,3} = p^2.$$

Thus, it will be enough to show,

$$\mathrm{Var}\left(\Phi\left(\frac{t_{p,\alpha}}{\sqrt{\sum \alpha_i Y_i^2}}\right)\right) \le 8\left(\frac{\|\alpha\|_4}{\|\alpha\|_2}\right)^4. \tag{13.29}$$

For $\sigma^2 > 0$ denote by G_{σ^2} a random variable with law $\mathcal{N}(0, \sigma^2)$. We then have

$$\Phi\left(\frac{t_{p,\alpha}}{\sqrt{\sum \alpha_i Y_i^2}}\right) = \mathbb{P}\left(G_{\sum \alpha_i Y_i^2} \le t_{p,\alpha}\right),$$

and

$$\mathrm{Var}\left(\Phi\left(\frac{t_{p,\alpha}}{\sqrt{\sum \alpha_i Y_i^2}}\right)\right) \le \mathbb{E}\left[\left(\Phi\left(\frac{t_{p,\alpha}}{\sqrt{\sum \alpha_i Y_i^2}}\right) - \Phi\left(\frac{t_{p,\alpha}}{\|\alpha\|_2}\right)\right)^2\right]$$

$$= \mathbb{E}\left[\left(\mathbb{P}\left(G_{\sum \alpha_i Y_i^2} \le t_{p,\alpha}\right) - \mathbb{P}\left(G_{\|\alpha\|_2^2} \le t_{p,\alpha}\right)\right)^2\right]$$

$$\le \mathbb{E}\left[\mathrm{TV}\left(G_{\sum \alpha_i Y_i^2}, G_{\|\alpha\|_2^2}\right)^2\right].$$

For the total variation distance between 2 Gaussian random variables we have the following bound (see Proposition 3.6.1 in [11], for example):

$$\mathrm{TV}\left(G_{\sigma_1^2}, G_{\sigma_1^2}\right) \le 2\frac{|\sigma_1^2 - \sigma_2^2|}{\max(\sigma_1^2, \sigma_2^2)}.$$

This implies

$$\mathrm{Var}\left(\Phi\left(\frac{t_{p,\alpha}}{\sqrt{\sum \alpha_i Y_i^2}}\right)\right) \le \frac{4}{\|\alpha\|_2^4}\mathbb{E}\left[\left(\sum \alpha_i Y_i^2 - \|\alpha\|_2^2\right)^2\right].$$

As $Y \sim \mathcal{N}(0, D_\alpha)$, it is immediate to check

$$\mathbb{E}\left[\sum \alpha_i Y_i^2\right] = \sum \alpha_i^2 = \|\alpha\|_2^2.$$

Hence

$$\mathbb{E}\left[\left(\sum \alpha_i Y_i^2 - \|\alpha\|_2^2\right)^2\right] = \sum \alpha_i^2 \operatorname{Var}\left(Y_i^2\right) = 2\|\alpha\|_4^4.$$

This establishes (13.29) and finishes the proof.

$\square$

Lemma 13.9 *Let* $p \in (0, 1)$*, then*

$$\mathbb{E}[\tau_{G(n,p,\alpha)}(1, 2, 3)\tau_{G(n,p,\alpha)}(1, 2, 4)] \leq 80\left(\frac{\|\alpha\|_4}{\|\alpha\|_2}\right)^4$$

Proof The proof is similar to Lemma 13.8 and uses the observation that if V_1 and V_2 are the random vectors corresponding to two vertices, then conditioned on their values, the random variables $\tau_{G(n,p,\alpha)}(1, 2, 3)$ and $\tau_{G(n,p,\alpha)}(1, 2, 4)$ are independent. Thus

$$\mathbb{E}[\tau_{G(n,p,\alpha)}(1, 2, 3)\tau_{G(n,p,\alpha)}(1, 2, 4)]$$
$$= \mathbb{E}\left[\mathbb{E}[\tau_{G(n,p,\alpha)}(1, 2, 3)\tau_{G(n,p,\alpha)}(1, 2, 4)|V_1, V_2]\right]$$
$$= \mathbb{E}\left[\mathbb{E}\left[\bar{A}_{1,3}\bar{A}_{2,3}|V_1, V_2\right]^2 \bar{A}_{1,2}^2\right] \leq \mathbb{E}\left[\mathbb{E}\left[\bar{A}_{1,3}\bar{A}_{2,3}|V_1, V_2\right]^2\right].$$

Lemma 13.8 implies $\mathbb{E}\left[\bar{A}_{1,3}\bar{A}_{2,3}\right] \leq 8\left(\frac{\|\alpha\|_4}{\|\alpha\|_2}\right)^4$, so that

$$\mathbb{E}[\tau_{G(n,p,\alpha)}(1, 2, 3)\tau_{G(n,p,\alpha)}(1, 2, 4)] \leq \mathbb{E}\left[\mathbb{E}\left[\bar{A}_{1,3}\bar{A}_{2,3}|V_1, V_2\right]^2\right]$$

$$\leq \operatorname{Var}\left(\mathbb{E}\left[\bar{A}_{1,3}\bar{A}_{2,3}|V_1, V_2\right]\right) + 64\left(\frac{\|\alpha\|_4}{\|\alpha\|_2}\right)^8.$$
$$(13.30)$$

Note that if $X \sim \mathcal{N}(0, D_\alpha)$,

$$\mathbb{E}\left[\bar{A}_{1,3}\bar{A}_{2,3}|V_1, V_2\right] = (1 - p)^2\mathbb{P}\left(\langle X, V_1\rangle \geq t_{p,\alpha}, \langle X, V_2\rangle \geq t_{p,\alpha}\right)$$
$$+ p^2\mathbb{P}\left(\langle X, V_1\rangle < t_{p,\alpha}, \langle X, V_2\rangle < t_{p,\alpha}\right)$$
$$- p(1 - p)\left(\mathbb{P}\left(\langle X, V_1\rangle \geq t_{p,\alpha}, \langle X, V_2\rangle < t_{p,\alpha}\right)\right.$$
$$\left.+ \mathbb{P}\left(\langle X, V_1\rangle < t_{p,\alpha}, \langle X, V_2\rangle \geq t_{p,\alpha}\right)\right).$$

For any $v, u \in \mathbb{R}^d$, denote by $\Sigma_{v,u}$ the matrix given by

$$\begin{bmatrix} \sum \alpha_i v_i^2 & \sum \alpha_i v_i u_i \\ \sum \alpha_i v_i u_i & \sum \alpha_i u_i^2 \end{bmatrix},$$

then the joint law of $\langle X, v \rangle, \langle X, u \rangle$ is $G_{v,u} := \mathcal{N}\left(0, \Sigma_{v,u}\right)$. The above can now be rewritten as

$$(1-p)^2 \mathbb{P}\left(G_{V_1, V_2} \in (t_{p,\alpha}, \infty) \times (t_{p,\alpha}, \infty)\right) + p^2 \mathbb{P}\left(G_{V_1, V_2} \in (-\infty, t_{p,\alpha})\right)$$
$$\times (-\infty, t_{p,\alpha})) - p(1-p)\left((G_{V_1, V_2} \in (-\infty, t_{p,\alpha}) \times (t_{p,\alpha}, \infty))\right)$$
$$+\mathbb{P}\left(G_{V_1, V_2} \in (t_{p,\alpha}, \infty) \times (-\infty, t_{p,\alpha}))\right). \tag{13.31}$$

In particular, if M_i are independent random Wishart matrices with law $\mathcal{W}_2(\alpha_i^2 I_2, 1)$ and $M = \sum M_i$ then the matrix Σ_{V_1, V_2} has the same law as M. In this case we regard (13.31) as a function h, of the covariance M. Using (13.30), we get

$$\mathbb{E}[\tau_{G(n,p,\alpha)}(1, 2, 3)\tau_{G(n,p,\alpha)}(1, 2, 4)] \le \mathrm{Var}\left(h(M)\right) + 16 \left(\frac{\|\alpha\|_4}{\|\alpha\|_2}\right)^8.$$

It is thus enough to establish an upper bound for $\mathrm{Var}(h(M))$. For a positive semi-definite matrix Σ, we denote $G_\Sigma \sim \mathcal{N}(0, \Sigma)$. As $h(M) \le 1$, we have the following inequality

$$\mathrm{Var}(h(M)) \le \mathbb{E}\left[\left(h(M) - h(\|\alpha\|_2^2 I_2)^2\right)\right] \le \mathbb{E}\left[\mathrm{TV}\left(G_M, G_{\|\alpha\|_2^2 I}\right)^2\right]$$
$$\le \mathbb{E}\left[\min\left(1, \mathrm{Ent}\left(G_M \| G_{\|\alpha\|_2^2 I_2}\right)\right)\right],$$

where we have used Pinsker's inequality (13.8) to bound the total variation. The relative entropy between the Gaussians (see [5]) is given by

$$\mathrm{Ent}\left(G_M \| G_{\|\alpha\|_2^2 I_2}\right) = \mathrm{Tr}\left(\frac{1}{\|\alpha\|_2^2} M\right) - \ln\left(\frac{\det(M)}{\det(G_{\|\alpha\|_2^2 I_2})}\right) - 2$$
$$= \mathrm{Tr}\left(\frac{M}{\|\alpha\|_2^2} - I_2\right) - \ln\left(\det\left(\frac{M}{\|\alpha\|_2^2}\right)\right).$$

For any $x \ge \frac{1}{2}$ we have the inequality $x - 1 - \ln(x) \le x^2$. So, if both eigenvalues of $\frac{M}{\|\alpha\|_2^2}$ are bigger than $\frac{1}{2}$

$$\mathrm{Ent}\left(G_M \| G_{\|\alpha_2^2\| I_2}\right) \le \left\| \frac{M}{\|\alpha\|_2^2} - I_2 \right\|_{HS}^2.$$

Otherwise, it is clear that

$$1 \leq 2 \left\| \frac{M}{\|\alpha\|_2^2} - I_2 \right\|_{HS}^2.$$

Combining the above, we have established

$$\text{Var}\,(h(M)) \leq 2\mathbb{E}\left[\left\| \frac{M}{\|\alpha\|_2^2} - I_2 \right\|_{HS}^2 \right]. \tag{13.32}$$

Recall that the diagonal elements of M are given by $\sum \alpha_i(V)_i^2$ where $V \sim \mathcal{N}(0, D_\alpha)$, thus

$$\mathbb{E}\left[M_{1,1}\right] = \mathbb{E}\left[\left(\sum \alpha_i (V_1)_i^2\right)\right] = \left(\sum \alpha_i^2\right) = \|\alpha\|_2^2,$$

and

$$\text{Var}(M_{1,1}) = \sum \text{Var}\left(\alpha_i (V_1)_i^2\right) = 2\,\|\alpha\|_4^4.$$

The off-diagonal element is given by $\sum \alpha_i (V_1)_i (V_2)_i$, for which we have

$$\mathbb{E}\left[M_{1,2}\right] = 0,$$

and

$$\text{Var}\left(M_{1,2}\right) = \mathbb{E}\left[\left(\sum \alpha_i (V_1)_i (V_2)_i\right)^2\right] = \sum \alpha_i^2 \mathbb{E}\left[(V_1)_i^2\right] \mathbb{E}\left[(V_2)_i^2\right] = \|\alpha\|_4^4.$$

Using these estimates in (13.32) we obtain

$$\text{Var}\,(h(M)) \leq 16 \left(\frac{\|\alpha\|_4}{\|\alpha\|_2}\right)^4.$$

plugging this into (13.30) gives the desired result. $\qquad\square$

Lemma 13.10 *Let $p \in (0, 1)$, then*

$$\mathbb{E}[\tau_{G(n,p,\alpha)}(1, 2, 3)\tau_{G(n,p,\alpha)}(1, 4, 5)] \leq 80 \left(\frac{\|\alpha\|_4}{\|\alpha\|_2}\right)^4$$

Proof Conditioned on the location of the vertex V_1, the random variables $\tau_{G(n,p,\alpha)}(1,2,3)$, $\tau_{G(n,p,\alpha)}(1,4,5)$ are independent and identically distributed, thus

$$\mathbb{E}[\tau_{G(n,p,\alpha)}(1,2,3)\tau_{G(n,p,\alpha)}(1,4,5)]$$
$$= \mathbb{E}\left[\mathbb{E}\left[\tau_{G(n,p,\alpha)}(1,2,3)\tau_{G(n,p,\alpha)}(1,4,5)|V_1\right]\right]$$
$$= \mathbb{E}\left[\mathbb{E}\left[\tau_{G(n,p,\alpha)}(1,2,3)|V_1\right]^2\right].$$

Using the tower property of conditional expectation

$$\mathbb{E}\left[\mathbb{E}\left[\tau_{G(n,p,\alpha)}(1,2,3)|V_1\right]^2\right]$$
$$= \mathbb{E}\left[\mathbb{E}\left[\mathbb{E}\left[\tau_{G(n,p,\alpha)}(1,2,3)|V_1,V_2\right]|V_1\right]^2\right]$$
$$\leq \mathbb{E}\left[\mathbb{E}\left[\tau_{G(n,p,\alpha)}(1,2,3)|V_1,V_2\right]^2\right] = \mathbb{E}\left[\mathbb{E}\left[\tau_{G(n,p,\alpha)}(1,2,3)|V_1,V_2\right]^2\right].$$

But in Lemma 13.9, using (13.30), we have essentially shown

$$\mathbb{E}\left[\mathbb{E}\left[\tau_{G(n,p,\alpha)}(1,2,3)|V_1,V_2\right]^2\right] \leq 80\left(\frac{\|\alpha\|_4}{\|\alpha\|_2}\right)^4,$$

thus the claim is proven.   □

Towards the proof of Theorem 13.2 we now estimate $\mathbb{E}\tau(G(n,p,\alpha))$. Note that since

$$\mathbb{E}\tau(G(n,p,\alpha)) = \binom{n}{3}\mathbb{E}\tau_{G(n,p,\alpha)}(1,2,3),$$

it is enough to estimate $\mathbb{E}\tau_{G(n,p,\alpha)}(1,2,3)$.

$$\mathbb{E}\tau_{G(n,p,\alpha)}(1,2,3) = \mathbb{E}\bar{A}_{1,2}\bar{A}_{1,3}\bar{A}_{2,3} = \mathbb{E}(A_{1,2}-p)(A_{1,3}-p)(A_{2,3}-p)$$
$$= \mathbb{E}A_{1,2}A_{1,3}A_{2,3} - p\left(\mathbb{E}A_{1,2}A_{2,3} + \mathbb{E}A_{1,2}A_{1,3} + \mathbb{E}A_{1,3}A_{2,3}\right)$$
$$+ p^2\left(\mathbb{E}A_{1,2} + \mathbb{E}A_{1,3} + \mathbb{E}A_{2,3}\right) - p^3$$
$$\geq \mathbb{E}A_{1,2}A_{1,3}A_{2,3} - p^3 - 24p\left(\frac{\|\alpha\|_4}{\|\alpha\|_2}\right)^4, \tag{13.33}$$

where the inequality follows from the fact that $\mathbb{E}A_{i,j} = p$ and Lemma 13.8. As

$$\left(\frac{\|\alpha\|_4}{\|\alpha\|_2}\right)^4 \leq \left(\frac{\|\alpha\|_3}{\|\alpha\|_2}\right)^3 \frac{1}{\|\alpha\|_2},$$

as long as $\|\alpha\|_2$ is large enough, the lower bound of Theorem 13.5 yields

$$\mathbb{E}\tau_{G(n,p,\alpha)}(1,2,3) \geq \delta_p \left(\frac{\|\alpha\|_3}{\|\alpha\|_2}\right)^3$$

for a constant $\delta_p > 0$, depending only on p. This shows

$$\mathbb{E}\tau(G(n,p,\alpha)) \geq \delta_p \binom{n}{3} \left(\frac{\|\alpha\|_3}{\|\alpha\|_2}\right)^3.$$

To bound from above the variance of $\tau(G(n,p,\alpha))$ we observe that $\tau_G(i,j,k)$ is independent from $\tau_G(i',j',k')$ whenever $\left|\{i,j,k\} \cap \{i',j',k'\}\right| = 0$, thus

$$\begin{aligned}
&\mathrm{Var}\left(\tau(G(n,p,\alpha))\right) \\
&= \sum_{\{i,j,k\}} \sum_{\{i',j',k'\}} \mathbb{E}\left[\tau_{G(n,p,\alpha)}(i,j,k)\tau_{G(n,p,\alpha)}(i',j',k')\right] \\
&\qquad - \mathbb{E}\left[\tau_{G(n,p,\alpha)}(i,j,k)\right]\mathbb{E}\left[\tau_{G(n,p,\alpha)}(i',j',k')\right] \\
&\leq \sum_{\{i,j,k\}} \mathbb{E}\left[\tau_{G(n,p,\alpha)}(i,j,k)\tau_{G(n,p,\alpha)}(i,j,k)\right] \\
&\qquad + \sum_{\{i,j,k,l\}} \mathbb{E}\left[\tau_{G(n,p,\alpha)}(i,j,k)\tau_{G(n,p,\alpha)}(i,j,l)\right] \\
&\qquad + \sum_{\{i,j,k,l,m\}} \mathbb{E}\left[\tau_{G(n,p,\alpha)}(i,j,k)\tau_{G(n,p,\alpha)}(k,l,m)\right] \\
&= \binom{n}{3}\mathbb{E}[\tau_{G(n,p,\alpha)}(1,2,3)\tau_{G(n,p,\alpha)}(1,2,3)] \\
&\qquad + \binom{n}{4}\binom{4}{2}\mathbb{E}[\tau_{G(n,p,\alpha)}(1,2,3)\tau_{G(n,p,\alpha)}(1,2,4)] \\
&\qquad + 5\binom{n}{5}\mathbb{E}[\tau_{G(n,p,\alpha)}(1,2,3)\tau_{G(n,p,\alpha)}(1,4,5)].
\end{aligned}$$

Noting that $\mathbb{E}[\tau_{G(n,p,\alpha)}(1,2,3)\tau_{G(n,p,\alpha)}(1,2,3)] \leq 1$, in conjunction with Lemmas 13.9 and 13.10 yields

$$\mathrm{Var}(\tau(G(n,p,\alpha))) \leq n^3 + 80n^5 \left(\frac{\|\alpha\|_4}{\|\alpha\|_2}\right)^4.$$

Combining all of the above

$$\mathbb{E}\left[\tau(G(n,p))\right] = 0, \quad \mathbb{E}[\tau(G(n,p,\alpha))] \geq \delta_p \binom{n}{3} \left(\frac{\|\alpha\|_3}{\|\alpha\|_2}\right)^3,$$

and

$$\max\{\mathrm{Var}(\tau(G(n, p, \alpha))), \mathrm{Var}(G(n, p))\} \leq n^3 + 80n^5 \left(\frac{\|\alpha\|_4}{\|\alpha\|_2}\right)^4.$$

Chebyshev's inequality implies that

$$\mathbb{P}\left(\tau(G(n, p, \alpha)) \leq \frac{1}{2}\mathbb{E}[\tau(G(n, p, \alpha))]\right) \leq 200\frac{\left(\frac{\|\alpha\|_2}{\|\alpha\|_3}\right)^6 n^3 + 80n^5 \frac{\|\alpha\|_2^2\|\alpha\|_4^4}{\|\alpha\|_3^6}}{\delta_p^2 n^6},$$

and also

$$\mathbb{P}\left(\tau(G(n, p)) \geq \frac{1}{2}\mathbb{E}[\tau(G(n, p, \alpha))]\right) \leq 200\frac{\left(\frac{\|\alpha\|_2}{\|\alpha\|_3}\right)^6 n^3 + 80n^5 \frac{\|\alpha\|_2^2\|\alpha\|_4^4}{\|\alpha\|_3^6}}{\delta_p^2 n^6}.$$

Note that due to the normalization (13.13) $\frac{\|\alpha\|_2^2\|\alpha\|_4^4}{\|\alpha\|_3^6} \leq \frac{\|\alpha\|_2^2}{\|\alpha\|_3^2}$. Putting the above expressions together we thus have:

$$\mathrm{TV}\left(\tau(G(n, p, \alpha)), \tau(G(n, p))\right) \geq 1 - C\frac{\left(\frac{\|\alpha\|_2}{\|\alpha\|_3}\right)^6}{n^3} - C\frac{\left(\frac{\|\alpha\|_2}{\|\alpha\|_3}\right)^2}{n},$$

for a constant C depending only on p. This concludes the proof of Theorem 13.3.

13.5 Proof of the Lower Bound

As stated in the introduction, we can view $G(n, p, \alpha)$ as a function of an appropriate random matrix, as follows. Let $\mathbb{Y}$ be a random $n \times d$ matrix with rows sampled i.i.d. from $\mathcal{N}(0, D_\alpha)$. Define $W = W(n, \alpha) = \mathbb{Y}\mathbb{Y}^T / \|\alpha\|_2 - \mathrm{diag}\left(\mathbb{Y}\mathbb{Y}^T / \|\alpha\|_2\right)$. Note that for $i \neq j$, $W_{ij} = \langle \gamma_i, \gamma_j \rangle / \|\alpha\|_2$, where γ_i, γ_j are the rows of $\mathbb{Y}$. Thus the $n \times n$ matrix A defined as

$$A_{i,j} = \begin{cases} 1 & \text{if } W_{ij} \geq t_{p,\alpha}/\|\alpha\|_2 \text{ and } i \neq j \\ 0 & \text{otherwise} \end{cases}$$

has the same law as the adjacency matrix of $G(n, p, \alpha)$. Denote the map that takes W to A by $H_{p,\alpha}$, i.e., $A = H_{p,\alpha}(W)$.

Similarly, we may view $G(n, p)$ as function of an $n \times n$ matrix with independent Gaussian entries. Let $M(n)$ be a symmetric $n \times n$ random matrix with 0 entries in the diagonal, and whose entries above the diagonal are i.i.d. standard normal random variables. If Φ is the cumulative distribution function of the standard Gaussian, then the $n \times n$ matrix B, defined as

$$
B_{i,j} = \begin{cases} 1 & \text{if } M(n)_{ij} \geq \Phi^{-1}(p) \text{ and } i \neq j \\ 0 & \text{otherwise} \end{cases}
$$

has the same law as the adjacency matrix of $G(n, p)$. Denote the map that takes $M(n)$ to B by K_p, i.e., $B = K_p(M(n))$.

Using the triangle inequality and by the previous two paragraphs, we have that for any $p \in (0, 1)$

$$
\begin{aligned}
\mathrm{TV}(G(n, p), G(n, p, \alpha)) &= \mathrm{TV}(K_p(M(n)), H_{p,\alpha}(W(n, \alpha))) \\
&\leq \mathrm{TV}(H_{p,\alpha}(M(n)), H_{p,\alpha}(W(n, \alpha))) \\
&\quad + \mathrm{TV}(K_p(M(n)), H_{p,\alpha}(M(n))) \\
&\leq \mathrm{TV}(M(n), W(n, \alpha)) + \mathrm{TV}(K_p(M(n)), H_{p,\alpha}(M(n))).
\end{aligned}
$$

The second term is of lower order and will be dealt with later. The first term is bounded using Pinsker's inequality, (13.8), yielding

$$
\mathrm{TV}(M(n), W(n, \alpha)) \leq \sqrt{\frac{1}{2}\mathrm{Ent}[M(n)\,||\,W(n, \alpha)]}.
$$

We'll use a similar argument to the one presented in [2] which follows an inductive proof using the chain rule for relative entropy. We observe that a sample of $W(n + 1, \alpha)$ may be constructed from $W(n, \alpha)$ by adjoining the column vector (and symmetrically the row vector) $\mathbb{Y}Y/\|\alpha\|_2$ where $Y \sim \mathcal{N}(0, D_\alpha)$ is independent of $\mathbb{Y}$. Thus, using the notation, Z_n for a standard Gaussian in $\mathbb{R}^n$, by (13.9), we obtain

$$
\begin{aligned}
\mathrm{Ent}\big[W(n + 1, \alpha)\,||\,M(n + 1)\big] &= \mathrm{Ent}\big[W(n, \alpha)\,||\,M(n)\big] \\
&\quad + \mathbb{E}_{\mathbb{Y}}\mathrm{Ent}\big[\mathbb{Y}Y/\|\alpha\|_2\,\big|\,W(n, \alpha)\,\big|\big|\,Z_n\big].
\end{aligned}
$$

Since $W(n, \alpha)$ is a function of $\mathbb{Y}$, standard properties of relative entropy (see [4], chapter 2) show

$$
\begin{aligned}
\mathbb{E}_{\mathbb{Y}}\mathrm{Ent}\big[\mathbb{Y}Y/\|\alpha\|_2\,\big|\,W(n, \alpha)\,\big|\big|\,Z_n\big] &= \mathbb{E}_{\mathbb{Y}}\mathrm{Ent}\big[\mathbb{Y}Y/\|\alpha\|_2\,\big|\,\mathbb{Y}\mathbb{Y}^T/\|\alpha\|_2\,\big|\big|\,Z_n\big] \\
&\leq \mathbb{E}_{\mathbb{Y}}\mathrm{Ent}\big[\mathbb{Y}Y/\|\alpha\|_2\,\big|\,\mathbb{Y}\,\big|\big|\,Z_n\big].
\end{aligned}
$$

Note that $\mathbb{Y}Y/\|\alpha\|_2\,|\,\mathbb{Y}$ is distributed as $\mathcal{N}(0, \frac{1}{\|\alpha\|_2^2}\mathbb{Y}D_\alpha\mathbb{Y}^T)$. The relative entropy between two n-dimensional Gaussians, (see [5]) $\mathcal{N}_1 \sim \mathcal{N}(0, \Sigma_1), \mathcal{N}_2 \sim \mathcal{N}(0, \Sigma_2)$

is given by

$$\text{Ent}\,[\mathcal{N}_1||\mathcal{N}_2] = \frac{1}{2}\left(\text{tr}\left(\Sigma_2^{-1}\Sigma_1\right) + \ln\left(\frac{\det\Sigma_2}{\det\Sigma_1}\right) - n\right).$$

In our case $\Sigma_2 = I_n$ and $\mathbb{E}_{\mathbb{Y}}\,\text{tr}(\mathbb{Y}D_\alpha\mathbb{Y}^T) = n\,\|\alpha\|_2^2$. Thus the following holds:

$$\mathbb{E}_{\mathbb{Y}}\,\text{Ent}\left[\frac{1}{\|\alpha\|_2}\mathbb{Y}Y\,|\mathbb{Y}||\,Z_n\right] = -\frac{1}{2}\left(\mathbb{E}_{\mathbb{Y}}\,\ln\det\left(\frac{1}{\|\alpha\|_2^2}\mathbb{Y}D_\alpha\mathbb{Y}^T\right)\right).$$

Theorem 13.4 is then implied by the following lemma:

Lemma 13.11 $-\mathbb{E}_{\mathbb{Y}}\,\ln\det\left(\frac{1}{\|\alpha\|_2^2}\mathbb{Y}D_\alpha\mathbb{Y}^T\right) \leq C\left(n^2\left(\frac{\|\alpha\|_4}{\|\alpha\|_2}\right)^4 + \sqrt{n\left(\frac{\|\alpha\|_4}{\|\alpha\|_2}\right)^4}\right)$ *for a universal constant $C > 0$.*

The proof will follow similar lines as Lemma 2 in [2]. Namely, we will decompose the expectation on the event that the smallest eigenvalue of $\frac{1}{\|\alpha\|_2^2}\mathbb{Y}D_\alpha\mathbb{Y}^T$, denoted by $\lambda_{\min}$, is larger than $\frac{1}{2}$. Lemma 13.11 will then follow by the following two claims:

Claim 13.3

$$-\mathbb{E}_{\mathbb{Y}}\left[\ln\det\left(\frac{1}{\|\alpha\|_2^2}\mathbb{Y}D_\alpha\mathbb{Y}^T\right)\mathbb{1}_{\left\{\lambda_{\min}\geq\frac{1}{2}\right\}}\right] \leq C\left(n^2\left(\frac{\|\alpha\|_4}{\|\alpha\|_2}\right)^4 + \sqrt{n\left(\frac{\|\alpha\|_4}{\|\alpha\|_2}\right)^4}\right),$$

for a universal constant $C > 0$.

Proof We first use the inequality $-\ln(x) \leq 1 - x + (1-x)^2$ for $x \geq \frac{1}{2}$:

$$-\mathbb{E}_{\mathbb{Y}}\left[\ln\det\left(\frac{\mathbb{Y}D_\alpha\mathbb{Y}^T}{\|\alpha\|_2^2}\right)\mathbb{1}_{\left\{\lambda_{\min}\geq\frac{1}{2}\right\}}\right]$$

$$\leq \mathbb{E}_{\mathbb{Y}}\left[\left|\text{tr}\left(I_n - \frac{\mathbb{Y}D_\alpha\mathbb{Y}^T}{\|\alpha\|_2^2}\right)\right| + \left\|I_n - \frac{\mathbb{Y}D_\alpha\mathbb{Y}^T}{\|\alpha\|_2^2}\right\|_{HS}^2\right], \qquad (13.34)$$

where $\|\cdot\|_{HS}$ denotes the Hilbert-Schmidt norm. Before proceeding, we first calculate several quantities. For $1 \leq j \leq n$ denote by A_j the jth row of $\mathbb{Y}\sqrt{D_\alpha}$ with entries $\{\sqrt{\alpha_i}\,y_{j,i}\}_{i=1}^d$.

1. The expected squared norm of A_j is given by $\mathbb{E}\,\|A_j\|^2 = \sum_i \mathbb{E}\,\alpha_i y_{j,i}^2 = \sum_i \alpha_i^2 = \|\alpha\|_2^2$. Since $y_{j,i}$ is a centred Gaussian with variance α_i.

2. When $j \neq k$, A_j and A_k are independent, and so $\mathbb{E} \left\| A_j \right\|^2 \left\| A_k \right\|^2 = \left(\sum_i \alpha_i^2 \right)^2 = \left\| \alpha \right\|_2^4$.

3. When $j \neq k$, the expected squared inner product between two rows is given by

$$\mathbb{E} \langle A_j, A_k \rangle^2 = \mathbb{E} \left(\sum_{i=1}^d \alpha_i \, y_{j,i} \, y_{k,i} \right)^2$$

$$= \sum_{i=1}^d \alpha_i^2 \mathbb{E} y_{j,i}^2 y_{k,i}^2 + \sum_{i_1 \neq i_2} \alpha_{i_1} \alpha_{i_2} \mathbb{E} y_{j,i_1} y_{k,i_1} y_{j,i_2} y_{k,i_2}$$

$$= \sum_{i=1}^d \alpha_i^4 = \left\| \alpha \right\|_4^4 \,.$$

4. The expected 4th power of the norm is given by

$$\mathbb{E} \left\| A_j \right\|^4 = \mathbb{E} \left(\sum_i \alpha_i \, y_{j,i}^2 \right)^2 = \sum_i \alpha_i^2 \mathbb{E} y_{j,i}^4 + \sum_{i \neq k} \alpha_i \alpha_k \mathbb{E} y_{j,i}^2 y_{j,k}^2$$

$$\leq 3 \sum_i \alpha_i^4 + \left(\sum_i \alpha_i^2 \right)^2 = 3 \left\| \alpha \right\|_4^4 + \left\| \alpha \right\|_2^4 \,,$$

when we remember that the 4th moment of a centred Gaussian with variance α_i is $3\alpha_i^2$.

We turn to bound each term of the sum (13.34):

$$\mathbb{E}_{\mathbb{Y}} \left| \operatorname{tr} \left(I_n - \frac{1}{\left\| \alpha \right\|_2^2} \mathbb{Y} D_\alpha \mathbb{Y}^T \right) \right|$$

$$\leq \sqrt{\mathbb{E}_{\mathbb{Y}} \operatorname{tr}^2 \left(I_n - \frac{1}{\left\| \alpha \right\|_2^2} \mathbb{Y} D_\alpha \mathbb{Y}^T \right)} = \sqrt{\mathbb{E}_{\mathbb{Y}} \left(\sum_{j=1}^n \left(1 - \frac{\left\| A_j \right\|^2}{\left\| \alpha \right\|_2^2} \right) \right)^2}$$

$$= \sqrt{\mathbb{E}_{\mathbb{Y}} \left(n^2 - \frac{2n}{\left\| \alpha \right\|_2^2} \sum_{j=1}^n \left\| A_j \right\|^2 + \frac{1}{\left\| \alpha \right\|_2^4} \sum_{j \neq k} \left\| A_j \right\|^2 \left\| A_k \right\|^2 + \frac{1}{\left\| \alpha \right\|_2^4} \sum_{j=1}^n \left\| A_j \right\|^4 \right)}$$

$$\leq \sqrt{n^2 - 2n^2 + 2 \binom{n}{2} + \frac{n}{\left\| \alpha \right\|_2^4} \left(3 \left\| \alpha \right\|_4^4 + \left\| \alpha \right\|_2^4 \right)} = \sqrt{3n \frac{\left\| \alpha \right\|_4^4}{\left\| \alpha \right\|_2^4}} \,.$$

Similarly, we may deal with the second term:

$$\mathbb{E}_{\mathbb{Y}} \left\| I_n - \frac{1}{\|\alpha\|_2^2} \mathbb{Y} D_\alpha \mathbb{Y}^T \right\|_{HS}^2 = \left(\sum_{k,j} \frac{1}{\|\alpha\|_2^4} \mathbb{E}_{\mathbb{Y}} \langle A_j, A_k \rangle^2 \right) - n$$

$$= \frac{1}{\|\alpha\|_2^4} \sum_{j=1}^n \mathbb{E}_{\mathbb{Y}} \|A_j\|^4 + \frac{1}{\|\alpha\|_2^4} \sum_{j \neq k} \langle A_j, A_k \rangle^2 - n$$

$$\leq \frac{n}{\|\alpha\|_2^4} (3 \|\alpha\|_4^4 + \|\alpha\|_2^4) + \frac{2}{\|\alpha\|_2^4} \binom{n}{2} \|\alpha\|_4^4 - n$$

$$= 3n \frac{\|\alpha\|_4^4}{\|\alpha\|_2^4} + (n^2 - n) \frac{\|\alpha\|_4^4}{\|\alpha\|_2^4} \leq 3n^2 \frac{\|\alpha\|_4^4}{\|\alpha\|_2^4}.$$

Combining (13.34) with the last two displays gives

$$-\mathbb{E}_{\mathbb{Y}} \left[\ln \det \left(\frac{1}{\|\alpha\|_2^2} \mathbb{Y} D_\alpha \mathbb{Y}^T \right) \mathbb{1}_{\{\lambda_{\min} \geq \frac{1}{2}\}} \right] \leq 3 \left(n^2 \left(\frac{\|\alpha\|_4}{\|\alpha\|_2} \right)^4 + \sqrt{n \left(\frac{\|\alpha\|_4}{\|\alpha\|_2} \right)^4} \right).$$

$\square$

Claim 13.4

$$\mathbb{E}_{\mathbb{Y}} \left[\ln \det \left(\frac{1}{\|\alpha\|_2^2} \mathbb{Y} D_\alpha \mathbb{Y}^T \right) \mathbb{1}_{\{\lambda_{\min} < 1/2\}} \right] < n \exp(-C \|\alpha\|_2^2),$$

for a universal constant $C > 0$.

Proof Observe that for any $\xi \in (0, \frac{1}{2})$:

$$-\mathbb{E}_{\mathbb{Y}} \left[\ln \det \left(\frac{\mathbb{Y} D_\alpha \mathbb{Y}^T}{\|\alpha\|_2^2} \right) \mathbb{1}_{\{\lambda_{\min} < 1/2\}} \right] \leq n \mathbb{E} \left(-\log(\lambda_{\min}) \mathbb{1}_{\{\lambda_{\min} < 1/2\}} \right)$$

$$= n \int_{\log(2)}^\infty \mathbb{P}(-\log(\lambda_{\min}) > t) dt$$

$$= n \int_0^{1/2} \frac{1}{s} \mathbb{P}(\lambda_{\min} < s) ds$$

$$\leq \frac{n}{\xi} \mathbb{P}(\lambda_{\min} < 1/2) + n \int_0^\xi \frac{1}{s} \mathbb{P}(\lambda_{\min} < s) ds.$$

$$\tag{13.35}$$

By allowing ξ to be some small constant, we'll need to bound $\mathbb{P}(\lambda_{\min} < 1/2)$ and $\mathbb{P}(\lambda_{\min} < s)$ for small s.

Recall that for any s, $\lambda_{\min} < s$ implies the existence of $\theta \in \mathbb{S}^{n-1}$ such that

$$\theta^T \frac{\mathbb{Y} D_\alpha \mathbb{Y}^T}{\|\alpha\|_2^2} \theta < s \text{ , or equivalently } \left\| \sqrt{D_\alpha} \mathbb{Y}^T \theta \right\|^2 < s \|\alpha\|_2^2 .$$

Also, if θ is such that $\left\| \frac{\sqrt{D_\alpha} \mathbb{Y}^T}{\|\alpha\|_2} \theta \right\| < \sqrt{s}$, then for any $\theta' \in \mathbb{S}^{n-1}$,

$$\left\| \frac{\sqrt{D_\alpha} \mathbb{Y}^T}{\|\alpha\|_2} \theta' \right\| < \sqrt{s} + \sqrt{\lambda_{\max}} \left\| \theta - \theta' \right\| ,$$

where $\lambda_{\max}$ is the largest eigenvalue of $\frac{\mathbb{Y} D_\alpha \mathbb{Y}^T}{\|\alpha\|_2^2}$.

We will first bound $\mathbb{P}(\lambda_{\min} < 1/2)$, using an ε-net argument. Note that for each θ, $\sqrt{D_\alpha} \mathbb{Y}^T \theta$ is distributed as $\mathcal{N}(0, D_\alpha^2)$. Consider the Euclidean metric on $\mathbb{S}^{n-1}$ and let $0 < \varepsilon < 1$. We may cover $\mathbb{S}^{n-1}$ with $\left(\frac{3}{\varepsilon} \right)^n$ balls of radius ε (see Lemma 2.3.4 in [14], for example) to achieve

$$\mathbb{P}\left(\lambda_{\min} < 1/2 \right) \leq \left(\frac{3}{\varepsilon} \right)^n \mathbb{P}\left(\left\| \mathcal{N}(0, D_\alpha^2) \right\| < \sqrt{\frac{1.1}{2} \|\alpha\|_2^2} \right) + \mathbb{P}\left(\sqrt{\lambda_{\max}} > \frac{0.1}{\sqrt{2}\varepsilon} \right).$$

$$(13.36)$$

To bound $\mathbb{P}\left(\sqrt{\lambda_{\max}} > \frac{0.1}{\sqrt{2}\varepsilon} \right)$ we will use another ε-net with $\varepsilon = \frac{1}{2}$. Along with the fact that $\|\theta - \theta'\| \leq \frac{1}{2}$ implies $\left\| \frac{\sqrt{D_\alpha} \mathbb{Y}^T (\theta - \theta')}{\|\alpha\|_2} \right\| \leq \frac{\sqrt{\lambda_{\max}}}{2}$, we may see that

$$\mathbb{P}\left(\sqrt{\lambda_{\max}} > \frac{0.1}{\sqrt{2}\varepsilon} \right) \leq 6^n \mathbb{P}\left(\left\| \sqrt{D_\alpha} \mathbb{Y}^T \theta \right\|^2 > \frac{0.01 \|\alpha\|_2^2}{4\varepsilon^2} \right)$$

$$= 6^n \mathbb{P}\left(\left\| \mathcal{N}(0, D_\alpha^2) \right\| > \sqrt{\frac{0.01}{4\varepsilon^2} \|\alpha\|_2^2} \right). \qquad (13.37)$$

But, for any $x > 0$:

$$\mathbb{P}\left(\left\| \mathcal{N}\left(0, D_\alpha^2\right) \right\| > \sqrt{x \|\alpha\|_2^2} \right) = \mathbb{P}\left(\sum_i \alpha_i^2 \chi_i^2 > x \|\alpha\|_2^2 \right),$$

where the χ_i^2 are i.i.d. Chi-squared random variables with 1 degree of freedom. Observe that $\mathbb{E}[\alpha_i^2 \chi_i^2] = \alpha_i^2$.

We may now utilize the sub-exponential tail of the χ^2 distribution and apply (13.5) with $v_i = \alpha_i^2$, noting that, by the normalization, (13.13), $\|\alpha\|_\infty = 1$. Thus, provided that $x > 3$

$$\mathbb{P}\left(\sum \alpha_i^2 \chi_i^2 > x \|\alpha\|_2^2\right)$$

$$\leq \mathbb{P}\left(\left|\sum \alpha_i^2 \chi_i^2 - \|\alpha\|_2^2\right| > (x-1)\|\alpha\|_2^2\right)$$

$$\leq 2\exp\left(-\min\left(\frac{x-1}{2}\|\alpha\|_2^2, \frac{(x-1)^2}{4}\|\alpha\|_2^2\right)\right) \leq 2\exp\left(-\|\alpha\|_2^2\right).$$
(13.38)

Substituting x for $\frac{0.01}{4\varepsilon^2}$ in (13.37) shows that when $\frac{0.01}{4\varepsilon^2} > 3$ then

$$\mathbb{P}\left(\sqrt{\lambda_{\max}} > \frac{0.1}{\sqrt{2}\varepsilon}\right) \leq 6^n \exp(-\|\alpha\|_2^2).$$

The exact same considerations as in (13.38) also show that

$$\mathbb{P}\left(\left\|\mathcal{N}(0, D_\alpha^2)\right\| < \sqrt{\frac{1.1}{2}\|\alpha\|_2^2}\right)$$

$$\leq \mathbb{P}\left(\left|\sum_i \alpha_i^2 \chi_i^2 - \|\alpha\|_2^2\right| > \frac{0.9}{2}\|\alpha\|_2^2\right)$$

$$\leq 2\exp\left(-\frac{0.9^2}{16}\|\alpha\|_2^2\right) \leq 2\exp\left(-\frac{\|\alpha\|_2^2}{20}\right).$$

Plugging the above two displays into (13.36), when ε is small enough, yields

$$\mathbb{P}(\lambda_{\min} < 1/2) \leq 2\left(\frac{3}{\varepsilon}\right)^n e^{-\frac{\|\alpha\|_2^2}{20}} + 2 \cdot 6^n e^{-\|\alpha\|_2^2} \leq 4\exp\left(\frac{3n}{\varepsilon} - \frac{\|\alpha\|_2^2}{20}\right).$$
(13.39)

For general $0 < s < 1/2$, in a similar fashion to (13.36), using an s-net gives the bound

$$\mathbb{P}\left(\lambda_{\min} < s\right) \leq \left(\frac{3}{s}\right)^n \mathbb{P}\left(\left\|\mathcal{N}(0, D_\alpha^2)\right\| < \sqrt{1.1s\|\alpha\|_2^2}\right) + \mathbb{P}\left(\sqrt{\lambda_{\max}} > 0.1/\sqrt{s}\right).$$
(13.40)

Now, $\mathcal{N}(0, D_\alpha^2)$ can be written as $D_\alpha Z_d$ where Z_d is a standard Gaussian d-dimensional vector. In [10, Proposition 2.6], it was shown that there exists universal constants $C_L, C' > 0$ such that for any $t < C'$:

$$\mathbb{P}\Big(\|D_\alpha Z\| < t\,\|D_\alpha\|_{HS}\Big) \le \exp\left(C_L \ln(t)\left(\frac{\|D_\alpha\|_{HS}}{\|D_\alpha\|_{op}}\right)^2\right) = \exp\Big(C_L \ln(t)\,\|\alpha\|_2^2\Big)$$

$$= t^{C_L\|\alpha\|_2^2},$$

with equality stemming from the facts that $\|D_\alpha\|_{HS} = \|\alpha\|_2$ and $\|D_\alpha\|_{op} = \|\alpha\|_\infty = 1$. Thus

$$\mathbb{P}\left(\Big\|\mathcal{N}(0, D_\alpha^2)\Big\| < \sqrt{1.1s\,\|\alpha\|_2^2}\right) \le 2s^{\frac{C_L}{2}\|\alpha\|_2^2}. \tag{13.41}$$

By revisiting (13.37) and replacing $\sqrt{2}\varepsilon$ with $\sqrt{s}$ we note that for small s

$$\mathbb{P}\Big(\sqrt{\lambda_{\max}} > 0.1/\sqrt{s}\Big) \le 6^n \mathbb{P}\left(\Big\|\mathcal{N}(0, D_\alpha^2)\Big\| > \sqrt{\frac{0.01}{2s}\,\|\alpha\|_2^2}\right)$$

$$\le 6^n \mathbb{P}\left(\Big|\sum_i \alpha_i^2 \chi_i^2 - \|\alpha\|_2^2\Big| > \left(\frac{0.01}{2s} - 1\right)\|\alpha\|_2^2\right).$$

And, provided that $s \le \frac{0.01}{4}$, (13.38) shows

$$\mathbb{P}(\sqrt{\lambda_{\max}} > 0.1/\sqrt{s}) \le 6^n \exp\left(-\frac{1}{2s}\left(\frac{0.01}{2} - s\right)\|\alpha\|_2^2\right) \le 6^n e^{-\frac{0.01\|\alpha\|_2^2}{4s}}. \tag{13.42}$$

By using (13.42) and (13.41) to bound (13.40) we obtain

$$\mathbb{P}(\lambda_{\min} < s) \le 2\left(\frac{3}{s}\right)^n s^{\frac{C_L}{2}\|\alpha\|_2^2} + \exp\left(2n - \frac{0.01\,\|\alpha\|_2^2}{4s}\right), \quad \forall s \le \frac{0.01}{4}.$$

We have thus shown, by combining (13.39), together with the last inequality into (13.35) and choosing ξ to be a small enough constant:

$$\frac{n}{\xi}\mathbb{P}(\lambda_{\min} < 1/2) + n\int_0^\xi \frac{1}{s}\mathbb{P}(\lambda_{\min} < s)ds$$

$$\le \frac{n}{\xi}12\exp\left(\frac{3n}{\varepsilon} - \frac{\|\alpha\|_2^2}{20}\right) + n\int_0^\xi 3^n s^{\frac{C_L}{2}(\|\alpha\|_2^2 - n - 1)} + \frac{1}{s}e^{\left(2n - \frac{0.01\|\alpha\|_2^2}{4s}\right)}ds.$$

Assuming that $\xi \leq \frac{1}{e}$ and that $\|\alpha\|_2^2 > n + 1$,

$$n \int_0^\xi 3^n s^{\frac{C_L}{2}(\|\alpha\|_2^2 - n - 1)} ds \leq n 3^n \xi^{\frac{C_L}{2}(\|\alpha\|_2^2 - n)} \leq n e^{\frac{C_L}{2}(n - \|\alpha\|_2^2) + 2n},$$

$$n \int_0^\xi \frac{1}{s} e^{\left(2n - \frac{0.01\|\alpha\|_2^2}{4s}\right)} ds \leq n e^{2n} \int_0^\xi e^{-\frac{0.01\|\alpha\|_2^2}{8s}} ds \leq n e^{2n} \xi e^{-\frac{0.01\|\alpha\|_2^2}{8\xi}}.$$

To obtain the desired result we observe that if $n^3 \left(\frac{\|\alpha\|_4}{\|\alpha\|_2}\right)^4 \to 0$ then $\left(\frac{\|\alpha\|_2}{\|\alpha\|_4}\right)^4 >> n^3$. the inequality $\|\alpha\|_2^2 \geq \left(\frac{\|\alpha\|_2}{\|\alpha\|_4}\right)^{4/3}$ implies $\|\alpha\|_2^2 >> n$, which shows the existence of a constant $C > 0$ for which

$$\mathbb{E}_{\mathbb{Y}}\left[\ln \det\left(\frac{1}{\|\alpha\|_2^2}\mathbb{Y} D_\alpha \mathbb{Y}^T\right) \mathbb{1}_{\{\lambda_{\min} < 1/2\}}\right] < n \exp(-C\|\alpha\|_2^2).$$

$\square$

To finish the prove of Theorem 13.2(b) we must now deal with TV $\left(K_p(M(n)), H_{p,\alpha}(M(n))\right)$.

Lemma 13.12 *Assume* $n^3 \left(\frac{\|\alpha\|_4}{\|\alpha\|_2}\right)^4 \to 0$, *then* TV$(K_p(M(n)), H_{p,\alpha}(M(n))) \to 0$.

Proof First, we again pass to relative entropy using (13.8), Pinsker's inequality:

$$\text{TV}(K_p(M(n)), H_{p,\alpha}(M(n)) \leq \sqrt{\text{Ent}\left[K_p(M(n))||H_{p,\alpha}(M(n))\right]}.$$

We note that both $K_p(M(n))$ and $H_{p,\alpha}(M(n))$ are simply Bernoulli matrices. The entries of $K_p(M(n))$ are i.i.d. Bernoulli(p), while the entries of $H_{p,\alpha}(M(n))$ are i.i.d. Bernoulli(p') where $p' = \Phi^{-1}\left(\frac{t_{p,\alpha}}{\|\alpha\|_2}\right)$. Defining Ent$[p||p'] :=$ Ent$\left[\text{Bernoulli}(p)||\text{Bernoulli}(p')\right]$ and using the chain rule (13.9) for relative entropy yields

$$\text{Ent}\left[K_p(M(n))||H_{p,\alpha}(M(n))\right] \leq n^2 \text{Ent}[p||p'].$$

One may verify that

$$\lim_{p' \to p} \frac{\text{Ent}[p||p']}{(p - p')^2} = \lim_{p' \to p} \frac{p \ln(\frac{p}{p'}) + (1 - p)\ln(\frac{1-p}{1-p'})}{(p - p')^2} = \frac{1}{2p - 2p^2}.$$

So, $\frac{\text{Ent}[p||p']}{(p-p')^2}$ is a continuous function on $(0, 1) \times (0, 1)$ and is bounded on every compact subset of its domain. Thus, there exists a constant C_p, depending on p such that

$$\text{Ent}[p||p'] \leq C_p(p - p')^2.$$

By Lemma 13.4, $|p - p'| \leq 3 \left(\frac{||\alpha||_3}{||\alpha||_2} \right)^3$, which affords the bound

$$\text{Ent}[p||p'] \leq 9C_p \left(\frac{||\alpha||_3}{||\alpha||_2} \right)^6.$$

But now, by Cauchy-Schwartz's inequality, $||\alpha||_3^3 = \sum_i \alpha_i \alpha_i^2 \leq \sqrt{||\alpha||_2^2 ||\alpha||_4^4}$. Combining all of the above

$$\text{TV} \left(K_p(M(n)), H_{p,\alpha}(M(n)) \right)^2 \leq \text{Ent}[K_p(M(n))||H_{p,\alpha}(M(n))]$$

$$\leq n^2 \text{Ent}(p||p') \leq 9C_p n^2 \left(\frac{||\alpha||_3}{||\alpha||_2} \right)^6 \leq n^2 \frac{||\alpha||_4^4 ||\alpha||_2^2}{||\alpha||_2^6} < n^3 \left(\frac{||\alpha||_4}{||\alpha||_2} \right)^4.$$

$\square$

Acknowledgements We would like to thank the anonymous referee for carefully reading this paper and for the thoughtful comments which helped improve the overall presentation.

References

1. V. Bentkus, A lyapunov-type bound in R^d. Theory Probab. Its Appl. **49**(2), 311–323 (2005)
2. S. Bubeck, S. Ganguly, Entropic CLT and phase transition in high-dimensional Wishart matrices. Int. Math. Res. Not. **2018**(2), 588–606 (2016)
3. S. Bubeck, J. Ding, R. Eldan, M.Z. Rácz, Testing for high-dimensional geometry in random graphs. Random Struct. Algoritm. **49**(3), 503–532 (2016)
4. T.M. Cover, J.A. Thomas, *Elements of Information Theory* (John Wiley & Sons, Hoboken, 2012)
5. J. Duchi, *Derivations for Linear Algebra and Optimization* (Berkeley, 2007). http://www.cs. berkeley.edu/~jduchi/projects/generalnotes.pdf
6. R. Durrett, *Probability: Theory and Examples* (Cambridge University Press, Cambridge, 2010)
7. M.L. Eaton, Chapter 8: The wishart distribution, in *Multivariate Statistics*, *Lecture Notes–Monograph Series*, vol. 53 (Institute of Mathematical Statistics, Beachwood, 2007), pp. 302–333
8. P. Erdős, A. Rényi, On the evolution of random graphs. Publ. Math. Inst. Hungar. Acad. Sci **5**, 17–61 (1960)
9. L. Hörmander, *The Analysis of Linear Partial Differential Operators I: Distribution Theory and Fourier Analysis* (Springer, Berlin, 2015)

10. R. Latala, P. Mankiewicz, K. Oleszkiewicz, N. Tomczak-Jaegermann, Banach-Mazur distances and projections on random subgaussian polytopes. Discret. Comput. Geom. **38**(1), 29–50 (2007)
11. I. Nourdin, G. Peccati, *Normal Approximations with Malliavin Calculus: from Stein's Method to Universality*, vol. 192 (Cambridge University Press, Cambridge, 2012)
12. V.V. Petrov, *Limit Theorems of Probability Theory* (Oxford University Press, Oxford, 1995)
13. N.G. Shephard, From characteristic function to distribution function: a simple framework for the theory. Economet. Theor. **7**(4), 519–529 (1991)
14. T. Tao, *Topics in Random Matrix Theory*, vol. 132 (American Mathematical Society, Providence, 2012)
15. R. Vershynin, Introduction to the non-asymptotic analysis of random matrices, in *Compressed Sensing* (Cambridge University Press, Cambridge, 2012), pp. 210–268

Chapter 14
On the Ekeland-Hofer-Zehnder Capacity of Difference Body

Efim Gluskin

Symplectic capacities form a wide class of the symplectic invariants. The first invariant of such kind was introduced by Gromov [10]. The notion symplectic capacity itself was introduced in the work of Ekeland and Hofer [5] where some different symplectic invariant was constructed via Hamiltonian dynamics. Shortly after this a variety of symplectic capacities was constructed reflecting different geometrical and dynamical properties. All these quantities play an important role in symplectic topology nowadays and are closely related with symplectic embedding obstructions on the one hand and with existence and behaviour of periodic orbits of Hamiltonian systems on the other. In [5] and [11], two symplectic capacities, known as the Ekeland-Hofer and Hofer-Zehnder capacities, were defined using a variational principle for the classical action functional from Hamiltonian dynamics. Moreover it was proved (see [5, 11, 15]), that for a smooth convex body this two capacities coincide and are given by the minimal action over all closed characteristics on the boundary of the body. It is possible to extend this definition to the class of all convex bodies (see e.g. [3]). In the following, we shall refer to the (coinciding) Ekeland-Hofer and Hofer-Zehnder capacities on the class of all convex bodies as the Ekeland-Hofer-Zehnder capacity and denote it C_{EHZ}.

Another important example of a symplectic capacity which is closely related to Gromov's non-squeezing theorem [10] is the cylindric capacity $\overline{C}$. It is the largest among all capacities. In particular, for any subset $V \subset \mathbb{R}^{2n}$ one has $C_{\mathrm{EHZ}}(V) \leq \overline{C}(V)$. In [7] the linearized version $C_{Sp(2n)}$ of the cylindric capacity was introduced. In general, it is much easier to compute $C_{Sp(2n)}$ rather that $\overline{C}$ but it is invariant only

E. Gluskin (✉)

School of Mathematical Sciences, Tel Aviv University, Tel Aviv, Israel

e-mail: gluskin@tauex.tau.ac.il

© Springer Nature Switzerland AG 2020

B. Klartag, E. Milman (eds.), *Geometric Aspects of Functional Analysis*,
Lecture Notes in Mathematics 2256,
https://doi.org/10.1007/978-3-030-36020-7_14

with respect to the linear symplectic transformations and translations. It is proved (see [7]) that for any centrally symmetric convex body $V \subset \mathbb{R}^{2n}$ one has:

$$\frac{1}{4} C_{Sp(2n)}(V) \leq C_{EHZ}(V) \leq \overline{C}(V) \leq C_{Sp(2n)}(V).$$

Recently [6] an example of a convex body $W \subset \mathbb{R}^{2n}$ satisfying

$$C_{EHZ}(W) \leq \frac{C}{\ln(n+1)} C_{Sp(2n)}(W)$$

(where C is a universal constant) was constructed. This example is based on the fact that for any convex body $V \subset \mathbb{R}^{2n}$ one has

$$C_{Sp(2n)}(V) \leq C_{Sp(2n)}(V + (-V)) \leq 6 C_{Sp(2n)}(V).$$

but for some convex body $W \subset \mathbb{R}^{2n}$ one has

$$C_{EHZ}(W + (-W)) \geq C \ln(n+1) C_{EHZ}(W)$$

with some positive universal constant C. In some sense, the last inequality is in contrast with the following difference body volume bound due to Rogers and Shephard [14]:

$$\mathrm{Vol}\big(V + (-V)\big)^{1/2n} \leq \left(\binom{4n}{2n} \mathrm{Vol}(V) \right)^{1/2n} \leq 4 \big(\mathrm{Vol}(V)\big)^{1/2n},$$

for any convex body $V \subset \mathbb{R}^{2n}$. In the current note we investigate the relation between Ekland-Hofer-Zehnder capacity of a general convex body V and its of difference body $V + (-V)$ and show that the example from [6] is close to optimal. Namely the following is true

Theorem 14.1 *Let $K \subset \mathbb{R}^{2n}$ be a convex body. Then the following inequality holds*

$$C_{EHZ}(K) \leq C_{EHZ}(K + (-K)) \leq C_1 \ln(n+1) C_{EHZ}(K),$$

where C_1 is an absolute constant.

The proof of this theorem is based on the following inequality

Theorem 14.2 *Let $A = (a_{ij})_{i,j=0}^{N-1}$ be the $N \times N$ matrix of rank k. Then with some universal constant C_2, $0 < C_2 < \infty$ the following inequality holds:*

$$\left| \sum_{i=0}^{N-1} \sum_{j=0}^{N-1} a_{ij} \, \mathrm{sign}(i-j) \right| \leq C_2 \ln(k+1) \sup \left\{ \left| \sum_{i=0}^{N-1} \sum_{j=0}^{N-1} a_{ij} x_i y_j \right| \; : \; |x_i|, \; |y_j| \right.$$

$$\left. \leq 1 \, , \; i, j = 0, \ldots, N-1 \right\}$$

This inequality is sharp (see proposition bellow) up to a multiplicative constant.

Organization of the Paper In Sect. 14.1 we reduce Theorem 14.1 to Theorem 14.2. Section 14.2 contains some preliminaries for the proof of theorem 14.2 which is presented in Sect. 14.3.

Notations We use the following (mostly standard) notations. For any real a we define

$$\text{sign}(a) := \begin{cases} +1 \text{ if } a > 0 \\ 0 \quad \text{if } a = 0 \\ -1 \text{ if } a < 0 \end{cases}$$

For a vector $X = (x_0, x_1, \ldots, x_{n-1})$, $X \in \mathbb{R}^n$, we write $X > 0$ if $x_j > 0$ for all $j = 0, 1, \ldots, n - 1$. $< \cdot, \cdot >$ is the standard inner product in $\mathbb{R}^n$. We equip the space $\mathbb{R}^{2n} = \mathbb{R}^n \oplus \mathbb{R}^n$ with the usual complex structure $J : \mathbb{R}^{2n} \to \mathbb{R}^{2n}$ given in coordinates by $J(x, y) = (-y, x)$. The space ℓ_p^n, $1 \le p \le \infty$, is $\mathbb{R}^n$ equipped with the norm

$$\|x\|_{\ell_p^n} = \left(\sum_{k=0}^{n-1} |x_k|^p \right)^{1/p} , \quad 1 \le p < \infty$$

and

$$\|x\|_{\ell_\infty^n} = \max \{|x_k| \ : \ k = 0, 1, \ldots, n - 1\},$$

for $x = (x_0, x_1, \ldots, x_{n-1}) \in \mathbb{R}^n$.

$$B_p^n = \left\{ x \in \ell_p^n \ : \ \|x\|_{\ell_p^n} \le 1 \right\},$$

its closed unit ball. For an absolutely continuous function γ defined on an interval we denote by γ' its derivative.

As usual for a Banach space X we denote by $\| \cdot \|_X$ its norm; X^* is its dual space and for two Banach spaces X and Y; $\mathcal{L}(X, Y)$ stands for the Banach space consisting of all bounded linear operator from X to Y equipped with the norm

$$\|T\|_{\mathcal{L}(X,Y)} = \sup \{\|Tx\|_Y \ : \ x \in X, \ \|x\|_X \le 1\}.$$

Given $T \in \mathcal{L}(X, Y)$ one denotes by $T^* \in \mathcal{L}(Y^*, X^*)$ its adjoins operator. We don't differ between an operator in finite dimensional spaces and its matrix. For $n \times n$ matrix $A = (a_{ij})_{i,j=0}^{n-1}$ its trace is defined as follows

$$\text{Tr } A = \sum_{i=0}^{n-1} a_{ii}.$$

More generally for an operator $T \in \mathcal{L}(X)$ of the finite rank

$$T x = \sum_{j=0}^{k} f_j(x) e_j \quad , \quad e_j \in X, \quad f_j \in X^*, \tag{14.0.1}$$

its trace defined by the formula

$$\mathrm{Tr}\ T = \sum_{j=0}^{k} f_j(e_j).$$

It is well known that trace doesn't depend on representation (14.0.1) and for two operators $T, S \in \mathcal{L}(X)$, rank $T < \infty$ one has

$$\mathrm{Tr}\ ST = \mathrm{Tr}\ TS$$

Recall that the polar body K^0 of a set $K \subset \mathbb{R}^n$ is defined as follows:

$$K^0 = \left\{ x \in \mathbb{R}^n \ : \ <x, y> \le 1 \ , \ \forall y \in K \right\}.$$

Certainly K^0 is a closed convex set and if 0 is an interior point of K then K^0 is bounded. It is well known that if K is a closed convex body and 0 is its interior point, then $\left(K^0\right)^0 = K$ (bipolar theorem).

14.1 Proof of Theorem 14.1

Let $K \subset \mathbb{R}^{2n}$ be a convex body with smooth boundary ∂K. A smooth closed curve γ on ∂K is called a closed characteristic of ∂K if γ' is parallel to $J\bar{n}$, where $\bar{n}$ is the outer normal to ∂K. It is known [13, 16] that for any smooth convex (or more generally starshaped) body, there exists at least one closed characteristic. In these terms the Ekeland-Hofer-Zehnder capacity of a smooth convex body K can be defined as follows:

$$C_{\mathrm{EHZ}}(K) = \frac{1}{2} \inf \left| \int_0^1 \langle \gamma(t), J\gamma'(t) \rangle \, dt \right|,$$

when infimum is taken over all closed characteristic $\gamma : [0, 1] \to \partial K$ on ∂K.

   Using Clarke's dual action principle (see [4]), Artstein-Avidan and Ostrover [2] gave certain dual description of the Ekeland-Hofer-Zehnder capacity. Recently, using similar methods, a slightly different description was obtained by Abbondandolo and Majer [1]. Namely let $V \subset \mathbb{R}^{2n}$ be a convex body. By $\mathcal{E}'_V$ we denote the set of all absolutely continuous vector-functions $\gamma : [0, 1] \to \mathbb{R}^{2n}$, such that

$\gamma(0) = \gamma(1)$ and $\gamma'(t) \in V$ for almost every $t \in [0, 1]$. Let $K \subset \mathbb{R}^{2n}$ be a convex body such that $0 \in \text{Int}(K)$, then its Ekeland-Hofer-Zehnder capacity of K can be computed by the following formula (see [1])

$$C_{\text{EHZ}}(K) = \frac{1}{2 \displaystyle\sup_{\gamma \in \mathcal{E}'_{K^0}} \left| \int_0^1 \langle \gamma'(t), J\gamma(t) \rangle \, dt \right|}. \tag{14.1.1}$$

Let's rewrite this equality in a slightly different form. Denote $\mathcal{E}_V$ the set of all measurable absolutely integrable vector functions $\varphi : [0, 1] \to \mathbb{R}^{2n}$ such that $\varphi(t) \in V$ a.e. and

$$\int_0^1 \varphi(t) dt = 0.$$

And let $\mathcal{E}_V^N$ be a set of all functions $\varphi \in \mathcal{E}_V$ such that φ is a constant on each interval

$$\left[\frac{k}{N}, \frac{k+1}{N} \right), k = 0, 1, \ldots, N - 1.$$

Certainly the set $\displaystyle\bigcup_{N=1}^{\infty} \mathcal{E}_V^N$ is dense in $\mathcal{E}_V$ in $L_2 \left([0, 1], \mathbb{R}^{2n} \right)$ topology. It is clear that $\gamma \in \mathcal{E}'_V$ iff $\gamma' \in \mathcal{E}_V$ and one has

$$\gamma(t) = \gamma(0) + \int_0^t \gamma'(s) ds.$$

Going back to (14.1.1) we obtain

$$\frac{1}{2C_{\text{EHZ}}(K)} = \sup_{\varphi \in \mathcal{E}_{K^0}} \left| \int_0^1 \left(\int_0^t \langle \varphi(t), J\varphi(s) \rangle \, ds \right) dt \right|.$$

By the denseness of $\displaystyle\bigcup_{N=1}^{\infty} \mathcal{E}_{K^0}^N$ in $\mathcal{E}_{K^0}$ and antisymmetry of J one has

$$\frac{1}{C_{\text{EHZ}}(K)} = \sup_N \sup_{\varphi \in \mathcal{E}_{K^0}^N} \left| \int_0^1 \int_0^1 \langle \varphi(t), J\varphi(s) \rangle \, \text{sign}(t - s) dt ds \right|. \tag{14.1.2}$$

The following simple consequence of this representation was firstly proved in [7] by a different method.

Corollary 14.1 *Let $V \subset \mathbb{R}^{2n}$ be a convex body, symmetric with respect to the origin* $(V = -V)$. *Then one has*

$$\frac{1}{\max\limits_{x,y \in V^0} \langle x, Jy \rangle} \leq C_{\text{EHZ}}(V) \leq \frac{4}{\max\limits_{x,y \in V^0} \langle x, Jy \rangle}.$$

Proof Let's denote by α

$$\alpha = \max\limits_{x,y \in V^0} \langle x, Jy \rangle = \max\limits_{x,y \in V^0} |\langle x, Jy \rangle|.$$

Then for any $\varphi \in \mathcal{E}_{V^0}^N$ one has

$$|\langle \varphi(t), J\varphi(s) \rangle| \leq \alpha, \quad \text{for every } t, s \in [0, 1].$$

So the following inequality holds

$$\left| \int_0^1 \int_0^1 \langle \varphi(t), J\varphi(s) \rangle \, \text{sign}(t - s) dt ds \right| \leq \alpha.$$

Taking supremum over all N and $\varphi \in \mathcal{E}_{V^0}^N$ one gets the left-hand inequality. To prove the right hand inequality take $u, v \in V^0$ such that

$$\langle u, Jv \rangle = \alpha$$

and define the function ψ as follows

$$\begin{aligned}
\psi(t) &= u & \text{for } t \in \left[0, \tfrac{1}{4}\right) \\
\psi(t) &= v & \text{for } t \in \left[\tfrac{1}{4}, \tfrac{1}{2}\right) \\
\psi(t) &= -u & \text{for } t \in \left[\tfrac{1}{2}, \tfrac{3}{4}\right) \\
\psi(t) &= -v & \text{for } t \in \left[\tfrac{3}{4}, 1\right]
\end{aligned}$$

It is clear that $\psi \in \mathcal{E}_{V^0}^4$ and by (14.1.2) one has

$$\frac{1}{C_{\text{EHZ}}(V)} \geq \left| \int_0^1 \int_0^1 \langle \psi(t), J\psi(s) \rangle \, \text{sign}(t - s) dt ds \right| = \alpha/4.$$

$\square$

Proof of Theorem 14.1 Without loss of generality let us assume that $0 \in \text{Int}(K)$. All symplectic capacities are monotonic with respect an inclusion, by definition (see e.g. [5]). In particular one has $C_{\text{EHZ}}(K) \leq C_{\text{EHZ}}(K + (-K))$. Now by (14.1.2) it is enough to prove that for a given N and a given $\varphi \in \mathcal{E}_{K^0}^N$ one has

$$\left| \int_0^1 \int_0^1 \langle \varphi(t), J\varphi(s) \rangle \operatorname{sign}(t-s) dt ds \right| \leq C_1 \frac{\ln(2n)}{C_{\text{EHZ}}(K + (-K))}. \qquad (14.1.3)$$

Denote by $v_j = \varphi(\frac{j}{N}) \subset \mathbb{R}^{2n}$, $j = 0, 1, \ldots, N - 1$. Since $\varphi \in \mathcal{E}_{K^0}^N$ one has $\varphi_j(t) = v_j$ for any $t \in \left[\frac{j}{N}, \frac{j+1}{N} \right)$,

$$v_j \in K^0 \qquad j = 0, 1, \ldots, N - 1$$

$$\sum_{j=0}^{N-1} v_j = N \int_0^1 \varphi(t) dt = 0. \qquad (14.1.4)$$

For a given $x = (x_0, x_1, \ldots, x_{N-1}) \in \mathbb{R}^N$, denote by $W(x)$ the vector

$$W(x) = \sum_{j=0}^{N-1} x_j v_j.$$

Let us verify that for $x \in \frac{1}{2N} B_\infty^N$ we have $W(x) \in K^0 \cap (-K^0)$. Indeed, in this case one has

$$\alpha_j := x_j + \frac{1}{2N} \geq 0, \qquad j = 0, 1, \ldots, N - 1;$$

$$\sum_{j=0}^{N-1} \alpha_j \leq 1$$

and by (14.1.4)

$$W(x) = \sum_{j=0}^{N-1} \alpha_j v_j \in \operatorname{conv}\{0, v_0, v_1, \ldots, v_{N-1}\} \subset K^0.$$

Similarly denoting $\beta_j := \frac{1}{2N} - x_j \geq 0$, we get $\sum_{j=0}^{N-1} \beta_j \leq 1$ and hence

$$-W(x) = \sum_{j=0}^{N-1} \beta_j v_j \in K^0.$$

It follows that

$$W(x) \in K^0 \cap (-K^0) \subset 2(K + (-K))^0.$$

By Corollary 14.1 one has

$$
\begin{aligned}
\frac{1}{C_{\mathrm{EHZ}}(K + (-K))} &\geq \frac{1}{4} \sup_{u,v \in (K+(-K))^0} |\langle v, Jv \rangle| \geq \\
&\geq \frac{1}{16} \left(\frac{1}{2N} \right)^2 \sup_{x,y \in B_\infty^N} |\langle W(x), JW(y) \rangle|.
\end{aligned}
\tag{14.1.5}
$$

Denote by $A = \left(a_{ij}\right)_{i,j=0}^{N-1}$ the $N \times N$ matrix with entries

$$a_{ij} = \langle v_i, Jv_j \rangle$$

It is clear that $rank(A) \leq 2n$. By the definition of $W(x)$ one has

$$\langle W(x), JW(y) \rangle = \langle Ax, y \rangle. \tag{14.1.6}$$

Denote by I the left hand side of (14.1.3). By the definition of a_{ij} one has

$$I = \frac{1}{N^2} \left| \sum_{i=0}^{N-1} \sum_{j=0}^{N-1} a_{ij} \, \mathrm{sign}(i - j) \right|.$$

By Theorem 14.2 we get

$$I \leq C_2 \ln \left(1 + rank(A)\right) \frac{1}{N^2} \cdot \sup_{x,y \in B_\infty^N} \langle Ax, y \rangle.$$

We conclude by (14.1.5) and (14.1.6).   □

14.2 Preliminaries for the proof of Theorem 14.2

To prove Theorem 14.2 we use some basic facts from the operator theory (see e.g. [8]). Let $T \in \mathcal{L}(H_1, H_2)$ be a compact operator which acts from one Hilbert space H_1 to another Hilbert space H_2. Let $\lambda_0 \geq \lambda_1 \geq \ldots$ be the sequence of all eigenvalues $\lambda_k = \lambda_k(T^*T)$ of the operator T^*T, taking account of their multiplicities. The k-singular value of T is defined as

$$s_k(T) = \sqrt{\lambda_k(T^*T)}$$

An operator $S \in \mathcal{L}(H)$ is called normal if $S^*S = SS^*$. For such operators $\lambda_k(S^*S) = |\lambda_k(S)|^2$ and so the sequence of the singular values of S coincides with the sequence of the absolute values of its eigenvalues. The following is a variant of the minimax principle

$$s_k(T) = \inf\{\|T - A\| : A \in \mathcal{L}(H_1, H_2), \text{rank } A \leq k\}.$$

Consequently

(a) if a sequence of the compact operators $T_n \in \mathcal{L}(H_1, H_2)$ converges to some operator T in the operator norm, then for any $k = 0, 1, 2, \ldots$ one has

$$s_k(T_n) \xrightarrow[n \to \infty]{} s_k(T).$$

(b) If H_0, H_1, H_2, H_3 are Hilbert spaces and $R \in \mathcal{L}(H_0, H_1)$, $T \in \mathcal{L}(H_1, H_2)$, $S \in \mathcal{L}(H_2, H_3)$ and T is a compact operator then STR is also a compact one and one has

$$s_k(STR) \leq \|S\| s_k(T) \|R\|.$$

Using spectral decomposition one gets (see e.g. [9], II.2.1)

$$\sum_{j=0}^{k-1} s_j(T) = \max |\text{Re Tr } TA|, \tag{14.2.1}$$

where maximum is taken over all operators $A \in \mathcal{L}(H_2, H_1)$ with rank $A \leq k$ and $\|A\| \leq 1$. Moreover if $H_1 = H_2$ and $T^* = \pm T$ then the maximum on the right hand side of (14.2.1) is attended on the operator A satisfying $A^* = \pm A$ correspondingly. A compact operator $T \in \mathcal{L}(H_1, H_2)$ is of Matsaev class $\mathbf{S}_\Omega$ if

$$\|T\|_{\mathbf{S}_\Omega} = \sup_{k \in \mathbb{N}} \frac{1}{\ln(1 + k)} \sum_{j=0}^{k-1} s_j(T) < \infty.$$

$\mathbf{S}_\Omega$ is a linear subspace of $\mathcal{L}(H_1, H_2)$ and $\| \cdot \|_{\mathbf{S}_\Omega}$ is one of its equivalent norms. Moreover, $\mathbf{S}_\Omega$ is a symmetrically normed ideal of operators. Particularly, for $R \in \mathcal{L}(H_0, H_1)$ and $S \in \mathcal{L}(H_2, H_3)$ one has

$$\|STR\|_{\mathbf{S}_\Omega} \leq \|S\| \|T\|_{\mathbf{S}_\Omega} \|R\|.$$

For a function $K \in L_2([0, 1] \times [0, 1])$ the Hilbert-Schmidt operator with the kernel K is defined by the following formula:

$$(Tf)(x) = \int_0^1 K(x, y)f(y)dy,$$

for $f \in L_2[0, 1]$.

It is well known (see e.g. [8]) that such operator is bounded. Moreover T is compact and one has

$$\|T\|_{\mathcal{L}(L_2[0,1])} \leq \|K\|_{L_2([0,1]^2)}.$$

For a given function $g \in L_2[0, 1]$ we define the operator $T_g \in \mathcal{L}(L_2[0, 1])$ by the following formula

$$\left(T_g f\right)(x) = \int_0^1 g(x)\,\mathrm{sign}(x - y)g(y)f(y)dy, \qquad f \in L_2[0, 1].$$

The following fact is well known to the experts (see e.g. [9]). For the reader convenience we reproduce its proof here.

Lemma 14.1 *For any function $g \in L_2[0, 1]$ the singular numbers of the operator T_g are bounded by*

$$s_k(T_g) \leq \frac{C_3}{k+2}\|g\|_{L_2[0,1]}^2$$

Where $0 < C_3 < \infty$ is some absolute constant. Particularly, $T_g \in \mathbf{S}_\Omega$ and

$$\|T_g\|_{\mathbf{S}_\Omega} \leq C_3\|g\|_{L_2[0,1]}^2.$$

Proof Denote by M the following multiplicator operator

$$(Mf)(x) = \mathrm{sign}\,(g(x))\,f(x).$$

It is clear that $M \in \mathcal{L}(L_2[0, 1])$ with $\|M\| = 1$ and one has $T_g = MT_{|g|}M$ and $T_{|g|} = M^* T_g M^*$. It follows

$$s_k(T_g) = s_k(T_{|g|}) \quad , \quad k = 0, 1, \dots$$

and so we can suppose that $g \geq 0$. Next, if a sequence of functions $g_n \in L_2[0, 1]$ converges to g in $L_2[0, 1]$, then the sequence of the kernels

$$K_n(x, y) = g_n(x)\,\mathrm{sign}(x - y)g_n(y)$$

converges in $L_2\left([0, 1]^2\right)$ sense to the kernel

$$K(x, y) = g(x)\,\mathrm{sign}(x - y)g(y).$$

Therefore $T_{g_n} \rightarrow Tg$ in the Hilbert Schmidt norm and in particular in the $\mathcal{L}(L_2[0, 1])$ norm. Now by the density of $C^\infty([0, 1])$ in $L_2[0, 1]$ it is enough to prove the Lemma for $g \in C^\infty([0,1])$ and $g > 0$. Let us point out that $T_g^* = -T_g$, in particular T_g is a normal operator. Hence $s_k(T_g) = |\lambda_k(T_g)|$. Let us find the eigenvalues $\lambda_k(T_g)$. If $\lambda \neq 0$ is an eigenvalue and f is the corresponding eigenfunction, then one has

$$
\begin{aligned}
f(x) &= \tfrac{1}{\lambda}(T_g f)(x) \\
&= \tfrac{1}{\lambda}\left[\int_0^x g(x)g(y)f(y)\,dy - \int_x^1 g(x)g(y)f(y)\,dy\right].
\end{aligned}
\tag{14.2.2}
$$

The right-hand of this equality is a continuous function, so $f \in C[0,1]$. By the Newton-Leibniz's formula we have $f \in C^1[0,1]$. Denote by φ the function

$$
\varphi(x) = \frac{f(x)}{g(x)}.
$$

Since $g > 0$ and $f, g \in C^1[0,1]$ one has $\varphi \in C^1[0,1]$. By (14.2.2) one has

$$
\varphi(x) = \frac{1}{\lambda}\left[\int_0^x g^2(y)\varphi(y)\,dy - \int_x^1 g^2(y)\varphi(y)\,dy\right].
\tag{14.2.3}
$$

By the Newton-Leibniz formula one gets

$$
\varphi'(x) = \frac{2}{\lambda}g^2(x)\varphi(x).
$$

Therefore

$$
\varphi(x) = C\exp\left(\frac{2}{\lambda}\int_0^x g^2(y)\,dy\right)
$$

with some constant C.

From (14.2.3) one sees that

$$
\varphi(0) = -\varphi(1).
$$

Hence

$$
1 = \exp(0) = -\exp\left(\frac{2}{\lambda}\|g\|_{L_2[0,1]}^2\right).
$$

That is

$$
\frac{2}{\lambda}\|g\|_{L_2[0,1]}^2 = i\,(2\ell + 1)\,\pi, \qquad \ell \in \mathbb{Z}
$$

In other words all the non-zero eigenvalues of T_g have the form

$$\lambda = \frac{2}{i\,(2\ell + 1)\,\pi}\,\|g\|^2_{L_2[0,1]}.$$

Goes back to (14.2.2) it is easy to check that all such numbers are indeed eigenvalues of T_g. Hence

$$s_k(T_g) = \begin{cases} \dfrac{2}{(k+1)\pi}\,\|g\|^2_{L_2[0,1]} & k \text{ is even} \\[2ex] \dfrac{2}{k\pi}\,\|g\|^2_{L_2[0,1]} & k \text{ is odd} \end{cases}$$

$\square$

For a given vector $\Lambda = (\lambda_0, \ldots, \lambda_{N-1})$ we denote by D_Λ the diagonal operator in ℓ_2^N:

$$D_\Lambda x = (\lambda_0 x_0, \lambda_1 x_1, \ldots, \lambda_{N-1} x_{N-1}),$$

where $x = (x_0, \ldots, x_{N-1}) \in \ell_2^N$.

In this notation we have the following consequence of the Lemma

Corollary 14.2 *Let* $S = (s_{ij})_{i,j=0}^{N-1}$ *be* $N \times N$ *matrix with the entries* $s_{ij} = \text{sign}(i - j)$. *Then for any vector* $\Lambda \in \ell_2^N$ *one has*

$$\|D_\Lambda S D_\Lambda\|_{\mathbf{S}_\Omega} \le C_3 \|\Lambda\|^2_{\ell_2^N}.$$

Proof Denote by σ the operator from ℓ_2^N to $L_2[0, 1]$ defined as follows:

$$(\sigma x)(t) = \begin{cases} x_j \sqrt{N} \text{ for } t \in \left[\frac{j}{N}, \frac{j+1}{N}\right), & j = 0, 1, \ldots, N - 1 \\ 0 \quad \text{for } t = 1 \end{cases},$$

where $x = (x_0, x_1, \ldots, x_{N-1}) \in \ell_2^N$. The operator σ is an isometric embedding of ℓ_2^N in $L_2[0, 1]$, hence one has $\|\sigma\| = 1$. Let g be $g = \sigma \Lambda$. We have

$$D_\Lambda S D_\Lambda = \sigma^* T_g \sigma.$$

From ideal property of Matsaev class and Lemma 14.1 we get

$$\|D_\Lambda S D_\Lambda\|_{S_\Omega} \le \|\sigma^*\| \|T_g\|_{S_\Omega} \|\sigma\|$$
$$\le C_3 \|g\|^2_{L_2[0,1]}$$
$$= C_3 \|\Lambda\|^2_{\ell_2^N}.$$

$\square$

We use the following form of the Grothedieck's Theorem (see [12])

Lemma 14.2 *Let T be a linear operator from ℓ_∞^N to ℓ_1^M. Then there exist two vectors $\Lambda_0 \in \ell_2^N$ and $\Lambda_1 \in \ell_2^M$ such that $\Lambda_j > 0$, $j = 0, 1$,*

$$\|\Lambda_0\|_{\ell_2^N} \leq 1 \quad , \quad \|\Lambda_1\|_{\ell_2^M} \leq 1$$

and one has

$$\|D_{\Lambda_1}^{-1} T D_{\Lambda_0}^{-1}\|_{\mathcal{L}(\ell_2^N, \ell_2^M)} \leq C_4 \|T\|_{\mathcal{L}(\ell_\infty^N, \ell_1^N)}.$$

The following corollary is well known to the experts

Corollary 14.3 *Let T be a linear operator from ℓ_∞^N to ℓ_1^N, then there exists a vector $\Lambda \in \ell_2^N$, such that $\Lambda > 0$, $\|\Lambda\|_{\ell_2^N} \leq 1$ and one has*

$$\|D_\Lambda^{-1} T D_\Lambda^{-1}\|_{\mathcal{L}(\ell_2^N)} \leq 4C_4 \|T\|_{\mathcal{L}(\ell_\infty^N, \ell_1^N)}.$$

Proof It is enough to put $\Lambda = \frac{1}{2}(\Lambda_0 + \Lambda_1)$, where Λ_0, Λ_1 are the vectors from Lemma 14.2. $\qquad\square$

14.3   Proof of Theorem 14.2

Proof of Theorem 14.2 By Corollary 14.3 there exists a vector $\Lambda \in \ell_2^N$, $\Lambda > 0$ such that $\|\Lambda\|_{\ell_2^N} \leq 1$ and

$$\|D_\Lambda^{-1} A D_\Lambda^{-1}\|_{\mathcal{L}(\ell_2^N)} \leq 4C_4 \|A\|_{\mathcal{L}(\ell_\infty^N, \ell_1^N)}.$$

Denote by R the operator $R = D_\Lambda^{-1} A D_\Lambda^{-1}$. Let us point out that rank $R =$ rank $A = k$ and

$$\|R\|_{\mathcal{L}(\ell_2^N)} \leq 4C_4 \|A\|_{\mathcal{L}(\ell_\infty^N, \ell_1^N)}.$$

Recall that S is the matrix with the entries $s_{ij} = \text{sign}(i - j)$. Then using (14.2.1) one gets

$$\left| \sum_{i=0}^{N-1} \sum_{j=0}^{N-1} a_{ij} \, \text{sign}(i - j) \right| = |\text{Tr } SA|$$
$$= |\text{Tr } SD_\Lambda R D_\Lambda|$$
$$= |\text{Tr } D_\Lambda S D_\Lambda R|$$
$$\leq \|R\|_{\mathcal{L}(\ell_2^N)} \sum_{j=0}^{k-1} s_j (D_\Lambda S D_\Lambda).$$

By definition of the operator R and the class $\mathbf{S}_\Omega$, the last expression is bounded by

$$4C_4 \ln(k+1) \|A\|_{\mathcal{L}(\ell_\infty^N, \ell_1^N)} \|D_\Lambda S D_\Lambda\|_{\mathbf{S}_\Omega}.$$

By Corollary 14.2 one has

$$\|D_\Lambda S D_\Lambda\|_{\mathbf{S}_\Omega} \le C_3 \|\Lambda\|_{\ell_2^N}^2 \le C_3.$$

The proof now follows from the above estimates, and from the fact that $\|A\|_{\mathcal{L}(\ell_\infty^N, \ell_1^N)}$ is exactly the expression on the right-hand side of Theorem 14.2's inequality. □

Proposition 14.1 *For any k, $4 \le k \le N$ there exists an operator $B \in \mathcal{L}\left(\ell_\infty^N, \ell_1^N\right)$ such that*

$$\text{rank } B \le k, \quad B = -B^* \text{ and}$$

$$\|B\|_{\mathcal{L}(\ell_\infty^N, \ell_1^N)} \le 1, \tag{14.3.1}$$

and such that

$$\text{Re Tr } SB \ge C_5 \ln k,$$

where $C_5 > 0$ is a universal constant. Moreover for a given subspace $F \subset \ell_\infty^N$ with

$$\dim F = m, \qquad 2m + k \le N/2$$

there exists an operator $B \in \mathcal{L}\left(\ell_\infty^N, \ell_1^N\right)$ satisfying (14.3.1) such that $F \subset \ker B$ and

$$\text{Re Tr } SB \ge C_6 \ln \frac{k}{m+1}.$$

where $C_6 > 0$ is a universal constant.

Proof Denote

$$L = \left\{ x = (x_0, \ldots, x_{N-1}) \in \ell_2^N : \sum_{j=0}^{N-1} x_j = 0 \right\}$$

and let P_0 be orthogonal projection on subspace L. It is easy to check that for $\zeta \in \mathbb{C}$, $\zeta^N = 1$ the vector

$$V_\zeta = \left(1, \zeta, \zeta^2, \ldots, \zeta^{N-1}\right) \in \mathbb{C}^N$$

is an eigenvector of the operator $T = P_0^* S P_0$ with the corresponding eigenvalue $\lambda_\zeta = \frac{1+\zeta}{1-\zeta}$ if $\zeta \neq 1$ and $\lambda_1 = 0$. It is clear that $T^* = -T$ and hence the set of its singular values coincides with the set

$$\left\{ |\lambda_\zeta| \ : \ \zeta^N = 1 \right\} = \left\{ 0, \ \left| \frac{\cos \frac{\pi \ell}{N}}{\sin \frac{\pi \ell}{N}} \right| \ : \ \ell = 1, 2, \dots, N-1 \right\}.$$

Denote by $F^\perp = \left\{ x \in \ell_2^N \ : \ <x, y> = 0, \forall y \in F \right\}$ the orthogonal complement of the subspace F, and by Q the orthogonal projection on $F^\perp$. Let T_1 be the operator $T_1 = Q^* T Q$. It is clear that $T_1^* = -T_1$ and $\mathrm{rank}\,(T - T_1) \leq 2 \dim F = 2m$. Consequently, for the singular numbers of T_1 one has

$$s_{j+2m}(T) = s_{j+2m}(T_1 + (T - T_1))$$

$$\leq s_j(T_1) + s_{2m}(T - T_1) = s_j(T_1).$$

Therefore

$$\sum_{j=0}^{k-1} s_j(T_1) \geq 2 \sum_{j=m+1}^{m+\left[\frac{k}{2}\right]} \frac{\cos \frac{\pi j}{N}}{\sin \frac{\pi j}{N}} \geq \sqrt{2} \frac{N}{\pi} \ln \left(1 + \frac{k-3}{2(m+1)} \right).$$

It follows from (14.2.1) that there exists an operator $B_0 \in \mathcal{L}(\ell_2^N)$ such that rank $B_0 \leq k$; $B_0^* = -B_0$ and

$$|\mathrm{Re}\,\mathrm{Tr}\,T_1 B_0| \geq \frac{\sqrt{2}}{\pi} N \ln \left(1 + \frac{k-3}{m+1} \right),$$

putting

$$B = \frac{1}{N} (P_0 Q) B_0 (P_0 Q)^*$$

finishes the proof of the remaining part of the proposition.   □

Remark The proposition shows that the logarithmic factor in Theorem 14.2 is sharp. Its particular case with $F = \mathrm{span}\,\{(1, 1, \dots, 1)\}$ was settled in [6] by an explicit example. It was the main technical step in the construction of a convex body with logarithmic gap between the Ekeland-Hofer-Zehnder capacity of it and of its difference body.

Acknowledgements The author is extremely grateful to the Creator for giving him the understanding presented in this paper, to Yaron Ostrover for many stimulating discussions, to the referee for valuable comments and suggestions and to Lev Buhovsky and the referee for the generous help with preparing this paper.

References

1. A. Abbondandolo, P. Majer, A non-squeezing theorem for convex symplectic images of the Hilbert ball. Calc. Var. Partial Differ. Equ. **54**(2), 1469–1506 (2015)
2. S. Artstein-Avidan, Y. Ostrover, A Brunn-Minkowski inequality for symplectic capacities of convex domains. Int. Math. Res. Not. IMRN (13), Art. ID rnn044, 31 (2008)
3. S. Artstein-Avidan, Y. Ostrover, Bounds for Minkowski billiard trajectories in convex bodies. Int. Math. Res. Not. IMRN **2014**(1), 165–193 (2014)
4. F.H. Clarke, A classical variational principle for periodic Hamiltonian trajectories. Proc. Amer. Math. Soc. **76**(1), 186–188 (1979)
5. I. Ekeland, H. Hofer, Symplectic topology and Hamiltonian dynamics. Math. Z. **200**(3), 355–378 (1989)
6. E.D. Gluskin, Symplectic capacity and the triangular truncation operator. Algebra i Analiz **30**(3), 66–75 (2018)
7. E.D. Gluskin, Y. Ostrover, Asymptotic equivalence of symplectic capacities. Comment. Math. Helv. **91**(1), 131–144 (2016)
8. I.C. Gohberg, M.G. Kreĭn, *Introduction to the Theory of Linear Nonselfadjoint Operators*, vol. 18. Translated from the Russian by A. Feinstein. Translations of Mathematical Monographs (American Mathematical Society, Providence, 1969)
9. I.C. Gohberg, M.G. Kreĭn, *Theory and Applications of Volterra Operators in Hilbert Space*, vol. 24. Translated from the Russian by A. Feinstein. Translations of Mathematical Monographs (American Mathematical Society, Providence, 1970)
10. M. Gromov, Pseudo holomorphic curves in symplectic manifolds. Invent. Math. **82**(2), 307–347 (1985)
11. H. Hofer, E. Zehnder, A new capacity for symplectic manifolds, in *Analysis, et cetera* (Academic Press, Boston, 1990), pp. 405–427
12. G. Pisier, Grothendieck's theorem, past and present. Bull. Amer. Math. Soc. (N.S.) **49**(2), 237–323 (2012)
13. P.H. Rabinowitz, Periodic solutions of Hamiltonian systems. Comm. Pure Appl. Math. **31**(2), 157–184 (1978)
14. C.A. Rogers, G.C. Shephard, The difference body of a convex body. Arch. Math. (Basel) **8**(3), 220–233 (1957)
15. C. Viterbo, Metric and isoperimetric problems in symplectic geometry. J. Amer. Math. Soc. **13**(2), 411–431 (2000)
16. A. Weinstein, Periodic orbits for convex Hamiltonian systems. Ann. Math. (2) **108**(3), 507–518 (1978)

LECTURE NOTES IN MATHEMATICS   🐎 Springer

Editors in Chief: J.-M. Morel, B. Teissier;

Editorial Policy

1. Lecture Notes aim to report new developments in all areas of mathematics and their applications – quickly, informally and at a high level. Mathematical texts analysing new developments in modelling and numerical simulation are welcome.

   Manuscripts should be reasonably self-contained and rounded off. Thus they may, and often will, present not only results of the author but also related work by other people. They may be based on specialised lecture courses. Furthermore, the manuscripts should provide sufficient motivation, examples and applications. This clearly distinguishes Lecture Notes from journal articles or technical reports which normally are very concise. Articles intended for a journal but too long to be accepted by most journals, usually do not have this "lecture notes" character. For similar reasons it is unusual for doctoral theses to be accepted for the Lecture Notes series, though habilitation theses may be appropriate.

2. Besides monographs, multi-author manuscripts resulting from SUMMER SCHOOLS or similar INTENSIVE COURSES are welcome, provided their objective was held to present an active mathematical topic to an audience at the beginning or intermediate graduate level (a list of participants should be provided).

   The resulting manuscript should not be just a collection of course notes, but should require advance planning and coordination among the main lecturers. The subject matter should dictate the structure of the book. This structure should be motivated and explained in a scientific introduction, and the notation, references, index and formulation of results should be, if possible, unified by the editors. Each contribution should have an abstract and an introduction referring to the other contributions. In other words, more preparatory work must go into a multi-authored volume than simply assembling a disparate collection of papers, communicated at the event.

3. Manuscripts should be submitted either online at www.editorialmanager.com/lnm to Springer's mathematics editorial in Heidelberg, or electronically to one of the series editors. Authors should be aware that incomplete or insufficiently close-to-final manuscripts almost always result in longer refereeing times and nevertheless unclear referees' recommendations, making further refereeing of a final draft necessary. The strict minimum amount of material that will be considered should include a detailed outline describing the planned contents of each chapter, a bibliography and several sample chapters. Parallel submission of a manuscript to another publisher while under consideration for LNM is not acceptable and can lead to rejection.

4. In general, **monographs** will be sent out to at least 2 external referees for evaluation.

   A final decision to publish can be made only on the basis of the complete manuscript, however a refereeing process leading to a preliminary decision can be based on a pre-final or incomplete manuscript.

   Volume Editors of **multi-author works** are expected to arrange for the refereeing, to the usual scientific standards, of the individual contributions. If the resulting reports can be forwarded to the LNM Editorial Board, this is very helpful. If no reports are forwarded or if other questions remain unclear in respect of homogeneity etc, the series editors may wish to consult external referees for an overall evaluation of the volume.

5. Manuscripts should in general be submitted in English. Final manuscripts should contain at least 100 pages of mathematical text and should always include

   - a table of contents;
   - an informative introduction, with adequate motivation and perhaps some historical remarks: it should be accessible to a reader not intimately familiar with the topic treated;
   - a subject index: as a rule this is genuinely helpful for the reader.
   - For evaluation purposes, manuscripts should be submitted as pdf files.

6. Careful preparation of the manuscripts will help keep production time short besides ensuring satisfactory appearance of the finished book in print and online. After acceptance of the manuscript authors will be asked to prepare the final LaTeX source files (see LaTeX templates online: https://www.springer.com/gb/authors-editors/book-authors-editors/manuscriptpreparation/5636) plus the corresponding pdf- or zipped ps-file. The LaTeX source files are essential for producing the full-text online version of the book, see http://link.springer.com/bookseries/304 for the existing online volumes of LNM). The technical production of a Lecture Notes volume takes approximately 12 weeks. Additional instructions, if necessary, are available on request from lnm@springer.com.

7. Authors receive a total of 30 free copies of their volume and free access to their book on SpringerLink, but no royalties. They are entitled to a discount of 33.3 % on the price of Springer books purchased for their personal use, if ordering directly from Springer.

8. Commitment to publish is made by a *Publishing Agreement*; contributing authors of multiauthor books are requested to sign a *Consent to Publish form*. Springer-Verlag registers the copyright for each volume. Authors are free to reuse material contained in their LNM volumes in later publications: a brief written (or e-mail) request for formal permission is sufficient.

Addresses:
Professor Jean-Michel Morel, CMLA, École Normale Supérieure de Cachan, France
E-mail: moreljeanmichel@gmail.com

Professor Bernard Teissier, Equipe Géométrie et Dynamique,
Institut de Mathématiques de Jussieu – Paris Rive Gauche, Paris, France
E-mail: bernard.teissier@imj-prg.fr

Springer: Ute McCrory, Mathematics, Heidelberg, Germany,
E-mail: lnm@springer.com

Printed in the United States
By Bookmasters